BIOGEOGRAPHY
AND
ADAPTATION

BIOGEOGRAPHY AND ADAPTATION

Patterns of Marine Life

Geerat J. Vermeij

HARVARD UNIVERSITY PRESS
CAMBRIDGE, MASSACHUSETTS
AND LONDON, ENGLAND
1978

Library of Congress Cataloging in Publication Data

Vermeij, Geerat J.   1946-
  Biogeography and adaptation.

  Bibliography: p.
  Includes index.
  1. Geographical distribution of animals and plants.
2. Adaptation (Biology).   3. Marine biology.
I. Title.
QH84.V47   575   78-3722
ISBN 0-674-07375-4

Aan Vader, Moeder, en Arie
and to Edith

Preface

FOR as long as I can remember, two subjects have fascinated me more than almost any others: geography and the study of organic forms. It thus seemed natural to combine these interests and to satisfy them on the world's seashores, where aesthetically pleasing animals and plants abound in one of my favorite environments. An intense pleasure in practicing this brand of natural history is the recognition of patterns and relationships, which then form the raw material needed for new ideas. The most intriguing patterns are always those for which no explanation is immediately apparent, or those which at first glance seem to contradict an earlier idea.

This book is an account of some patterns in the sizes and shapes of marine organisms in relation to geography and time. It is, first and foremost, an empirical statement. Some of the inferred relationships between form and environment are based on experimental tests of specific hypotheses, but many others are speculative explanations of patterns that will demand closer scrutiny in the future.

A secondary objective is the construction of a synthesis and predictive theory in which all the empirical patterns and observations are related to one another. To this end I have asked how plants and animals that differ in the expression of predation-related traits differ in their susceptibility to speciation and to extinction. I have come to believe that biological factors in the environment are a source of evolutionary change and of instability, and that this change is most evident in the tropics and in other environments where biochemical reactions are least constrained by cold or by other physiological limitations.

One of the basic principles underlying this book is that biogeographical patterns and the process of evolution at the ecological level can be best understood from interpretation of the traits of individual species or populations. Natural selection, after all, acts on phenotypic characters, and we can recognize evolution only because we see a change in phenotypic or genotypic traits. I have chosen therefore to pursue a geography of traits rather than a biogeography of names or of numbers. Thus, secondary importance has been placed on the delimitation of biogeographical provinces, on patterns in the number of species, and on the incidence of endemism in the biota of a given region. Instead, I have emphasized patterns in the distribution of adaptations.

Of course, I have been somewhat selective in the patterns I have chosen to treat. Several important topics remain largely or completely neglected: life-history phenomena, the biology of larvae, physiological mechanisms of adaptation to stress, and symbioses between plants and animals. Inclusion of these topics would, I think, have diffused the main theme of the book.

Although I should like to think that some of my conclusions have general applicability to animals and plants in the sea, if not to all organisms, the reader must be cautioned that I have examined critically only a few groups in a few accessible environments. Most of my experience has been with skeletonized invertebrates, especially molluscs and the larger decapod crustaceans. I have explored far more thoroughly the intertidal zone as compared to the subtidal, and the benign tropics have been emphasized at the expense of what are euphemistically called the "temperate" zones. Regrettably, I have little to say about fishes, soft-bodied animals, the plankton, or the deep sea. These omissions reflect inexperience, not indifference or lack of interest. I feel that it is better to rely on groups and habitats that are familiar to me than to make uneducated guesses about things I know less well. Even so, I have made an effort to apply the insights of investigators working on other groups and in other habitats, and I have not hesitated to stray onto land or into fresh water. Only additional work will tell which of my claims will stand the test of time and which clams will not.

This book could not have been written without the kind help of many people and many institutions. Numerous friends and colleagues have discussed various parts of the book with me, read portions of the manuscript, or all of it , or permitted me to read their unpublished works; many have given valuable advice, pointed to crucial references, or offered useful criticisms. I am deeply grateful to these colleagues and friends, among whom I may mention J. D. Allan, C. E. Birkeland, P. K. Dayton, A. G. Fischer, D. E. Gill, P. W. Glynn, S. J. Gould, J. B. Graham, J. P. Grime, N. G. Hairston, R. C. Highsmith, R. S. Houbrick, J. B. C. Jackson, A. J. Kohn, E. G. Leigh, J. Lubchenco (J. L. Menge), D. B. Macurda, R. B. Manning, B. A. Menge, D. H. Morse, M. E. Nicotri, the late A. A. Olsson, R. T. Paine, A. R. Palmer, J. W. and K. G. Porter, the late G. E. Radwin, M. Reaka, R. Robertson, J. Rosewater, S. Smith-Gill, S. M. Stanley, R. R. Strathmann, C. W. Thayer, L. Van Valen, D. Vink, A. B. Williams, S. A. Woodin, W. P. Woodring, M. Yamaguchi, and E. J. Zipser.

My field work has been carried out at various marine laboratories and other institutions: Marine Biological Association, Plymouth, England; Discovery Bay Marine Laboratory, Discovery Bay, Jamaica; Caraibisch Marien-Biologisch Instituut, Piscadera Baai, Curaçao; Smithsonian Tropical Research Institute, Panama Canal Zone; Organization of Tropical Studies, Costa Rica; Instituto Oceanográfico (now Laboratório de Ciencias Marinhas), Recife, Brazil; Instituto Oswaldo Cruz, Rio de Janeiro, Brazil; Estación de Biología Marina, Montemar, Chile; Instituto del Mar del Perú, Callao, Peru; Instituto Nacional de Pesca del Ecuador, Guayaquil, Ecuador; Darwin Biological Station, Academy Bay, Galápagos, Ecuador; Hawaii Institute of Marine Biology. Coconut Island (Moku o Loe), Hawaii; University of Guam Marine Laboratory, Guam; Micronesian Mariculture Demonstration Center, Koror, Palau; University of the Philippines, Quezon City, Philippines; Department of Zoology, University of Singapore, Singapore; Office de Recherche Scientifique et Technique d'Outre Mer, Nosy-Be, Madagascar, and Abidjan, Ivory Coast; Hans Steinitz Memorial Marine Laboratory, Eilat, Israel; Department of Zoology, Hebrew University, Jerusalem, Israel; Institut Fondamental d'Afrique Noir, Dakar, Senegal; Institute of Oceanography and Marine Biology, Freetown, Sierre Leone; Department of Zoology, University of Ghana, Legon, Ghana; Florida State University Marine Laboratory, Turkey Point, Florida; and the Friday Harbor Laboratories, Friday Harbor, Washington. Numerous individuals have acted as helpers, guides, and assistants in these institutions, and their help is gratefully acknowledged.

I have been privileged to use the extensive molluscan and other collections at the U.S. National Museum of Natural History, Washington, D.C.; Academy of Natural Sciences, Philadelphia; Museum of Comparative Zoology, Harvard University, Cambridge, Massachusetts; British Museum of Natural History, London; Rijksmuseum voor Natuurlijke Historie, Leiden; and Zoologisch Museum, Amsterdam. Several libraries and librarians have been enormously helpful in providing me with important literature: J. Marquardt and others at the main and divisional libraries of the U.S. National Museum of Natural History, Washington, D.C.; the library at the Marine Biological Association, Plymouth, England; and the personal libraries of L. G. Eldredge and A. J. Kohn.

The photographs in this book have been ably taken by F. Dixon, M. Montroll, J. W. Porter, and L. Reed, and I am deeply grateful to them. I acknowledge the kindness of *Nature* in permitting reproduction of one of my figures that had previously appeared in that publication. I also wish to thank Mr. and Mrs. S. Zipser for providing me with a reliable typewriter, and the late R. E. Dickinson for teaching me how to swim.

The research upon which this book is based has been generously funded by grants from the Society of Sigma Xi, Organization of Tropical Studies, Geological Society of America, National Geographic Society, University of Maryland, and the Oceanography Section, Program of Biological Oceanography, National Science Foundation. The book was begun while I held a fellowship from the John Simon Guggenheim Memorial Foundation, which also funded various aspects of my research. I am extremely grateful for the generosity and support of these institutions, societies, and foundations.

For the past five years, I have enjoyed the scientific assistance, editorial criticism, and friendship of Bettina Dudley. Her help has immeasurably improved and expanded my work and has been essential in writing this book.

My deepest thanks are reserved for those closest to me and are perhaps best expressed in my native tongue: Tenslotte ben ik bijzonder dankbaar aan mijn ouders, de heer en mevrouw J. L. Vermeij, aan mijn broer Arie P. Vermeij, en aan mijn vrouw Edith Zipser. Zonder de steun, de nieuwsgierigheid in planten en dieren, en vooral de zo noodzakelijke aanmoediging van mijn ouders en broer, zou ik de richting van de wetenschap, en van de natuurlijke historie in het bijzonder, nooit zijn ingeslagen. Edith is in alle opzichten onmisbaar, als collega, als raadgeefster, als redactrice, als dierbare echtgenote. Deze mensen hebben mij doen beseffen, hoe afhankelijk wij zijn van de vriendschap en weldadigheid van anderen. [Finally, I am especially grateful to my parents, Mr. and Mrs. J. L. Vermeij, to my brother, Arie P. Vermeij, and to my wife,

Edith Zipser. Without the support and encouragement of my parents and brother, and their fascination with plants and animals, I would not have chosen science or natural history as a career. Edith is in all respects indispensable: as colleague, adviser, editor, and as beloved and esteemed wife. These people have made me realize how dependent we are on the friendship and beneficence of others.]

Contents

BIOGEOGRAPHY
AND
ADAPTATION

CHAPTER 1

Climate and Limitations of Form

ORGANISMS are adapted to the conditions in which they live. The morphology, behavior, and life history of a plant or animal species not only incorporate its adaptations, but reflect the historical idiosyncrasies and limitations of its ancestors. In the same way the habitat range and geographical distribution of a species are products both of historical events and of the physiological limitations set by the environment within which the species' adaptations are suitable. Ultimately these limitations of environment depend on the amount and availability of energy, and on the way in which this energy is apportioned among such competing functions as reproduction, growth, feeding, and defense.

The availability of energy varies systematically over the earth's surface, within regions from habitat to habitat (microgeographically) as well as from one region to another (geographically). As conditions in the physical environment change along a gradient, so do the associated organisms. Not only do the names of species and genera change, but, more importantly, the architecture (the geometry and mechanical design) of organisms varies with latitude, shore level, sediment and water depth, or altitude.

Two processes are capable of bringing about the observed correlation between architecture and environment. First, such environmental

Table 1.1   Simplified scheme of biogeographical provinces and regions in the sea.

Province	Region
Arctic	Arctic Ocean south to Labrador and Newfoundland, Aleutians, and northern Scandinavia
Cold-temperate North Atlantic	Newfoundland and Gulf of St. Lawrence to Cape Cod, Massachusetts
	Scandinavia and Baltic Sea to southwest England and Britanny
	Transitional area between Cape Cod and Cape Hatteras, North Carolina
Cold-temperate North Pacific	Siberia, northern Japan, Aleutians to Puget Sound
	Transitional area from Puget Sound to Point Conception, California
Warm-temperate Northwest Atlantic	Cape Hatteras to Cape Canaveral or Palm Beach, Florida
	Northern Gulf of Mexico
Warm-temperate East Atlantic	Brittany and southwest England to Mediterranean Sea and Mauritania
	Azores, Canary Islands
Warm-temperate Northwest Pacific	Japan, Korea, and China to about 27° N
Warm-temperate Northeast Pacific	Point Conception to Magdalena Bay, Baja California
Tropical Western Atlantic	Palm Beach, Bermuda, Bahamas, east coast of Yucatan Peninsula through West Indies south to near Rio de Janeiro, Brazil
Tropical Eastern Atlantic	Mauritania or Senegal to southern Angola
	Cape Verde Islands
Tropical Indo-West-Pacific	East coast of Africa from northern Red Sea to Natal, South Africa
	East through Indian Ocean, Indonesia, Philippines, northern Australia, Melanesia, and Micronesia to Hawaii, Marquesas, southeastern Polynesia, and Easter Island
	North to southern Japan
	South to near Perth, Western Australia, and southern Queensland
Tropical Eastern Pacific	Magdalena Bay and Gulf of California to northern Peru
	Rivellaguigedos, Clipperton, Cocos, and Galapagos
Warm-temperate Southwest Atlantic	Southern Brazil to northern Argentina

Province	Region
Warm-temperate South Africa	South and southeast coasts of South Africa
Warm-temperate Australia	Coast between Perth and eastern South Australia
	Southern Queensland and New South Wales
Warm-temperate New Zealand	Most of North Island
Warm-temperate Southeast Pacific	Northern Peru to central Chile
Cold-temperate South America	South of Chiloe, Chile, and most of Argentinian coast
Cold-temperate southern Africa	West coast of South Africa and most of Namibia
Cold-temperate Australia	Eastern South Australia, Victoria, and Tasmania
Cold-temperate New Zealand	Most of South Island and adjacent islands
Sub-Antarctic and Antarctic	Kerguelen, MacQuarie, islands south and southeast of New Zealand
	Tristan da Cunha
	Antarctica

characteristics as temperature and salinity may directly determine the rates of internal physiological processes and thus directly impose constraints on form. Second, organisms exhibit a wide variety of adaptations that take advantage of or ameliorate the effects of their external environment. An adaptation may be thought of as a characteristic that allows an organism to live and reproduce in an environment where it probably could not otherwise persist.

Much of this book will be devoted to the description and interpretation of geographical and microgeographical patterns of adaptation in marine organisms. If we are to recognize what constitutes an adaptive pattern, however, it is necessary to draw an unambiguous distinction between adaptation on the one hand, and nonadaptive response to an environmental gradient on the other; that is, we must identify the limitations that various types of environment impose on the range of possible adaptations. This is the mission of the first chapter. In it I shall briefly review existing climatic conditions and biogeographical provinces, then summarize how temperature and other physical characteristics affect metabolism, growth, calcification, and overall body form. Many geographical patterns will be shown to arise in whole or in part from direct environmental control of physiological processes and body shape.

Climate and Geography

The steep latitudinal gradients in the earth's climate today, together with the dispersed distribution of continents and seas, have created a rather large number of biogeographically distinct regions. These have been discussed at length by Ekman (1953) and Briggs (1974a) and are summarized in Table 1.1. For convenience I shall recognize four major tropical shallow-water provinces. Each of these regions may be variously subdivided according to the occurrence of endemic species (Briggs, 1974a), but the number and the geographical limits of the subdivisions vary both with the systematic group under study and with the environment being investigated. For example, a Brazilian subprovince of the tropical Western Atlantic province can be recognized on the basis of open-surface rocky-shore gastropods, light-loving corals and gorgonians, and perhaps soft-bottom bivalves; but it loses all distinctiveness for cryptic intertidal molluscs, mangrove-associated animals, and shade-tolerant corals (Laborel, 1969; Kempf, 1970; Vermeij and Porter, 1971). In the Eastern Pacific, more biogeographical subdivisions can be recognized using rocky-shore fishes than if soft-bottom fishes were relied upon (Rosenblatt, 1963, 1967).

As Hall (1964) has noted, most tropical waters do not fall much below 18° C in temperature for more than one month out of the year; usually they are far warmer. In general, the temperature range on the tropical west coasts of continents is greater than on the east coasts because of seasonal upwelling of cold, nutrient-rich waters (Lawson, 1966; Glynn, 1972; Dana, 1975). Some representative temperature ranges from various localities around the world are given in Table 1.2.

The list of temperate biogeographical provinces in Table 1.1 is a simplification of the scheme outlined by Briggs (1974a) and reflects geographical convenience as much as biological reality. Particularly in the Southern Hemisphere, many distinct biotic provinces have been recognized (see Stephenson, 1944, for southern Africa; Knox, 1960, 1963, for New Zealand and the sub-Antarctic islands; and Bennett and Pope, 1960, for Australia), but here they are grouped into only a few cold-temperate and warm-temperate regions. Moreover, the geographical limits of temperate-zone provinces are somewhat arbitrary and subject to varying interpretations. Provincial limits generally are coastal points where a large number of species reach their northern or southern distributional limits. Still, it must be emphasized that range limits also occur on many other points along the coast and tend to be somewhat variable over the years. For example, normally tropical species of the crab genus *Callinectes* occur in sheltered bays as far north as Los Angeles in warm-temperate California (Garth, 1960). The mussel *Mytilus edulis* often settles as much as 500 km south of its normal southern limit in the Western Atlantic at Cape Hatteras, North Carolina, but it cannot survive there for longer than a few months (Wells and Gray, 1960). Although the blue crab *Callinectes sapidus*, the oyster *Crassostrea virginica*, and the drilling moon snail *Polinices duplicatus* typically are found in the Western Atlantic south of Cape Cod, Massachusetts, in warm years they may extend as far north as Nova Scotia, or they may maintain local populations in warm bays (Dexter, 1962; Williams, 1974).

In most of the temperate zone, the west coasts of the continents and the shores of oceanic islands are characterized by comparatively small annual fluctuations in temperature (Table 1.2). Currents of cold high-latitude water, often associated with seasonal upwelling, flow toward the equator on the coasts of Peru and Chile, California, and southwestern Africa, while the warm Gulf Stream moderates the shores of Western Europe. The east coasts of continents in temperate latitudes usually are characterized by widely fluctuating sea surface temperatures and by continental climates. Nowhere is this fluctuation so great as on the Atlantic coast of the United States, where winter temperatures may be as much as 26° C lower than summer readings.

Temperature and Metabolic Rate

Most chemical reactions are temperature dependent. For every 10° C rise in temperature, the rate of reaction in the absence of temperature-compensating mechanisms increases twofold to threefold; that is, values of the so-called Q_{10} of the reaction lie between 2 and 3. Thus the rate of aerobic metabolism, and therefore food intake per unit of body weight, increase with a rise in temperature for an organism whose body temperature is closely tied to ambient temperature.

An immediate and important consequence of this relationship is that activities requiring large and rapid expenditures of metabolic energy, such as jumping, flying, running, and rapid burrowing, are difficult for organisms to generate and sustain at low temperatures unless the organisms have evolved special mechanisms for raising their body temperature above that of their immediate surroundings. For example, rapid-burrowing donacid bivalves and hippid anomuran crabs, so characteristic of tropical and warm-temperate sandy beaches, are absent from cold-temperate and polar shores. Ansell and Trevallion (1969), studying the mobility of sandy-beach molluscs in southwest India, showed that the speed of burrowing of *Donax incarnatus* at 30° C is about 2 seconds, three times

Table 1.2   Temperature regimes at the surface of the sea in selected localities.

Locality	Latitude	Range of monthly mean temperatures (° C)
Arctic		
Proven, northwest Greenland	72° N	−2 to 7°
Eastern North America		
Bay of Fundy	44° N	2 to 11°
Woods Hole, Massachusetts	41° N	3 to 20°
Ocean City, Maryland	38° N	2 to 23°
Beaufort, North Carolina	34° N	9 to 29°
Tampa Bay, Florida	27° N	15 to 33°
Western Europe		
Trelleborg, Sweden	60° N	1 to 15°
Southwest England	50° N	7 to 18°
Mediterranean		
Western Mediterranean	38 to 42° N	12–13 to 20–25°
Israel	31° N	16 to 28°
East Asia		
Southwest Kamtchatka	53° N	3 to 7°
Misaki, Honshu	35° N	3 to 30°
Western North America		
Neah Bay, Washington	49° N	6 to 13°
Pacific Grove, California	37° N	11 to 15°
Northern Gulf of California	31° N	15 to 31°
Tropical Western Atlantic		
North coast of Jamaica	17° N	24 to 28°
La Parguera, Puerto Rico	17° N	26 to 30°
Canal Zone	7° N	24 to 31°
West Africa		
Tenerife, Canary Islands	28° N	18 to 23°
Sierre Leone	9° N	20 to 27°
Ghana	4° N	19 to 29°
Cameroons	1° S	24 to 28°
Western Indian Ocean		
Eilat, Israel	29° N	18 to 27°
Dar es Salaam, Tanzania	7° S	25 to 28°
Mahé, Seychelles	4° S	28 to 32°
Nosy-Be, Madagascar	14° S	25 to 29°
Tulear, Madagascar	22° S	23 to 27°
Mauritius	20° S	22 to 27°
South Asia		
Okha, India	23° N	20 to 30°
Singapore	1° N	27 to 31°
Tropical Western Pacific		
South coast of Oahu, Hawaii	22° N	21 to 28°
Kona, Hawaii	20° N	22 to 29°
Marshall Islands	4 to 15° N	26–27 to 28–29°
Low Isles, Australia	17° S	23 to 28°

Locality	Latitude	Range of monthly mean temperatures (° C)
Tropical Western America		
Canal Zone	7° N	15 to 33°
Galapagos	0°	15 to 28°
Southern Atlantic		
Buenos Aires, Argentina	40° S	6 to 22°
Warm-temperate South Africa	30 to 35° S	14–16 to 20–22°
Australia and New Zealand		
Western South Australia	33 to 35° S	14–16 to 18–20°
Tasmania	40° S	9 to 19°
Piha, New Zealand	37° S	15 to 20°
Southwest South America		
Callao, Peru	12° S	16 to 21°
Montemar, Chile	33° S	12 to 17°
Sub-Antarctic and Antarctic		
MacQuarie Island	55° S	3 to 7°
Signy Island	61° S	−2 to 1°

Source: Data have been compiled from the works of Madsen (1940), Hutchins (1947), Stephenson and Stephenson (1954a, b), Wulff et al. (1968), Dexter (1969), Houbrick (1974a), Lowenstam (1954), Lewis (1964), Ekman (1953), Lipkin and Safriel (1971), Gislen (1943), Ricketts and Calvin (1968), Paine (1966a), Goreau (1959), Glynn (1968), Porter (1972a), Lawson (1955, 1956, 1957, 1969), Lawson and Norton (1971), Fishelson (1971), Taylor (1968), Pichon (1971), Gopalakrishnan (1970), Vohra (1971), Long (1974), Hobson (1974), Hiatt and Strassburg (1960), Stephenson et al. (1958), Abbott (1966), Olivier and Penchaszadeh (1968), Wommersley and Edmonds (1958), Bennett and Pope (1960), Paine (1971), Guiler (1959b), Vegas (1963), Kenny and Hayson (1962), and Walker (1972).

less than that required for Scottish *D. vittatus* at 14° C. For the Indian species, as well as for other tropical beach molluscs, such speeds are sufficient for reburial in the sand after the animal is dislodged by a wave; but the rate of burrowing of *D. vittatus* is too slow to permit this species to live intertidally on the beach. Indeed, *D. vittatus* is restricted to sands below low water in the British Isles. The retarding effect of temperature might have been even greater for *D. vittatus* and other temperate species of *Donax* were it not for the distinctly more streamlined shell as compared to that of tropical species (Figure 1.1).

Some rapid-burrowing bivalves do occur at or just below the low-water mark on temperate beaches. These include solenid razor clams (*Ensis*, *Solen*, and *Siliqua*) and various members of the Mactridae (Yonge, 1952; Holme, 1954). Again, however, these temperate species are very much slower than their tropical relatives. The Scottish *Mactra corallina* at 14° C completes a digging cycle in about 5 seconds; the same species at Naples

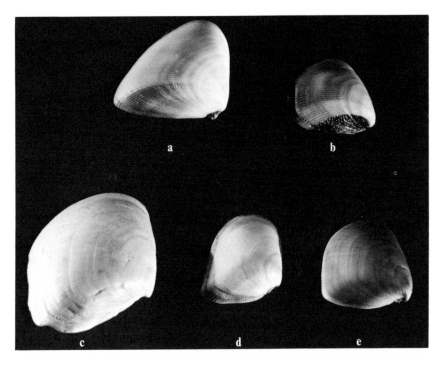

Figure 1.1   Tropical and temperate species of the beach clam *Donax*.
        a. *Donax rugosus*, Lumley Beach, Sierre Leone
        b. *D. denticulatus*, Cahuita, Costa Rica
        c. *D. vittatus*, Scheveningen, Netherlands
        d. *D. variabilis*, Boca Grande, west coast of Florida
        e. *D. trunculus*, Rimini, Italy
Temperate species of *Donax* tend to have shells that are more compressed and slender than tropical species. (Photograph by F. Dixon.)

at 21° C takes only 2.5 seconds to complete the cycle, while *M. olorina* in southwest India, bathed in 30° C water, manages to perform all the functions of the digging cycle in 1.5 seconds (Ansell and Trevallion, 1969).

Other active ectotherms, such as flying fishes, jumping strombid gastropods, fungiid corals able to right themselves when overturned, and fast-running grapsid and ocypodid crabs, are generally limited to tropical and warm-temperate waters. Stephenson (1959, 1972) has noted that the Carininae, a subfamily of swimming crabs (Portunidae) with a bipolar distribution, are slower and more benthic than other free-living members of the predominantly warm-water family. Warm-water scallops (Pectinidae) may be more adept swimmers than individuals or related species from cooler waters (Thayer, 1971).

Strict dependence of activity and metabolic rates on temperature confines most organisms to regions of warm or comparatively constant temperatures. Near or below the freezing point of water the hydrophobic bonds holding protein subunits together decrease in strength, while the ionic bonds within subunits become stronger (see Hochachka and Somero, 1973). As a result, it is difficult to design complex enzymes that are flexible enough to catalyze biological reactions at low temperatures and at the same time prevent denaturation. If an organism is exposed to a wide range of temperatures during its life cycle, temperature-induced conformational changes in both enzyme and substrate molecules may limit the thermal range over which a single enzyme can effectively catalyze a given reaction. To overcome this obstacle a series of isozymes, whose optimal activities lie at different temperatures, may be needed to catalyze the same reaction in different parts of the thermal range.

Hochachka and Somero (1973) give an excellent account of these biochemical limitations and distinguish two general mechanisms by which organisms can overcome strict temperature dependence: (1) they may maintain a constant (usually high) internal body temperature; or (2) they may regulate rates of reaction by various means of temperature compensation so that the Q_{10} of a reaction over a commonly encountered temperature interval is well below 2, even though body temperature is still closely coupled to that of the external medium.

The maintenance of a constant body temperature (homeothermy) usually is associated with high metabolic rates and with the production of large amounts of metabolic heat (endothermy), as in mammals, birds, some large insects, and a number of fast-swimming pelagic fishes. Except at high latitudes, endothermy is rare in the sea compared to its prevalence among large animals on land, perhaps because temperature fluctuations in water are far less dramatic than those on land or in the air.

Metabolic rate compensation, on the other hand, is widespread in fishes and marine invertebrates (for reviews see Bullock, 1955; Vernberg, 1962; Newell, 1970). For example, individuals of the California mussel *Mytilus californianus* and the limpet *Acmaea limatula* living near the upper limit of their intertidal distribution have been shown to have pumping rates and heartbeat rates respectively nearly identical to the rates of individuals at lower shore levels, despite the somewhat higher average temperatures experienced by the higher individuals (Rao, 1953; Segal et al., 1953; Segal, 1956a, b). The rate at which the cirri of Arctic *Chthamalus dalli* beat while this barnacle is feeding differs little from the rate in more southern populations of this species (Southward and Southward, 1967).

A third and probably minor mechanism of temperature control in marine animals is the raising of internal body temperature by absorption of sunlight, a process enhanced by an organism's dark coloration (Hamilton, 1973). Many intertidal snails, chitons, bivalves, sea urchins, and barnacles are dark in color, a characteristic that allows them to achieve higher temperatures and therefore probably more efficient catalysis than animals of lighter color. Dark coloration seems to be particularly common among intertidal animals of cooler shores. One indication of the prevalence of this phenomenon is the specific names of some Chilean intertidal molluscs: *Perumytilus purpuratus*, *Tegula atra*, *Prisogaster niger*, *Enoplochiton niger*. At the highest shore levels many species are light in color so as to prevent overheating (Vermeij, 1971b).

Rate compensation as a rule cannot entirely eliminate the effects of temperature. For example, Arctic fishes swim faster than expected relative to tropical fishes if swimming speed doubles with a $10°$ C rise in temperature (Scholander et al., 1953), but they are slow and inactive next to fishes from warmer waters (Arnaud and Hureau, 1966; Holeton, 1974). Moreover, temperature compensation is often limited to a relatively small segment of the thermal range encountered by an individual (see, for example, Davies, 1966).

One ecological consequence of this incomplete compensation is that many shore ectotherms on temperate and polar coasts hibernate in winter, either by burying themselves in sediment as do some hermit crabs (Rebach, 1974) or by migrating to deeper water where freezing is less likely. Seasonal migrations of this kind are known in limpets (Lewis, 1954; Frank, 1965; Walker, 1972), predatory snails (Connell, 1970; Luckens, 1970), sea stars (Paine, 1969; Seed, 1969b), and decapod crabs (Glynn, 1965; Crothers, 1968). Even when no migration occurs, as in some species of *Thais* and some sea stars, feeding and mobility may almost cease in winter (Paine, 1969, 1974; Christensen, 1970; Feare, 1971b; Spight, 1973). Even high-latitude endotherms are often seasonal in their occurrence or activity, since their ectothermic food supply is likely to be affected by temperature. Thus, most temperate and polar birds make seasonal migrations, while many ducks, pipits, and other year-round residents of higher latitudes often restrict their marine feeding to the winter months (Gibb, 1956; Olney, 1963, 1965).

Growth and Size

Many aspects of biological form are intimately related to, and perhaps controlled by, the rate of growth. Thus, geographical and microgeographical patterns in such features may not have an adaptive explanation, but could result from a direct dependence on growth rate, which is itself a function of age, size, temperature, availability of food, and other factors.

Among most ectotherms, for example, individuals or species at low temperatures grow more slowly at any given size than do animals at higher temperatures under otherwise equivalent conditions. This commonly known fact has been carefully documented in such diverse marine animals as bivalves, barnacles, snails, and reef corals (Connell, 1961a, 1970; Frank, 1965; Gaillard, 1965; Pannella and MacClintock, 1968; Seed, 1969a; Rhoads and Pannella, 1970; Glynn and Stewart, 1973; Spight, 1973). Intertidal diatoms in Oregon and a number of subtidal perennial algae in Nova Scotia, including *Hijikia, Desmarestia, Cystoseira,* and kelps of the genus *Laminaria,* constitute curious exceptions to this rule (Castenholz, 1961; Mann, 1973). The perennial algae apparently grow at near-freezing winter temperatures by tapping food reserves accumulated in summer. Annual subtidal algae, such as *Chorda,* and intertidal fucoid algae cease to grow in winter and have no mechanisms for storing nutrients.

At the cellular level, growth rate is determined both by the rate of cell division and by the increase in size of individual cells between episodes of mitosis. Slow rates of growth or of cell division in amphibians, protists, and other animals go hand in hand with larger cell size and could in some cases account for the increase in cell and body size observed with increasing latitude among many ectotherms (Ray, 1960). Presumably, DNA replication and membrane-mediated processes crucial to cell division are strongly temperature dependent. S. Smith-Gill has further suggested to me that differentiation and sexual maturation may be even more strongly temperature dependent than is somatic growth. If this is so, then the very commonly observed retardation in the time of onset of reproduction toward the poles (Vernberg, 1962; Frank, 1975) may not require an adaptive explanation. Much more work needs to be done, however, before adaptation can be ruled out as an explanation of latitudinal differences in first age of reproduction (Frank, 1975). Moreover, the poleward increase in body size within many species of mammals and birds (Bergmann's Rule) seems to be an adaptive response to lower temperature (see, for example, Brown and Lee, 1969) and to reduced competition with other endotherms (McNab, 1971); it is unlikely to be related to temperature-dependent growth rates.

The retarding effects of low temperatures on growth rates may be masked or even reversed by other factors. The slower growth rates of upper-shore as compared to lower-shore organisms, well known in algae, molluscs, barnacles, and echinoderms (Dehnel, 1956; Connell, 1961a, b, 1970; Seed, 1969a; Zaneveld, 1969; Branch, 1974; Lewis and Bowman,

1975), are not associated either with lower temperatures or with larger adult body size. In fact, temperatures during the growing season are on the average somewhat higher on the upper shore than at levels submerged under water for longer periods (Southward, 1958a; Fraenkel, 1968; Vermeij, 1971b; Wolcott, 1973). The reduction in growth rate in an upshore direction seems to result largely from shorter feeding periods and lower food availability and is usually associated with smaller adult body size. Excellent examples of these correlated trends may be seen both within and between species of barnacles (Dayton, 1971; Achituv, 1972), mussels (Seed, 1968, 1969a; Lavallard et al., 1969), limpets (Shotwell, 1950; Sutherland, 1970; Vermeij, 1972b; Branch, 1974, 1975b; Lewis and Bowman, 1975), and temperate predatory snails (Connell, 1970; Luckens, 1975).

Certain animals, having overcome the feeding restrictions at high shore levels, are able to attain sizes similar to or larger than those of related species lower on the beach. Pulmonate limpets (Siphonariidae), for example, often feed while exposed to the air (Abe, 1941); in New Zealand, western tropical America, and the Palau Islands they exhibit an upshore increase in adult body size (Borland, 1950; Morton and Miller, 1968; Vermeij, 1973b). Fiddler crabs (genus *Uca*) feed only when out of water and thus in principle could attain larger body sizes as submergence time decreases in an upshore or inshore direction (von Hagen, 1970; Crane, 1975). Tropical muricid gastropods typical of the middle and high intertidal zones often attain larger sizes than their low intertidal kin (Vermeij, 1973c; Taylor, 1976), possibly because their barnacle and mussel food is shared with fewer competitors and therefore is more abundant. Other reversals of the general size reduction toward the land are found in littorinids, neritids, and certain other gastropod groups exhibiting special adaptations to life on the upper shore (Vermeij, 1973b). By reducing their metabolic rate to very low levels when out of water, these snails have partially overcome problems of energy consumption in an environment of low productivity. In many cases they can support a larger volume of tissue than would be expected for a marine animal.

Growth and Shape

The intensity of allometry (change in shape with increasing size) is another feature that in many organisms seems to be closely tied to growth rate. The various features of gastropod shells and crab claws that we shall be considering are indicated in Figure 1.2. Pedomorphic individuals of the Bermudian land snail genus *Poecilozonites* were found by Gould (1968) to

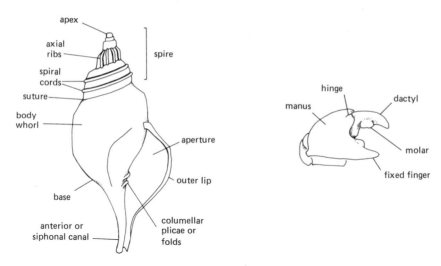

Figure 1.2   Terminology for the gastropod shell *(left)* and the crab claw *(right)*. (Illustrations by E. Zipser.)

exhibit faster growth rates, more diffuse color patterns, thinner shells, and relatively lower spires than snails growing at normal rates. The lower spire resulted from a less intense doming—that is, from a reduced positive allometry of spire height relative to shell diameter. In many snails, including this one, the intensity of individual doming increases as growth slows down during ontogeny, so that spire height is strongly dependent on growth. Frank (1975) has noted that the relatively tall spires of the trochid snail *Tegula funebralis* in British Columbia and Washington as compared to individuals in California result in part from the slower growth rates of the northern populations. Similar trends probably occur within members of the genus *Thais* and among intertidal littorinids (Spight, 1973; Vermeij, 1973b).

R. Highsmith has pointed out to me that the observed increase in doming with reduction in growth rate may result from a purely geometrical effect. Portions of the growing edge farthest from the axis of coiling must add more calcium carbonate per unit of time than points of the growing edge closer to the axis. Hence, as growth slows down, these outer points may be affected more drastically and accrete more slowly than points closer to the axis. Such differential growth reduction will result in doming.

Features dependent on growth rate in some cases are so numerous that populations differing only in growth rate have been claimed to belong to distinct species. Bandel (1974), for example, recognizes two West Indian species of the littorinid snail genus *Nodilittorina*: (1) a high-spired, small-

apertured species with indistinct columellar shelf and closely set beaded
sculpture (N. *tuberculata*); and (2) a lower-spired form with larger aperture, distinct columellar shelf, and more widely spaced beads (N. *dilatata*).
The second form has a generally more northern distribution than the first.
Most or all of the shell characteristics that distinguish these two forms are
sensitive to differences in growth rate, and even the purported radular differences might be dependent on growth rate or on rate of wear. More evidence is needed to prove that these snails belong to two distinct species.

In bivalve mussels (Mytilidae) and brachiopods the shell increases in
relative obesity (the distance between valves perpendicular to the commissure) as growth slows down during ontogeny (Coe and Fox, 1942; Rudwick,
1962; Seed, 1968; Lavallard et al., 1969). Slow-growing *Mytilus edulis* at
high shore levels and high latitudes often are markedly obese, while fast-
growing subtidal individuals and warm-temperate populations belonging
to the subspecies *diegensis* in California and *galloprovincialis* in southern
Europe are characterized by flat, fan-shaped shells (Coe, 1946; Soot-Ryen,
1955; Hepper, 1957; Seed, 1968, 1971).

S. Mitchell recently has found that bivalves and brachiopods from warm
waters have a larger number of radial ribs or riblets on their shells than do
individuals of the same or closely related species from colder waters. In
somewhat the same fashion, Johnson and Barnett (1975) showed that
various mesopelagic fishes have larger numbers of fin rays, vertebrae, and
other structures when food supply and hence growth rate of postmetamor-
phosed juveniles are low. Latitudinal differences in these and other meristic characters of fishes have been summarized by Barlow (1961), who
concluded that the degree of temperature dependence is often not the
same for different organs. It would appear from these data that genetic
differences among populations exaggerate or modify any direct effect that
environmental factors may have in determining the number of serially
repeated body parts (Barlow, 1961; Rosenblatt, 1963).

In spite of the possible control of allometric intensity by growth rate and
related environmental factors, it must not be concluded that allometry is
nonadaptive, or always a mere by-product of variation in growth rate.
Gould (1966) and others have shown that allometry is not only adaptive,
but mechanically necessary for species which during their lifetime traverse
a large range in body size. If no allometry were to take place, an organism
would remain geometrically similar to itself at all stages of growth; the
result would be a steady decrease in the surface-to-volume ratio and a
change in many other mechanical properties. Surface-dependent processes such as respiration, filter feeding, and crawling would decrease in
effectiveness as size increases; so would locomotion and other functions

dependent on the cross-sectional area of muscles. My main intent here is to point out that the mere existence of a latitudinal or microgeographical pattern in allometric intensity cannot be taken as prima facie evidence for an adaptive trend.

Calcification

Many geographical patterns involve the form and robustness of the skeleton, which in the majority of marine organisms is external and composed of calcium carbonate. For example, tropical bivalves often have large, thick, heavily sculptured shells; but species living in the cold waters of the polar regions and the deep sea are almost always small, thin-shelled, and devoid of color and external sculpture (Nicol, 1964, 1965, 1966, 1967). Bivalves cemented to the substratum by one valve are common in warm waters, but are unknown in the deep sea or near the poles, and only a few (such as *Hinnites multirugosus* and some oysters) are found in cold-temperate waters (Clarke, 1962; Nicol, 1964).

Similarly, polar and deep-sea gastropods are poorly calcified compared to their warm-water relatives. Graus (1974) has shown that temperate gastropods from the Woods Hole (Massachusetts) region (41° N) produce more internal living space per unit of deposited shell material than do snails from warmer, more southern waters. This is exemplified not only by the thinner shell walls of Woods Hole gastropods, but also by differences in shell shape. A relatively low-spired gastropod shell lacking sculpture and possessing a circular or broadly elliptical aperture builds more internal volume per unit of deposited material (and therefore has a higher calcification efficiency) than does a snail with a higher spire, more sculpture, or a markedly elongate aperture. Using these and other criteria of calcification efficiency, Graus was able to show that the demands of calcium carbonate conservation imposed by physiological limitations restrict the range of shell shapes available to temperate snails. Accordingly, shells with low calcification efficiencies (such as those with tall spires, narrow apertures, and well-developed external nodes or ribs) become increasingly more common toward the equator and are almost unknown among larger shells on cold-temperate shores. This trend, best exemplified among the gastropods living on sandy or muddy bottoms, is rather less evident in rocky habitats. In fact, as we shall see in Chapter 2, several rocky-shore gastropod families exhibit an overall decrease in spire height toward the equator (Vermeij, 1977b).

Calcification efficiency should also be high among snails on land, in fresh water, and in the deep sea, since calcium carbonate is relatively less available in these environments than in most shallow marine waters. Indeed, snails in these calcium-poor habitats tend to have thin shells, and sculpture and narrow apertures are rare (Vermeij and Covich, 1978). On the other hand, high-spired snails have evolved many times in both terrestrial and fresh-water habitats, and almost half of the fifteen known gastropods with open coiling occur on land and in fresh water (Rex and Boss, 1976). Open coiling is also known in several deep-sea snails (Rex, 1976), even though this geometry would seem to fly in the face of calcium conservation! In short, the efficient use of calcium carbonate in building shells may restrict the diversity of form among gastropods in calcium-poor environments, but the advantages of certain calcium-squandering geometries may outweigh or compromise the demands of conservation (Graus, 1974).

Arnaud (1974) has shown that shelled molluscs are not the only group whose calcification is impaired in the cold waters of the Antarctic. Most Antarctic taxa with calcareous skeletons (foraminifers, shelled molluscs, brachiopods) are fragile and small in body size. The echinoderms, with internal skeletons made of calcite, are relatively large in Antarctic waters, but their skeletons are poorly calcified. Benthic brachyuran and anomuran crabs, whose chitinous exoskeleton usually is rather well calcified, are wholly unknown from Antarctica; the only decapod crustaceans found there are planktonic shrimp-like forms with very thin carapaces (Yaldwyn, 1965; Arnaud, 1974). Groups using silica in their skeletons (diatoms, hexactinellid sponges) and those lacking substantial hard parts (nemerteans, nudibranchs, octopods, tunicates, sea anemones) are highly diversified and impressively large in body size at high southern latitudes (Dayton et al., 1970, 1974; Arnaud, 1974).

These contrasts strongly suggest a physiological limitation on calcification at low temperatures. Graus (1974) concluded that this limitation results mainly from the increased solubility product of calcium carbonate and the lower solubility of carbon dioxide at low temperatures, both of which render calcification more costly in terms of energy (see also Revelle and Fairbridge, 1957). Such energy expenditures would be particularly high for crustaceans, which must periodically shed their calcium-infused skeleton and then recalcify a completely new one. I have suggested that the reduction in relative claw size among benthic brachyuran crabs toward the poles could be a direct consequence of this difficulty (Vermeij, 1977a).

At least part of the problem of depositing calcium carbonate in cold waters may be that shell material is resorbed during anaerobiosis under near-freezing conditions. Below about 4° C, anaerobic metabolism seems to replace aerobiosis in the eastern North American salt-marsh mussel *Geukensia* (= *Modiolus*) *demissa*, a species that can survive temperatures as low as -12° C in winter (Murphy and Pierce, 1975; Murphy, 1977a, b). The reasons for conversion to anaerobiosis at low temperatures are not well understood, but may be connected with increasing tolerance to freezing. Anaerobic metabolism leads to the production of alanine, proline, and possibly other acids in the blood, whose neutralization is effected by calcium ions (Crenshaw and Neff, 1969; Hochachka and Somero, 1973). Under anaerobic conditions at near-freezing temperatures, or while the valves are tightly closed during prolonged tidal exposures, this calcium can be resorbed from the inner surface of the shell valves (Pannella and Mac-Clintock, 1968; Crenshaw and Neff, 1969), giving the valves a chalky rather than a glossy appearance. The widespread occurrence of chalky valve interiors among deep-burrowing and cold-water clams (Nicol, 1967; Rhoads and Morse, 1971) may mean that the conversion to anaerobiosis at low temperatures is common. If so, it follows that calcification at low temperatures becomes difficult even if calcium levels in the external medium are high. It is instructive that the largest Antarctic bivalve, the scallop *Adamussium colbecki* (Nicol, 1967; Arnaud, 1974), belongs to a group of relatively active clams with a propensity for swimming. The valve interiors of scallops are rarely chalky, and Nicol finds that certain Arctic scallops are among the few cold-water bivalves with bright coloration and well-developed shell sculpture. It is conceivable that these clams, as well as members of the actively burrowing and jumping Cardiidae (cockles), cannot or do not metabolize anaerobically, or do not resorb calcium carbonate from their shells. Cardiids always have shiny valve interiors, and Jackson (1973) finds them to be among the bivalves least tolerant to low oxygen levels.

Gastropods may also convert to anaerobic metabolism and resorb calcium carbonate at near-freezing temperatures, but this suggestion has not been experimentally verified. Most coiled archaeogastropods cannot or do not resorb calcium from their shells (Vermeij, 1977c), and it may not be coincidental that they are absent from intertidal rocks on coasts annually exposed to freezing. Thus, middle intertidal trochids are not found north of British Columbia and southwest England, even though lowest intertidal and subtidal species persist to polar seas (Fretter and Graham, 1962; Arnaud, 1974).

In the deep sea, low oxygen levels as well as low temperatures may promote anaerobiosis and limit calcification. Rhoads and Morse (1971) have argued that well-calcified organisms cannot exist, as a rule, if the dissolved oxygen concentration falls below about one milliliter of oxygen per liter of water. Such dysaerobic conditions occur in many deep basins, including those in the Black Sea and off southern California, where bottom circulation is restricted.

Gastropods found in salt marshes and mangrove swamps have a high incidence of secondary shell resorption and may be limited in calcification because of the acid soils on which they live (Vermeij, 1973c). In several groups (Potamididae, Truncatellidae, Planaxidae), species living highest in the swamps lose the tip of the spire as a result of decalcification and thinning of the apical whorl partitions. Other families (Neritidae and Ellobiidae) completely resorb the inner whorl partitions so as to make a shell cavity from which any trace of spirality has been removed (see also Woodward, 1892; Harry, 1951; Morton, 1955).

From the above it is clear that restrictions on calcification limit the shapes and thicknesses of skeletons under conditions of low temperature, low pH, and low oxygen levels. With the gradual easing of such limitations toward the tropics or toward other physiologically unstressed areas, adaptations unrelated to calcium conservation broaden the variety of shell types encountered. The nature of these adaptations will be explored in the next chapter.

Before leaving the subject of calcification, however, I shall briefly mention a curious latitudinal pattern in shell mineralogy first noticed by Lowenstam (1954). In nature, calcium carbonate occurs in any of three crystal forms: calcite, vaterite, and aragonite. Calcite is the most compact and stable form and has a slightly lower solubility in water than does aragonite. Vaterite is rare, and among living organisms it is known only in the internal skeletons or spicules of certain chordates (Lowenstam and Abbott, 1975). Lowenstam observed that a number of animals precipitate both aragonite and calcite in their skeletons (see also Waller, 1972; Adegoke, 1973). In the mussel *Mytilus edulis*, whose outer shell layer is calcitic but whose inner shell layer is composed of aragonite, the calcite-to-aragonite ratio decreases as mean annual sea temperature rises (Lowenstam, 1954). Similar mineralogical trends are found within species of the gastropod family Littorinidae and among species of mytilid and pectinid bivalves, as well as in serpulid polychaetes. In the littorinids, mytilids, and serpulids, aragonite precipitation predominates at temperatures above 15 to 18° C. Scleractinian corals, whose skeleton is composed almost exclusively of aragonite, only form reefs in waters in which the temperature does not fall much below 18 to 20° C (Wells, 1957; Veron, 1974). By contrast, brachiopods are today a predominantly cold-water phylum (Valentine, 1969) and have entirely calcitic shells.

Several untested explanations present themselves to account for the clinal variation in skeletal mineralogy. In bimineralic shells aragonite may grow more quickly at high temperatures than calcite, either by marginal accretion or by general shell thickening on the inner surface (Dodd, 1964). Differential absorption rates in some cases could account for the temperature dependence of mineralogy. If, for example, the aragonite on the inner shell surface of *Mytilus edulis* is being resorbed at low temperatures in conjunction with anaerobic metabolism, and if the outer calcitic layer is not subject to such resorption, then the calcite-to-aragonite ratio will increase as water temperature decreases, an expectation consistent with Lowenstam's (1954) trend. Other explanations are possible, and the whole topic deserves careful experimental investigation.

A large number of animals do not conform to Lowenstam's temperature-dependent trend in skeletal mineralogy. Articulate brachiopods in the past were almost certainly not so confined to high latitudes as they are today, yet all possessed calcitic shells. Most bivalves, including such typical cold-water families as the Cyamiidae, Hiatellidae, Nuculidae, Astartidae, and Thraciidae, have entirely aragonitic shells (Kennedy et al., 1969). The same is true of most gastropods regardless of latitude. Certain groups that are well represented in tropical waters, such as echinoderms, calcareous Foraminifera, and oysters (Ostreidae), have shells largely or entirely made of calcite. The significance of shell mineralogy and the influence of environmental factors on it remain largely unexplored.

Energy, Climate, and Diversity

A central axiom of ecology is that the energy available to an individual organism is limited, and that this energy must be allocated to such competing functions as feeding, growth, locomotion, defense, reproduction, and general body maintenance. Since biochemical reaction rates in ectotherms tend to increase as the temperature rises, the cost of body maintenance also increases with temperature (Hochachka and Somero, 1973); the costs associated with locomotion, calcification, and perhaps other functions may decrease. Natural selection is a prime determinant of how available energy is apportioned among competing body functions. As temperature rises, the range of possibilities of this allocation also increases: not only are certain modes of life and methods of feeding possible only at high temperatures and high metabolic rates, but some extreme types of energy allocation, in which one function (such as defense or reproduction) receives the lion's share of the available resources, would be impossible if

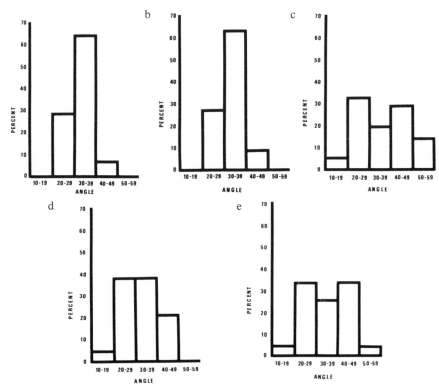

Figure 1.3 Morphological diversity of temperate and tropical assemblages of gastropods with coiled shells. Assemblages were collected from open rocky surfaces in the intertidal zone. The trend toward greater morphological diversity in the tropics as compared to the temperate zone, illustrated here with respect to spire height (apical half-angle of the spire), is equally apparent for other determinants of shell shape, such as expansion rate of the whorls and form of the aperture. Each species of gastropod in an assemblage is weighted equally in constructing the histograms. (Illustrations by E. Dudley.)

a. Botanical Beach, west coast of Vancouver Island, British Columbia (cold temperate). Of twenty-one species of gastropod collected, seven were limpets and fourteen had coiled shells. Actual range of apical half-angles 22° to 42°.

b. Montemar, Chile (temperate). Of twenty-five species of gastropod collected, fourteen were limpets and eleven had coiled shells. Actual range of apical half-angles 23° to 49°.

c. Paitilla Point, Panama City, Panama (tropical Eastern Pacific). Of twenty-nine species of gastropod collected, six were limpets, two were cowry like, and twenty-one had coiled shells. Actual range of apical half-angles 19° to 57°.

d. Fort Point, near Discovery Bay, north coast of Jamaica (tropical Western Atlantic). Of thirty-one species of gastropod collected, seven were limpets and twenty-four had coiled shells. Actual range of apical half-angles 15° to 46°.

e. Cahuita, Costa Rica (tropical Western Atlantic). Of twenty-seven species of gastropod collected, two were limpets, one was a cowry, and twenty-four had coiled shells. Actual range of apical half-angles 18° to 52°.

metabolic rates were very low, since other vital functions would be compromised too much.

On metabolic grounds alone, then, we should expect an increase in morphological or adaptive diversity toward regions of higher temperatures; this increase should be independent of any trend in taxonomic diversity. Such an equatorward adaptive diversification has been documented in land mammals (Fleming, 1973), night-flying moths (Ricklefs and O'Rourke, 1975), and gastropods (Graus, 1974; Vermeij, 1977b) (see Figure 1.3). Most temperate crabs are relatively slow, but no very slow xanthids or lightning-fast portunids occur in cold-temperate or polar waters.

A second expectation from the equatorward increase in heat energy is that adaptations to extreme habitats, probably involving unusual patterns of resource allocation, are more likely to occur under conditions of high temperature (in the tropics) than where temperatures are low or variable. Hutchinson (1959), for example, observed that many groups which in temperate regions are restricted to more or less normal marine salinity often invade fresh waters in the tropics. These include various bivalves (mytilids, donacids, teredinid shipworms), neogastropods (Marginellidae and Buccinidae—see Coomans and Clover, 1972), opisthobranchs (Acochlidiacea), decapod crabs, elasmobranchs, and various teleost fishes (see Roberts, 1972). The biochemical mechanisms involved in osmoregulation are apparently highly sensitive to temperature (see also Hochachka and Somero, 1973). Many Arctic forms (copepods and fishes) are also evolutionarily euryhaline, presumably because Arctic seas are characterized by low or strongly fluctuating salinity (Hutchinson, 1967).

Conclusions

In this chapter I have tried to show that many geographical and microgeographical patterns in behavior and form can be explained, in whole or in part, by the dependence of organic shape and metabolic activity on temperature and other external factors. It is therefore unnecessary to invoke natural selection as the primary determinant of these patterns, which include: (1) the restriction of metabolically active ectotherms to warm regions; (2) reduction in the degree of calcification under conditions of low temperature, low pH, and low oxygen; (3) various latitudinal and shore-level gradients of shape within many species; and (4) an increase in morphological diversity toward areas of higher temperature. Adaptive patterns, of course, have been superimposed on these trends, and in some cases may well have exaggerated or modified them. Such patterns are the subject of the next three chapters.

PART ONE

Patterns of Adaptation along Gradients

CHAPTER 2

Shelled Gastropods

GEOGRAPHICAL gradients in body form, behavior, and life history result not only from the direct dependence of the rates of life processes on temperature and other external factors, but they also reflect a differential adaptive response to environmental gradients. Sometimes the adaptation is a behavioral or architectural modification that tempers the potentially harmful effects of such conditions as excessive dryness (desiccation), high temperature, wave stress, osmotic stress, and the like. At other times, however, a geographical or microgeographical pattern is best interpreted as an adaptive pattern related not to the physical gradient directly, but to the changing biological surroundings. The particular adaptive responses of species along a given physical gradient are dictated by the nature and intensity of biological relationships such as competition and predation, and therefore cannot be fully understood without an awareness of these relationships.

In this chapter and the next two, I shall describe and interpret geographical and microgeographical patterns of adaptation in selected groups of marine animals and plants. I shall try to show that, as environmental conditions become less limiting to life processes along gradients, predation becomes a progressively more pervasive selective force. Gastropods constitute excellent animals with which to begin this inquiry; not only are they among my favorite animals, they are geometrically simple and ecologically well understood.

Adaptations to Physical Stress

Certain gradients in gastropod shell form seem primarily to reflect differential adaptation to physical stress. This is particularly well illustrated among snails living in the high intertidal zone of rocky shores (Vermeij, 1973b). Three types of snail, each with a different way of approaching the stresses of temperature and desiccation, are most commonly seen on the high rocky shore: periwinkles of the family Littorinidae, nerites (Neritidae), and limpets (Patellidae, Acmaeidae = Tecturidae, and Siphonariidae).

Periwinkles extend to higher levels than do other shore snails. They have small conical shells, which during ebb tides are attached to the substratum not by the foot (as in the other two groups and most other snails) but by a sheet of mucus at the outer shell lip (Vermeij, 1971a, b; Bingham, 1972). The soft parts are withdrawn in the shell behind the tightly fitting operculum. As a consequence, water loss is minimized, and the animal cannot use evaporative cooling as a means of temperature regulation, again in contrast to limpets and nerites. In fact, the only options available to littorines for preventing excess heat uptake and for maximizing heat loss during periods of aerial exposure to high temperatures are to minimize shell contact with the underlying hot substratum and to maximize reflectance from the shell surface. Within the Littorinidae, there is a general trend for tropical species to have relatively higher spires and therefore smaller apertures (reduced contact with the substratum) than species at higher latitudes. This trend, demonstrated in Figure 2.1, is accompanied toward lower latitudes by an increase in external sculpture. Thus, tropical periwinkles have strong spiral ridges or spirally arranged rows of small tubercles or beads on their shells, while most temperate littorines are smooth or at best weakly ridged. Compared with a smooth surface, the beaded external surface provides a larger area for reflectance of heat, but leaves the area for absorption of direct solar radiation unaffected.

Nerites are typical of most tropical and many warm-temperate shores. As shown in Figure 2.2, they have globose, low-spired shells with the internal cavity remodeled into a large, domed space that lacks the spiral structure characteristic of most shells (Woodward, 1892; Vermeij, 1973b). The remodeled interior seems to function as a water reservoir while the snail is exposed to the air (see Lewis, 1963; Vermeij, 1973b). Species of

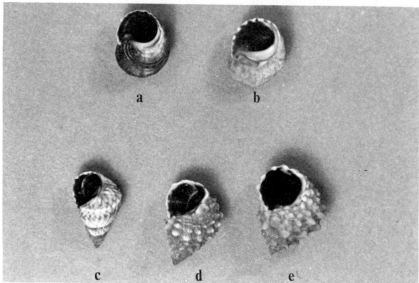

Figure 2.1   Dorsal *(above)* and apertural *(below)* views of some temperate and tropical periwinkles (Littorinidae).

      a. *Littorina sitkana*, False Bay, San Juan Island, Washington

      b. *L. planaxis*, near Carmel, California

      c. *L. lineata*, Pointe des Chateaux, Guadeloupe

      d. *Nodilittorina tuberculata*, Pointe des Chateaux

      e. *Echininus nodulosus*, Pointe des Chateaux

Temperate littorines from the highest zones of the shore are smoother and lower spired than their tropical counterparts. (Photographs by M. Montroll.)

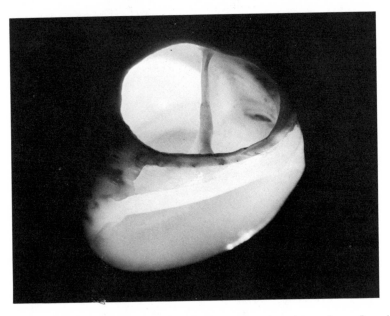

Figure 2.2 Internal shell cavity of *Nerita polita* (specimen from Inarajan, Guam). The side of the shell away from the aperture has been removed to show that the shell cavity is not spiral, as in most snails, but is incompletely divided by a septum into two chambers. (Photograph by F. Dixon.)

Nerita restricted to the high shore in the warmest parts of the tropics have relatively more globose shells, smaller apertures, and larger water reservoirs than species from the same zone at higher latitudes (Figure 2.3).

Cap-shaped limpets, distributed throughout the world, are most prominent in abundance and number of species on temperate shores (Knox, 1963; Vermeij, 1973b). Like nerites, they normally attach to the substratum by the sole of the foot, but a mucus film has been reported in the high intertidal *Acmaea digitalis* in western North America (Wolcott, 1973). Water may be stored in the space between the mantle and the inner shell surface (Segal, 1956b; Seapy and Hoppe, 1973); hence, a high-conic shell has a larger water reservoir than a lower shell of the same basal area. In spite of this, no clear latitudinal pattern is discernible in limpet shell form. However, for any given shell volume, nerites can store more water while exposed to air than can limpets; the equatorward decline in limpets therefore is in some respects counteracted by the appearance of nerites, which are better adapted to withstand desiccation.

The latitudinal patterns observed in nerites and periwinkles are also to be seen on a microgeographical scale. As shown in Figure 2.4, high-shore littorines on tropical and many temperate coasts have a taller spire and more pronounced shell sculpture than do sympatric lower-shore species

Figure 2.3 Dorsal *(above)* and apertural *(below)* views of some tropical and warm-temperate species of *Nerita*.

      a. *N. plicata*, Ngurukthapel, Palau
      b. *N. picea*, Coconut Island, Oahu, Hawaii
      c. *N. spengleriana*, Ngurukthapel
      d. *N. atramentosa*, northern New Zealand

Tropical species of *Nerita* (a and c) are more globose and often more heavily sculptured than are subtropical (b) or warm-temperate (d) species. (Photographs by F. Dixon.)

Figure 2.4 Dorsal *(left)* and apertural *(right)* views of high-shore and low-shore littorinids at Fort Point, Jamaica.

a. *Tectarius muricatus*, highest zone

b. *Echininus nodulosus*, next zone to seaward

c. *Littorina lineata*, highest intertidal, co-occurring with *Echininus* and with *Nodilittorina*

d. *L. lineolata*, below and to seaward of the preceding two species

e. *L. ziczac*, below *L. lineolata*

Tropical high-shore littorinids are commonly higher spired and more sculptured than species found lower on the same shore. The lowest occurring species at Fort Point, *L. meleagris*, is a very small species (5 mm) with a smooth exterior and a short spire. (Photographs by F. Dixon.)

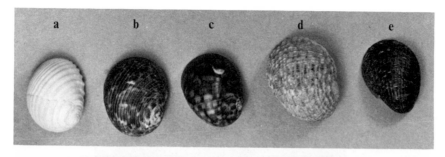

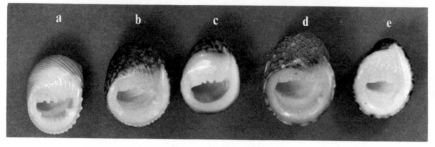

Figure 2.5   Dorsal (*above*) and apertural (*below*) views of high-shore and low-shore species of *Nerita* in the Palau Islands.

a. *N. plicata*, Ngurukthapel; highest zone on both limestone and volcanic rock
b. *N. undata*, Ngerdis Cove, Arakabesan; below *N. plicata* on volcanic rock; the Palauan form differs in color from *N. undata*, found elsewhere in the Indo-West-Pacific
c. *N. maxima*, Ngemelis; below *N. plicata* on limestone
d. *N. squamulata*, Ngerdis Cove; high middle intertidal
e. *N. albicilla*, Ngerdis Cove; middle intertidal

Species of *Nerita* high on the shore are more globose than those in the lower zones. (Photographs by F. Dixon.)

(Vermeij, 1973b; Heller, 1976). Among species of *Nerita* there is an upshore increase in relative globosity (Figure 2.5), which implies a decrease in relative aperture size and a larger water reservoir. Tropical limpets tend to be high-conic on the upper shore and low-conic or flattened at middle and low intertidal levels (Figure 2.6). In most cases high-shore snails are able to maintain body temperatures that are lower relative to rock and air temperatures than species at lower levels on the shore (Vermeij, 1971b). For high intertidal gastropods, then, moving upshore is somewhat comparable to approaching the equator.

On some temperate shores, however, shore-level patterns in littorine and limpet form are difficult to discern—or they may even be reversed. Spire height among periwinkle species actually decreases upshore in the

Figure 2.6   High-shore and low-shore limpets from Botanical Beach, ocean coast of Vancouver Island, British Columbia.

        a. *Acmaea mitra*; low intertidal
        b. *A. scutum*; middle intertidal
        c. *A. pelta*; middle intertidal
        d. *A. persona*; shaded high intertidal
        e. *A. digitalis*; open high intertidal

Most high-shore limpets have high-conic shells, while lower-shore species have a flatter profile. The low-shore A. *mitra* is a conspicuous exception. (Photograph by F. Dixon.)

eastern and southern Mediterranean and in California (Vermeij, 1973b). High-shore limpets in the Puget Sound area *(Acmaea digitalis* and A. *persona)* are more high-conic than such lower-shore species as A. *strigatella*, A. *pelta*, and especially A. *scutum*; but the low intertidal and subtidal A. *mitra* has an embarrassingly high-peaked shell whose height may be as much as 0.7 times the mean basal diameter. Similarly confusing shore-level patterns have been observed among limpets in Chile and South Africa (Marincovich, 1973; Vermeij, 1973b; Branch, 1975a). It may be significant, however, that the high-shore limpets achieve their high-conic shape in a different way from the lower-shore forms. In the high intertidal species, such as the northeast Pacific A. *digitalis* and A. *persona*, the Chilean A. *orbignyi*, and the South African *Patella granularis* and *Helcion pectunculus*, relative shell height increases allometrically with increasing size, and the apex is displaced toward the anterior of the shell. Tall low-shore species, such as A. *mitra*, the Chilean *Scurria viridula*, and the South African *Patella argenvillei*, are high-conic even at small sizes and have the apex located centrally (see also Branch, 1975a).

Wolcott (1973) has stressed that shore-level differences in shape or size among species of limpet are not sufficient to explain the much greater tolerance to desiccation of upper-shore as compared to lower-shore forms. In central California, for instance, most individuals of the middle intertidal A. *scutum* (ratio of height to mean diameter, 0.20 to 0.30) perish after less than 36 hours of exposure to air, while the higher-shore A. *digitalis* (ratio 0.35 to 0.45) can withstand desiccation for up to a week. The water reservoir of A. *digitalis*, however, is only about 25 percent greater than that of A. *scutum* and is therefore not the only feature determining the tolerance of limpets to desiccation; nor is there a large difference in the water loss that can be tolerated by species in the different zones (70 percent for the mid-shore A. *pelta* and A. *scutum* and for the shade-loving high-shore A. *persona*, 80 percent for A. *digitalis*). The principal mechanism that permits A. *digitalis* to withstand prolonged periods of exposure to the air is the formation of a sheet of mucus that effectively seals the soft parts from the outside air (Wolcott, 1973). Thus, shore-level changes in shell shape are only a small portion of the physiological adaptation of limpets to the desiccation and temperature gradients in the intertidal zone.

These temperate patterns, and many other snail-shell adaptations, may be related to gradients in wave exposure. Most wave-tolerant snails have cap-shaped (limpet-like) or low-spired coiled shells with large apertures and a low center of gravity, usually located over the foot of the animal. Several polymorphisms have been described in littorines (Struhsaker, 1968; Heller, 1975, 1976), muricids (Kitching et al., 1966; Kitching and Lockwood, 1974), and limpets (Bastida et al., 1971; Lewis and Bowman, 1975), in which wave-exposed individuals have lower spires (or flatter shells), a relatively larger aperture and foot, and often a smoother shell than animals found in more sheltered surroundings. Many wave-tolerant species, however, have well-sculptured shells whose rough surface may function to break up and distribute the force of the waves.

The limitations placed on shell form by severe wave action may be appreciated by noting the much greater diversity of shell types among snails living on mud, in sand, or under stones than in snails from open

Figure 2.7   Cryptic *(left)* and open-surface *(right)* snails from West Indian rocky shores, dorsal views. (Photographs by F. Dixon.)

       a. *Mitra nodulosa*, Discovery Bay, Jamaica
       b. *Tegula hotessieriana*, Piscadera Baai, Curaçao
       c. *Bailya parva*, Rio Bueno, Jamaica
       d. *Nitidella ocellata*, Vieux Habitants, Guadeloupe
       e. *Conus mus*, Rio Bueno
       f. *Columbella mercatoria*, Discovery Bay
       g. *Tegula excavata*, Vieux Habitants
       h. *Cittarium pica*, Playa Chikitu, Curaçao

rocky surfaces (Vermeij, 1973a, c) (see Figure 2.7). On most open hard bottoms, the apical half-angle of the coiled shell never falls below 14 degrees and rarely below 16 degrees; that is, there is an upper limit to how tall the spire can be. In cryptic environments and on soft bottoms, a much greater variety of shell geometries is found, including gravitationally unstable, high-spired forms in which the apical half-angle may be as low as 5 degrees.

    A few authors (Vermeij, 1973b; Branch, 1976) have argued that maximum sustained wave force is greater in middle and high latitudes than on most tropical coasts. This may further help to explain why the gravitationally stable limpets are so conspicuous and diverse in temperate latitudes, whereas the somewhat less stable nerites become progressively more prominent on warmer shores. The restriction to warm waters of high-spired cerithiids, narrow-apertured mitrids and cones, and other rocky-shore snails with shells incapable of resisting wave stress may be a further manifestation of the generally lighter wave action in the tropics. In the case of narrow-apertured snails, however, many species are characteristically found on algal ridges and on benches exposed to the full force of the tropical ocean (see, for example, Kohn, 1959a; Salvat, 1970). Even though a latitudinal gradient in wave force may partly explain the poleward reduction in geometrical diversity of snail shells, it gives little insight into the adaptations of the tropical forms.

Adaptations to Predation

Many features of gastropod shell architecture seem to be related more to predation than to physical factors directly. I shall try to show that predation involves an increasing emphasis on shell destruction toward the tropics, and that the intensity of this and of other modes of predation increases from high to low latitudes. These gradients in predation are reflected in turn by a greater shell sturdiness toward the tropics and along certain microgeographical gradients.

Shelled gastropods have numerous predators whose methods of attack and capture vary widely. The major mechanisms are summarized below, with an indication of the predators who use them and the principal means by which they are thwarted.

Entering the aperture to extract the soft parts without damaging the shell: asteroids, brachyuran and perhaps anomuran crabs, fishes, birds, octopods, other gastropods; prevented by long narrow aperture, toothed aperture, strong retractor muscles, inflexible tightly fitting operculum.

Breaking the shell: brachyuran crabs, stomatopods, lobsters, fishes, birds, sea otters(?), leatherback turtle, *Dirona* (opisthobranch); prevented by thick shell, reinforced outer lip, tight coiling, strong external sculpture, short spire.

Drilling through the shell: muricid and naticid gastropods, octopods; prevented by thick shell, strong sculpture(?).

Swallowing the prey whole, shell and all: sea anemones, asteroids, fishes, birds, some gastropods; prevented by large size, high-relief sculpture, tightly fitting operculum(?).

Predators that invade the aperture or otherwise extract the soft tissues without swallowing or assaulting the shell are potentially impeded by a

Figure 2.8   Strong columellar folds on the inner lip of *Turbinella angulata* (specimen taken near Santa Marta, Colombia). Retractor muscles, by which the soft parts are withdrawn into the shell, are attached to the columellar folds. (Photograph by F. Dixon.)

number of characteristics commonly found in marine gastropods. These features include highly toothed or narrowly elongate apertures; a large, thick, tightly fitting, inflexible operculum; and large, powerful retractor muscles, as reflected in extensive areas of columellar muscle attachment (Figure 2.8).

Forcipulate asteroids (sea stars) are common predators of limpets and coiled gastropods in temperate regions. Most of the larger hard-bottom sea stars kill their gastropod prey by extruding the stomach into the aperture of the prey and digesting the soft parts externally (Mauzey et al., 1968; Paine, 1969; Rosenthal, 1971; Menge, 1972a, b; Simpson, 1976). A large number of shore snails are known to have well-developed and effective running responses to predatory sea stars (Feder, 1963; Margolin, 1964b; Mauzey et al., 1968; Ansell, 1969; Phillips, 1976) and do not possess morphological characteristics that could be interpreted as thwarting asteroid predators; but other species, such as the western North American limpet *Acmaea mitra* and the keyhole limpet *Diodora aspera*, do not run away from sea stars (Margolin, 1964a, b). In the case of *D. aspera* contact with, or proximity to, sea stars elicits a mantle response in which the middle fold of the mantle is extended to cover virtually the whole shell exterior, thereby repelling the sea star or perhaps rendering the suction of the tube feet of the predator ineffective. *A. mitra* responds to the presence of *Pisaster* by clamping its shell tightly to the rock. A third gastropod that does not run from *Pisaster*, the buccinid *Searlesia dira*, has been seen to rob food from this predator and for unknown reasons is not usually attacked by it. Obviously, there is yet much to be learned about gastropod adaptations that discourage predation by sea stars.

Predatory crabs occasionally remove flesh by inserting one or both claws into the aperture without otherwise damaging the shell. This can occur only when the claws are thin or small, or when the prey's aperture is so large that even claws of quite substantial size can reach the soft parts. Extraction of soft parts without injury to the shell has been observed for the British crab *Carcinus maenas* (Ebling et al., 1964; Kitching et al., 1966; Gibson, 1970) as it feeds upon large individuals of the wide-apertured, open-coast form of *Thais* (= *Nucella*) *lapillus*. This method is ineffective on small individuals of the wide-apertured form, or on specimens of the more sheltered narrow-apertured form of any size. In Guam, I have observed the xanthid *Eriphia sebana* extract the soft parts of large wide-mouthed *Nerita albicilla* in spite of the latter's thick calcareous operculum, but the method fails on all other snails tested in that area because of their relatively small apertures. In Panama, the common intertidal *Eriphia squamata* can extract *Thais triangularis* out of its wide-mouthed shell, but is unable to do so with other species (Zipser and Vermeij, 1978). Several grapsids have been seen to dislodge and scoop out limpets (Chapin, 1968), abalones (Shepherd, 1973), and coiled *Tegula* snails (Hiatt, 1948), but how widespread this behavior is cannot be properly evaluated until more systematic experiments are conducted on these mostly herbivorous crabs.

Many gastropods feed on other snails by inserting the proboscis into the prey's aperture and immobilizing the victim with a toxin or anesthetic; they are listed below.

Cymatiidae: *Cymatium nicobaricum* and other species (Houbrick and Fretter, 1969)

Thaididae: *Thais armigera, Neothais scalaris, Haustrum haustorium, Nassa serta* (Menge, 1973; Luckens, 1975)

Melongenidae: *Busycon* spp., *Melongena corona* (Magalhaes, 1948; Carriker, 1951; Paine, 1963a)

Fasciolariidae: *Pleuroploca* spp., *Fasciolaria tulipa* (Wells, 1958b; Paine, 1963a, 1966b; Maes, 1967)

Vexillidae: *Thala floridana* (Maes and Raeigle, 1975)

Volutidae: *Aulica vespertilio* (Gonor, 1966)

Conidae: *Conus marmoreus, C. textile, C. dalli, C. pennaceus,* and related species (Kohn, 1959a; Nybakken, 1968; Kohn and Nybakken, 1975)

In spite of their narrow, slit-like apertures, adult cowries (Cypraeidae) are common victims of cones and of *Cymatium nicobaricum* in the Indo-West-Pacific and seem to comprise (along with trochids) a large component of the diet of *Nassa serta* in Guam. Even species of *Conus* with very narrowly elongate apertures fall prey to *Cymatium* (Kohn, 1959a; Houbrick and Fretter, 1969), as well as to such other species of *Conus* as *C. marmoreus* in the Indo-West-Pacific and *C. dalli* in the Eastern Pacific (Kohn, 1959a; Nybakken, 1968; Kohn and Nybakken, 1975). The large, inflexible opercula of *Nerita albicilla* and *Vasum turbinellus* render these common reef-flat species immune from predation by *Cymatium nicobaricum* in Guam. At Eniwetok a species of *Thais* (probably *T. armigera*) feeds almost exclusively on the limpet *Siphonaria*, but one was observed eating a specimen of *N. albicilla* (Menge, 1973).

Little is yet known about the prevalence of extraction from the aperture as a means of gastropod predation by fishes. Eales (1949) noted that the dogfish *Scyliorhinus caniculus* has many soft parts of *Buccinum undatum* in its stomach, but no trace of the shell; she speculated that this British elasmobranch might wrench the soft parts out of the shell without ingesting the latter. Similar modes of feeding have been postulated by Randall (1964) for the West Indian sting ray *Dasyatis americana*, the snapper *Lutjanus griseus*, and the grunts *Haemulon plumieri* and *H. sciurus* when these fishes eat conchs (*Strombus gigas*). Palmer (1977) reports that the kelp greenling (*Hexagrammos decagrammos*) in the Puget Sound region will bite the foot off *Calliostoma annulatum* and *Fusitriton oregonensis* after these snails have been dislodged from their site of attachment. In fact, Palmer has argued that *Ceratostoma foliatum*, a muricid snail common on subtidal rock walls, is adapted to the potential attack of this fish by the possession of three sharp varices per whorl; these render the probability of falling with the aperture downward and with the foot unavailable for biting much higher (35 to 70 percent) than it is for gastropods that lack the varices but otherwise are of comparable shape (almost zero probability).

If wrenching or biting occurs while the foot is extruded outside the aperture, then apertural defenses will be unimportant as adaptations to wrenching unless they are accompanied by a very rapid reflex for the soft parts to withdraw. This, in turn, might be accomplished by large columellar retractor muscles. It is interesting and puzzling that none of the known or suspected prey of the kelp greenling have large columellar folds. Equally puzzling is the fact that fresh-water snails in the Rift Valley lakes of Africa lack such folds, even though certain cichlid fishes have become specialized to bite the foot or to wrench the soft parts out of the shell (Corbet, 1961; Fryer and Iles, 1972; Greenwood, 1974).

Among birds, oystercatchers *(Haematopus)* have been seen by Feare (1967, 1971a) to wrench the foot from *Thais lapillus* without breaking or ingesting the shell. Gulls *(Larus)*, oystercatchers, and probably other birds are known to feed extensively on rocky-shore limpets, again by dislodging them and without ingesting the larger shells (Webster, 1941; Harris, 1965; Feare, 1971a; Hartwick, 1976; Simpson, 1976). Sometimes the shell will be cracked if the blow with the bill is hard enough (Feare, 1971a), but generally the shell is uninjured. Feeding without ingestion of the shell is also known among some fresh-water birds, including the limpkin *Aramus guarauna* (Snyder and Snyder, 1969) and some ducks (Bartonek and Hickey, 1969).

Sea otters *(Enhydra lutris)* are known to take abalones *(Haliotis)* off rocks without ingesting or injuring the shell (see, for example, Lowry and Pearse, 1973). As with birds, however, very little is known about predation success in relation to shell form.

Several species of *Octopus* have been shown to feed on gastropods (Pilson and Taylor, 1961; Arnold and Arnold, 1969; Wodinsky, 1969). Experiments by Wodinsky have elegantly proved that *Octopus* usually attempts to pull the body out of the shell before it resorts to drilling and then applying a muscle relaxant. In the Bahamas most species of gastropod must be drilled by *Octopus*, but the wide-apertured *Tonna maculosa* often can be pulled out without drilling (Arnold and Arnold, 1969). It is not yet known whether a very narrow aperture could prevent extraction of the soft parts by *Octopus* even after drilling and anesthesis. In the case of the cowry *Cypraea zebra*, however, extraction aided by drilling is successful in spite of the slit-like aperture (Wodinsky, 1969). It would be interesting to know if powerful retractor muscles might prevent successful attacks by *Octopus*.

It is uncertain whether hermit crabs (Paguridea) are predators of adult gastropods. Randall (1964) reported that the large West Indian *Petrochirus diogenes* could attack and eat young *Strombus gigas*, and in Guam I have seen the large red hairy *Dardanus megistos* attack possibly moribund *Lambis chiragra*. *Clibanarius corallinus*, a rather aggressive large hermit crab in Guam, was seen to attack and partially eat specimens of *Trochus niloticus* up to 5 cm in shell diameter in the laboratory. Near Alligator Harbor, on the Gulf coast of northern Florida, I collected an apparently living *Melongena corona* whose foot had been severely chewed and whose shell was shared with a large *Clibanarius vittata*. Magalhaes (1948), Greenwood (1972), Rossi and Parisi (1973), and Caine (1975) all suggest that hermit crabs may eat gastropods. I think it likely that hermits may attack adult snails injured by other potential predators. In any case, narrow apertures (or those occluded by teeth or thick opercula) would seem to be effective defenses against hermit crabs, which normally insert their chelae into the opening to extract the soft parts. These suggestions should be investigated experimentally, particularly since hermit crabs are a numerically conspicuous component of many shallow-water communities.

Spight (1976b) has shown in laboratory experiments that several species of *Pagurus* in the Puget Sound region can crush newly hatched *Thais* whose shells are less than 1.4 mm long. If this behavior occurs in the field, predation by pagurids is potentially an important source of hatchling mortality, especially in locations where hermit crabs are common.

Predators known to crush or otherwise break gastropod shells include various crustaceans, fishes, and birds. The leatherback turtle (*Caretta caretta*) is said to crush and eat large conchs (*Strombus gigas*) in the Bahamas (see Randall, 1964) and could be a locally important predator about which, however, virtually nothing is known. Robilliard (1971) reports that the northeast Pacific aeolid nudibranch *Dirona albolineata* can crush thin-shelled prosobranchs (*Lacuna* and *Margarites*) up to 7 mm in shell diameter with its jaws.

Most gastropod-eating crabs (Brachyura) have morphologically differentiated right and left claws and crush shells in only one of them (usually the right or master claw); but cancrids have subequal claws that are both used in shell destruction (Vermeij, 1977a). Each finger of the crusher claw normally has a proximal molariform tooth and a pointed, rather than spoon-shaped or hollowed-out, tip (Figure 2.9). Such a crushing surface is most spectacularly developed in tropical xanthids like *Carpilius*, *Eriphia*, *Ozius*, and *Menippe*, and in the parthenopid *Daldorfia*.

When attacking the shells of snails or hermit crabs, the tropical Pacific species of *Carpilius*, *Eriphia*, and *Daldorfia* often will attempt first to break off the apex or spire (Zipser and Vermeij, 1978). This method is highly successful when applied to such high-spired prey as *Cerithium*, but is less effective against *Drupa*, *Morula*, and other low-spired snails. On a high-spired shell, where opposite sides of the spire meet in an acute angle at the apex, the force between the closing fingers of the claw is mostly perpendicular to the shell surface and is therefore highly effective; but in a low-spired shell, opposite sides of the spire

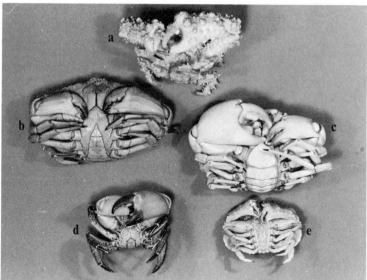

Figure 2.9   Dorsal (*above*) and ventral (*below*) views of some crushing crabs.
(Photographs by M. Montroll.)

       a. *Daldorfia horrida*, Cocos Island, Guam
       b. *Cancer productus*, Friday Harbor, Washington
       c. *Carpilius maculatus*, Pago Bay, Guam
       d. *Eriphia sebana*, Pago Bay
       e. *Calappa hepatica*, Pago Bay

come together at an obtuse angle, and the force applied by the claw on the spire has a large component parallel to rather than perpendicular to the surface to be crushed.

Only when the apex or body whorl cannot be crushed, as in low-spired shells or in large shells of almost any shape, will the massively clawed tropical xanthids and parthenopids resort to breaking the outer lip and peeling the outer whorl back spirally to expose the soft parts. Thus, low-spired or large snail shells need to possess adequate apertural defenses (outer-lip dentition, narrowly elongate opening, and so on) in addition to a sturdy spire and body whorl.

In many archaeogastropods, such as trochids and turbinids, the dual demands of a low spire (large apical half-angle) and a dentate or small aperture (low expansion rate of the whorl) cannot usually be accommodated in the same shell, since a low spire in these snails is typically accompanied by looser coiling, a wider umbilicus, and a larger aperture. For example, I have shown that the low-spired, widely umbilicate, large-mouthed West Indian trochid *Cittarium pica* (apical half-angle from 36° to 42°, whorl expansion rate 1.9 to 2.1) is vulnerable to predation by adult *Carpilius maculatus* crabs 11 cm to 14 cm wide up to a shell diameter of at least 5.7 cm; while the higher-spired, more tightly coiled *Trochus niloticus* (apical half-angle 30° to 34°, expansion rate 1.4) achieves immunity from crab predation at shell diameters greater than 4.0 cm (Vermeij, 1976). Since the two species have approximately the same shell thickness (1.1 mm to 3.8 mm for *C. pica*, 1.3 to 2.5 mm for *T. niloticus*), the difference in their vulnerability to predation by crabs seems to be directly tied to their overall geometry. In fact, the wide umbilicus of *C. pica* was often used by the crabs as a point of leverage for the crusher claw (Figure 2.10).

In conids, fasciolariids, and at least some muricid genera, however, apertural size and whorl expansion rate seem to be independent of, or positively correlated with, spire height; and extensive outer-lip dentition often secondarily restricts an otherwise wide aperture even in such very low-spired shells as *Drupa* (apical half-angle greater than 45°, whorl expansion rate greater than 2.0). These mostly neogastropod taxa combine a number of morphological features to make them sturdy and remarkably resistant to crushing by crabs and other predators. This is perhaps most impressive in short-spired members of the genus *Conus* (apical half-angle greater than 45°). The aperture of their shells is typically very narrow (length-to-width ratio usually between 6.0 and 9.5 except in a few Indo-West-Pacific fish-eating species), and the walls of the spire are greatly thickened. Species with taller spires (apical half-

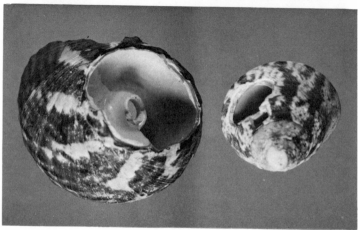

Figure 2.10 Two views of *Cittarium pica* from Jamaica *(left)* and *Trochus nilo-ticus* from Guam *(right)*, crushed by *Carpilius maculatus* in the laboratory. (Photographs by F. Dixon; from Vermeij, 1976. Reproduced with permission of *Nature.*)

angle from 39° to 45°) often have thinner shells and relatively broader apertures, with a length-to-breadth ratio never greater than 7.2 (Vermeij, 1977a).

Habitual spire-breaking in the fashion of *Carpilius, Eriphia,* and *Daldorfia* seems to be largely a tropical phenomenon, although it has also been described in the North Atlantic portunid *Carcinus maenas* (see Ebling et al., 1964; Kitching et al., 1966; Feare, 1970; Pettitt, 1975; Zipser and Vermeij, 1978). This mode of shell attack is not typical of such other temperate crabs as *Cancer* (Feare, 1970; Zipser and Vermeij,

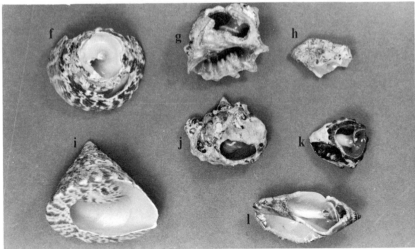

Figure 2.11 Examples of gastropod shells crushed by *Carpilius maculatus* (*above*) and by *Daldorfia horrida* (*below*) in the laboratory at Guam. (Photographs by F. Dixon.)

a. *Cerithium nodulosum*   g. *Drupa morum*
b. *Cypraea caputserpentis*   h. *Strombus mutabilis*
c. *Drupa morum*   i. *T. niloticus*
d. *Conus flavidus*   j. *D. morum*
e. *Trochus niloticus*   k. *Morula granulata*
f. *Trochus niloticus*   l. *Nassa serta*

Figure 2.12   Shells crushed by *Cancer productus* at Friday Harbor, Washington. (Photograph by F. Dixon.)

  a. *Fusitriton oregonesis*
  b. *Thais lamellosa*, sculptured form from Garrison Bay
  c. *T. lamellosa*, sturdier form from False Bay
  d. *T. lamellosa*, False Bay
  e. *F. oregonensis*
  f. *Ceratostoma foliatum*

1978), North American *Callinectes* (Hamilton, 1976), or the Mediterranean *Eriphia verrucosa* (Rossi and Parisi, 1973). Although there is some variation in the method of prey attack among members of the same species, these crabs mostly assault the prey shell at the outer lip (see Figure 2.11).

The most elaborate method of lip peeling, however, has been developed by sand-dwelling tropical crabs of the family Calappidae (Shoup, 1968). These crabs use a tooth on the external face of the crusher claw that closes on a corresponding notch in the fixed finger; a jagged spiral cut results, which may extend as much as two whorls back from the outer-lip margin (Figure 2.12). Thickened outer lips, a tall spire, and certain types of sculpture may aid sand-dwelling snails in resisting attacks by *Calappa*. On the Pacific coast of Panama I found that a small (27.1 mm wide) *Calappa convexa* successfully peeled *Nassarius luteo-*

Figure 2.13   Shells peeled by *Calappa convexa* in the laboratory at Naos Island, Panama. (Photograph by M. Montroll.)

    a. *Cantharus elegans*   c. *Cymia tectum*

    b. *Triumphis distorta*   d. *Terebra elata*

stoma 2 cm in length, as long as the latter's outer lip was not more than 1.3 mm thick. My observations and those of Miller (1975) further suggest that some high-spired snails, especially *Terebra*, can retract so far into the shell that the *Calappa* can rarely peel enough to expose the soft parts. Peeling also seems to be inhibited by the narrow apertures of many cones and miters (Mitridae and Vexillidae), and by the evenly spaced collabral sculpture (parallel to the outer lip) present in many sand gastropods (*Harpa, Rhinoclavis, Terebra, Vexillum, Otopleura,* and others—see Figures 2.13 and 2.19).

    Nephropid lobsters of the North Atlantic genus *Homarus* have crusher and cutter claws which, as in crabs, are used in fragmenting echinoid tests and the shells of gastropods and other prey (Squires, 1970; Mann, 1973). Spiny lobsters (Palinuridae), lacking claws, break prey shells with powerful molariform mandibles that come together in the vertical plane of symmetry of the animal (Figure 2.14). In the West

Indies Randall (1964) and Herrnkind and colleagues (1975) have seen *Panulirus argus* peel the outer part of the body whorl of *Strombus*, *Cittarium*, and other shells; as shown in Figure 2.14, shells attacked by spiny lobsters resemble those broken by *Calappa* (Figure 2.13). *P. gracilis* on the Pacific coast of Panama has similar behavior, and it is therefore not surprising that this lobster is thwarted by snail shells with heavily thickened outer lips. Thus, a *P. gracilis* 306 mm in total length was able to peel *Thais melonis*, a species with a simple outer lip, up to a length of 36 mm and a shell thickness of 3.7 mm; it also peeled adult *Cymia tectum* (49 mm long, 2.3 mm thick), *Opeatostoma pseudodon* (53 mm long, 2.3 mm thick), *Tegula pellisserpentis* (44 mm high, 3.7 mm thick), and small *Muricanthus radix* (up to 61 mm long, 3.7 mm thick). A hermited 24.6 mm long *Drupa ricinus* from Guam, reincarnated with a local hermit crab, sustained some damage to the outer lip, but would probably not have been killed by the lobster had the shell been occupied by the snail that built it. Similarly, although a 23.6 mm *Morula granulata* from Guam had a hole punched in the dorsal wall of the body whorl, the highly reinforced outer lip remained intact. The shell wall of

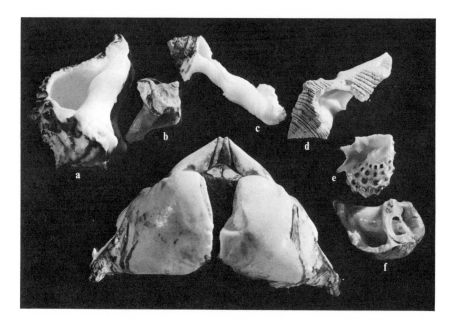

Figure 2.14   Shells crushed by the mandibles of a Pacific Panamanian spiny lobster *Panulirus gracilis*, shown at bottom. (Photograph by F. Dixon.)
a. *Muricanthus radix*   d. *Cymia tectum*
b. *Thais melonis*   e. *Drupa ricinus* (imported from Guam)
c. *Opeatostoma pseudodon*   f. *T. melonis*

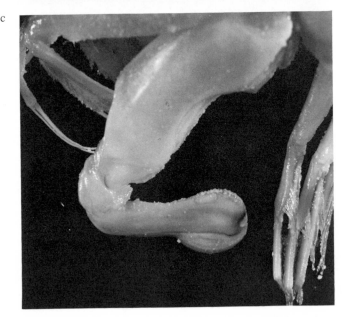

Figure 2.15 Stomatopods. (Specimens collected by M. Reaka; photographs by F. Dixon.)

     a. *Chloridopsis scorpio*, east coast of Thailand
     b. *Gonodactylus ternatensis*, near Phuket, Thailand
     c. Close-up of hammer-like second maxilliped of *G. ternatensis*

the Guamanian *D. ricinus* and *M. granulata* never becomes more than 2.0 mm thick.

Stomatopods employ the dactyls of their highly specialized second maxillipeds to feed (Figure 2.15). In *Gonodactylus*, *Odontodactylus*, and related genera, the base of this dactyl is modified into a hammer-like expansion, used to pound the shell of a prey snail or other animal until the shell fails (Burrows, 1969; Caldwell and Dingle, 1975). In other stomatopods limited to less well-armored prey, the tip of the dactyl is modified into a sharp spear that is often armed with lateral spines; this spear may be thrust into a thin-shelled or soft prey animal (Caldwell and Dingle, 1975; Dingle and Caldwell, 1975). Little is known about the impact or selective influence of stomatopods as predators of gastropods, but it may be imagined that many prey species would, if they could, readily agree with Bigelow when he so aptly named one of the hammering stomatopods, "O don't, O dactylus!"

Both elasmobranch and teleost fishes are well known as shell crushers (see Hiatt and Strassburg, 1960; Warmke and Erdman, 1963; Randall, 1964, 1967; Randall and Warmke, 1967; Hobson, 1968, 1974; Vivien,

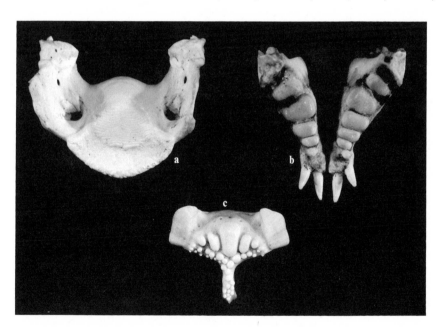

Figure 2.16   Crushing organs of some tropical fishes. (Photograph by F. Dixon.)
a. Jaw, *Diodon hystrix*, Guam (collected by F. Cushing)
b. Jaw, *Monotaxis grandoculis*, Palau (collected by J. Cochi)
c. Pharyngeal bone, *Coris aygula*, Guam (collected during fishing contest, South Pacific games, Umatac, 1975)

1973; Capape, 1976). These include rays (especially myliobatid eagle rays and rhinopterid cow-nosed rays), sharks (*Heterodontus* and *Galeocerdo*), filefishes (Monacanthidae), triggerfishes (Balistidae), trunkfishes (Ostraciontidae), porcupine fishes or spiny puffers (Diodontidae), porgies and their allies (Sparidae, Monotaxidae), grunts (Pomadasyidae), and wrasses (Labridae). Figure 2.16 shows some of the crushing organs. Rays crush large shells in their huge powerful jaws but, unlike most other fishes, rarely ingest any shell fragments. Puffers, porgies, trunkfishes, triggerfishes, and filefishes also use their jaws to crush hard-shelled prey, whereas wrasses, grunts, fresh-water cichlids, and many other fishes have hypertrophied pharyngeal teeth in the throat, which grind prey shells after they have been swallowed. In genera such as the wrasse *Coris*, only the very smallest shells (less than 3 to 5 mm in the largest dimension) are ingested without being broken, so that successful resistance to predation by these common reef fishes must involve a crush-resistant morphology even at relatively small shell sizes.

Several elegant experiments by Palmer (1978) have demonstrated the effectiveness of strong spines or other sculpture as adaptations against crushing. Using the puffer *Diodon hystrix* on the Pacific coast of Panama, he showed that the sharp knobs of *Thais kiosquiformis* and *Cymia tectum* render large representatives of these snails (longer than 35 mm) immune from being crushed in the powerful jaws of the puffer. When the knobs were filed off, these same individuals were successfully dismembered. Without spines the critical length of *Thais* would be increased by about 4 mm with respect to crushing by puffers. Not only does the sculpture effectively increase the size of the prey animal, but it also localizes the pressure applied by a predator to a small number of points where the shell is generally thickest (Vermeij, 1974b).

Strong sculpture has been implicated also in reducing the vulnerability of some shore muricid and thaidid snails to predation by cancrid and grapsid crabs (Fotheringham, 1971; Kitching and Lockwood, 1974). In Guam I often saw *Carpilius* and other crabs break away the tips of the knobs and spines of *Drupa*, *Vasum*, and *Chicoreus* without doing structural damage to the shell; in such cases the spines may prove to be useful defensive weapons. Spines or knobs may be less effective against peeling calappid crabs or spiny lobsters. When these animals attack such prey as *Thais*, *Leucozonia*, *Vasum*, and *Cymia*, the spines are broken off one by one.

Some birds also break the shells of their prey, but (at least for larger prey) achieve this with the aid of gravity. In North Carolina the herring gull (*Larus argentatus smithsonianus*) and ring-billed gull (*L. delawaren-*

sis) pick up *Busycon* whelks with their bills, fly to an appropriate site, then drop these snails and other prey on hard surfaces, where the shells are shattered (Magalhaes, 1948). Similar behavior is known in some land birds.

Gastropods of two families, the sand-dwelling Naticidae and the ecologically more diversified Muricidae (from which the Thaididae are often separated as a distinct family), prey on other molluscs by drilling the prey shell with an accessory boring organ (for reviews see Carriker and Yochelson, 1968; Sohl, 1969; Carriker and Van Zandt, 1972). Luckens (1975) suggests that the thick shell of *Nerita atramentosa* (= *N. melanotragus*) prevents *Neothais* and other muricids in New Zealand from drilling this species successfully; and Taylor (1976) reports the thick-shelled *Nerita textilis* to be one of the few potential prey species at Aldabra that is consistently unexploited by drilling muricids. Other than these casual observations, no work seems to have been done to explore gastropod features that might deter drilling gastropods and octopods. Gonor (1965, 1966) and various other workers have shown that gastropods often have well-marked escape responses to naticids, muricids, and other relatively slow predatory gastropods (see also Kohn and Waters, 1966; Ansell, 1969).

Many predators of gastropods are limited to relatively small prey, but this limitation is probably most severe for predators that swallow their catch whole. Thus even most wrasses, which must swallow the prey before these are crushed in the pharyngeal mill, tend to take only small molluscs, crabs, and hermit crabs. In the Antarctic, the nototheniid fish *Trematomus bernacchii* can accept snails of up to 10 mm in shell length, while *Notothenia neglecta* can take snails whose shells are not more than 5 mm in length (Arnaud and Hureau, 1966). Sand-dwelling sea stars in the genera *Luidia* and *Astropecten* are limited to small prey that are swallowed whole and slowly digested internally in the nonextruding stomach. For example, the largest European species of *Astropecten* (*A. arantiacus*), found at a depth of 4 to 50 m on the Mediterranean coast of France and with a mean radius of 136 mm, has a varied invertebrate diet of small shelled forms; only 9 percent of ingested snails, and 0.5 percent of ingested bivalves, are greater than 20 mm in their longest dimension (Massé, 1975). In Denmark, *A. irregularis* greater than 30 mm in radius have a similar diet, but 75 percent of the prey individuals are less than 3 mm long, and only a little more than 3 percent are greater than 5 mm long (Christensen, 1970; see also Hunt, 1925, for observations on this species off the south coast of England).

Similar size limitations apply to ducks and the majority of wading birds (Olney, 1963, 1965; Olney and Mills, 1963; Canton et al., 1974; Pettitt, 1975). In Great Britain, for example, turnstones (*Arenaria interpres*) can only take *Littorina littorea* of 8 mm or less in shell length; purple sandpipers (*Calidris maritima*) prey on *L. littorea* and *Thais lapillus* up to 5 to 8 mm in shell length; and the rock pipit *Anthus spinoletta* is limited to littorinid prey whose shells are at most 4 mm long (see Gibb, 1956; Feare, 1967, 1971a; Burton, 1974; Pettitt, 1975). In South Wales, the gull *Larus argentatus* can successfully swallow limpets (*Patella* spp.) of 5 to 38 mm in shell length, and it breaks about 5.2 percent of them (Harris, 1965). In contrast, animals that extract the soft parts externally and do not ingest the hard parts can take proportionally larger prey. This is true of forcipulate sea stars, gastropods, octopods, puffers, rays, crabs, lobsters, stomatopods, and some birds. Thus, oystercatchers (*Haematopus ostralegus*) in South Wales take limpets 12 to 59 mm in shell length and break about 2.9 percent of these (Harris, 1965). In Britain, *Carcinus maenas* crabs 75 mm wide can crush the open-coast form of *Thais lapillus* of up to 25 mm in length (Kitching et al., 1966; Gibson, 1970), and the oystercatcher takes adult *T. lapillus* and *L. littorea* greater than 20 mm long (Feare, 1967, 1971a; Pettitt, 1975). Large predatory snails in Florida (*Pleuroploca gigantea* and *Fasciolaria tulipa*) take prey gastropods that are as much as 85 percent of their own shell length (Paine, 1963a). In Guam, I have seen *Cymatium nicobaricum* kill snails more than 1.2 times its own length.

Because of their own large size, some predators that swallow their prey whole can take sizable victims. The gigantic northeast Pacific sea star *Pycnopodia helianthoides* (up to 1 or even 1.3 m across) can apparently ingest chitons 10 cm long (Mauzey et al., 1968). The Antarctic nototheniid *Trematomus hansoni* can ingest snails (such as *Neobuccinum eatoni*) up to 50 mm in shell length (Arnaud and Hureau, 1966). Some sea anemones may engulf and ingest large gastropods (Dayton, 1973a; Sebens, 1976). Members of the gastropod families Olividae and Volutidae envelop their prey of small gastropods and bivalves in the foot and thereby are able to capture animals that would be too large if the latter were to be ingested in the stomach, shell and all (Gonor, 1966; Olsson and Crovo, 1968). On the south coast of Curaçao, for instance, I have seen a 38.0 mm *Voluta musica* feeding on a 20.9 mm long keyhole limpet (*Fissurella barbadensis*) in the field; another *V. musica*, 29.4 mm long, was found feeding on a 12.7 mm *Nassarius albus*. (I thank R. Hensen for bringing the latter to my attention).

Large size may be attained by rapid growth through the vulnerable phase and may be enhanced by the development of spines, varices, and other protruding though not necessarily sturdy sculptural elements. Such features, spectacularly developed in deeper-water muricids, may also prevent shells from being swallowed. Biologists working with plankton have long appreciated the importance of spines, shields, and other structures in reducing the rate at which small planktonic organisms are consumed by larger copepods, rotifers, and fishes (see, for example, Zaret, 1972; Porter, 1973; Zaret and Kerfoot, 1975). The effectiveness of these devices depends on which predators are present; large size, achieved by the presence of spines or shields, is an advantage when the predators are copepods or rotifers, but may be disadvantageous against fishes that detect their zooplankton prey by vision. Thus, Brooks and Dodson (1965) found large species of *Daphnia* dominating lakes in which the fish *Alosa pseudoharengus* (alewife) is absent, while smaller species of *Bosmina* and other cladocerans dominated lakes occupied by alewives.

Other adaptations against swallowing predators include tightly closing opercula that prevent the predator's digestive enzymes from reaching the prey's soft parts (see Christensen, 1970; Massé, 1975). Inflexible calcareous opercula are typical of turbinids, neritids, and some naticids (*Natica*, *Lunatia*, and related genera); and thick or tightly fitting horny opercula are found in many families. In his study of the Californian predatory opisthobranch *Navanax inermis*, Paine (1963b) noted that the small operculate mesogastropod *Barleeia* was usually not digested during its passage through the gut of *Navanax*.

Noxious or toxic secretions undoubtedly play an important role in prey defense, particularly for those snails whose shell is partly or wholly internal or altogether absent. Cowries (Cypraeidae) are known to secrete acid, which may deter some fishes from ingesting them (Thompson, 1969); however, I have seen *Cypraea helvola* in the stomachs of Guamanian *Coris aygula* wrasses, and Rehder and Randall (1975) have found cowries in the stomachs of this fish collected at Ducie Atoll in Polynesia. Nudibranchs and other opisthobranchs (*Phyllidia*, for instance) also secrete a variety of acids and other noxious substances that, especially among dorids, often have a strong odor of iodine. Thompson (1960) and Harris (1973) suggest that these secretions might have an antipredatory function, and Paine (1963b) has shown that *Navanax inermis*, which feeds almost exclusively on opisthobranchs, avoids the acid-secreting *Pleurobranchus* species and *Acteon punctocaelatus* as prey; *Rostanga* may be avoided because of unknown noxious secretions; and still other potential prey may be immune because of their hard spicular integument.

Members of the thaidid genus *Purpura* give off a purple dye with an unpleasant odor of garlic that is not easily removed from the human hand. Randall (1967) does not record any instance of *P. patula* being eaten by any of the 212 West Indian species of reef fish whose stomachs he examined. The Caribbean intertidal limpet *Acmaea pustulata* exudes a remarkably sticky and persistent mucus that could be unpleasant or harmful to prospective predators; but in this instance Randall has found the limpet in the stomachs of nine fishes: the holocentrids *Holocentrus rufus* and *H. vexillarius*, the pomadasyid grunts *Anisotremus surinamensis* and *Haemulon carbonarium*, the labrid wrasses *Halichoeres bivittatus* and *H. radiatus*, the clinid *Labrisomus nucipennis*, the ostraciontid trunkfish *Lactophrys trigonus*, and the diodontid puffer *Diodon hystrix*.

The reader will appreciate that much remains to be learned about predation on gastropods and the adaptations of snails to predation. Already, however, it seems evident that features that deter some predators may be quite ineffective against other possible attackers. For example, narrow apertures seem to discourage crabs, but not predatory gastropods. Behavioral escape is effective against slow predators such as asteroids and other gastropods, but is usually futile when faster crustaceans or fishes are involved. Occasionally snails on a boulder-strewn beach may retract and fall off the exposed rock surface into a crevice or other hiding place where a fish or crab cannot reach them. I have seen this behavior in *Nerita doreyana* in northern Madagascar, and it has also been reported in some fresh-water snails (Snyder and Snyder, 1971). Usually, however, we may expect adaptations of gastropods to rapidly moving predators to be morphological rather than behavioral. In all cases the characteristics of the snail must reflect some compromise between the often conflicting adaptive demands imposed by different types of predators and by the physical environment.

Geographical Gradients in Predation

To any marine naturalist it is evident that the various predators of gastropods are not distributed evenly throughout the world or over all shallow-water habitats. Most molluscivorous groups (teleost and elasmobranch fishes, crabs, lobsters, stomatopods, octopods, and gastropods) are richest in species in the tropics. Although birds and mammals also increase in species numbers and trophic specialization toward the lower latitudes (Cook, 1969; Fleming, 1973), they are most important as molluscivores in temperate and polar regions. Asteroid echinoderms (sea stars) occur at all latitudes, and the intraorally digesting *Astropecten* is a large, predominantly warm-water genus; but extraorally digesting forcipulate sea stars known to feed on molluscs are nearly all found in the middle and high latitudes. Most tropical forcipulates feed on detritus, filamentous algae, or sessile corals and sponges (Yamaguchi, 1975a). The many-armed *Heliaster*, reported by Paine (1966a) to be a top predator in the northern Gulf of California and to take many gastropods as prey, is found throughout the tropical Eastern Pacific, but is exceedingly rare in most of Central America. The only Indo-West-Pacific asteroid which appears to prey extraorally on molluscs is *Acanthaster brevispinus*, a species known from the Philippines to northern Australia (Lucas and Jones, 1976).

The distribution of molluscivores agrees well with the views of Paine (1966a), MacArthur (1972), Levin (1975, 1976), and many others that predation pressure increases toward the tropics and toward conditions of reduced physiological stress generally. It is well to remember, however, that virtually no direct, incontrovertible evidence exists for increased predation pressure toward the tropics. Moreover, since conditions may vary greatly from year to year, especially in temperate regions, the impact of physical and biological sources of mortality is not necessarily constant over the years. Indeed, it is clear that predation on temperate populations may from time to time be very intense. For example, Feare (1967, 1971a) finds that the purple sandpiper *(Calidris maritima)* accounts for nearly all the mortality of young (less than 8 mm long) *Thais lapillus* on the Yorkshire coast, while the oystercatcher *Haematopus ostralegus* is an important source of adult mortality for this species; individuals of intermediate size die largely through the depredations of *Carcinus maenas*. Moore (1975) estimated from data collected in May 1974 that the gull *Larus californicus* accounts for at least 10 percent of the mortality of the common central Californian chiton *Nuttallina californica*. In the Gulf of St. Lawrence, Canton and associates (1974) found that the eider duck *(Somateria mollissima americana)* removes about 25 percent of the population of littorines in summer. About 10 percent of the mortality of *Thais lamellosa* individuals longer than 15 mm at Shady Cove on San Juan Island in the state of Washington results from predation by *Cancer productus* (Spight, 1976c). No quantitative estimates of predation intensity are available for tropical snails, but the impact of predation is likely to be strong.

Even if no direct evidence exists for an equatorward increase in overall intensity of predation, there can be little doubt that shell crushing becomes relatively more important as a mechanism of predation toward lower latitudes. Among brachyuran crabs the mostly temperate Cancridae have relatively smaller chelae than do most large, tropical crushing xanthids or parthenopids (Vermeij, 1977a). Within the genus *Cancer* the southern Californian *C. antennarius* has a larger claw (claw height to carapace width ratio 0.319, claw thickness to carapace width ratio 0.189, for males) than does the more northerly distributed but similarly large (15 cm broad) *C. productus* (ratios 0.244 and 0.136 respectively). Similarly, the crusher claw of the average male *Carcinus maenas*, a cold-temperate North Atlantic portunid (ratios 0.269 and 0.176) is only about 80 percent as high and 86 percent as thick as that of an equal-sized Mediterranean *C. mediterraneus* (ratios 0.337 and 0.205). The Mediterranean *Eriphia verrucosa* (ratios 0.415 and 0.290) has proportionally smaller crusher claws than other species in the genus, all of which are otherwise tropical.

Moreover, temperate crabs are limited to smaller prey than are their tropical counterparts of equal size (Vermeij, 1976). In Ireland, *Macropipus puber* 80 mm broad can crush thick-shelled *Thais lapillus* up to 20 mm in shell length and thin-shelled open-coast forms up to 25 mm long; *Carcinus maenas* 66 to 75 mm in carapace width take the thick-shelled form up to 12 mm and the open-coast form up to 25 mm in shell length (Kitching et al., 1966). However, trials with *Eriphia sebana* in Guam showed that crabs of this species 56 mm broad can crush hermited shells of the thick-shelled form of *T. lapillus* up to at least 27 mm in length (Vermeij, 1976).

The same latitudinal pattern in crushing ability may be discerned in other groups. The southern cold-temperate spiny lobster *Jasus frontalis* has proportionally smaller crushing mandibles than do the warm-temperate and tropical species of *Panulirus*; and within the latter genus, the Californian *P. interruptus* has smaller molars than the tropical Eastern Pacific *P. gracilis* (Kent, 1978). Hammering stomatopods are nearly all restricted to tropical coasts; only two of the eighteen Western Atlantic gonodactylids listed by Manning (1969) extend as far north as North Carolina in the warm-temperate Western Atlantic. Stomatopods that spear rather than hammer their prey are also predominantly tropical in distribution, but *Squilla empusa* is found as far north as cold-temperate Maine, while several species of *Hemisquilla* are shown off southern South America and New England (Manning, 1969).

Most fishes that crush shelled invertebrates are tropical or warm-temperate animals. For example, the wrasses (Labridae) are represented by only two species in the temperate Northwest Atlantic (*Tautoga onitis* and *Tautoglabrus adspersus*), while in the tropical West Indies at least fifteen species are found. Although shell-crushing rays often migrate to temperate waters as far north as southern New England in summer, they are not year-round residents there (Bigelow and Schroeder, 1953). Most polar fishes are comparatively slow species, which swallow their prey whole (see Arnaud and Hureau, 1966; Moore and Moore, 1974; Beurois, 1975). This habit does occur among some tropical fishes, such as the goatfishes (Mullidae), squirrelfishes (Holocentridae), and the clinid *Labrisomus* (Randall, 1967), but it is rare in comparison to crushing.

Further evidence for the increased importance of crushing predators toward the equator comes from the frequency of repaired injuries on the shells of living snails. These injuries, generally made by prospective predators who are unsuccessful at killing their prey, affect the outer shell lip and may be seen as jagged interruptions on the outer surface of the shell (Figure 2.17). Table 2.1 presents some preliminary data which suggest that the incidence of repaired injuries steadily increases from temperate to tropical latitudes. Thus, even if attempts at predation are unsuccessful, it is evident that shell features promoting sturdiness and resistance to crushing are more advantageous to tropical than to temperate snails. This conclusion holds even if it could be shown that the injuries result from physical rather than from biological causes, since it is the injuries rather than the sources of the injuries that impose the cost of repair.

It is therefore not surprising that the incidence and degree of development of gastropod shell characters that discourage crushing should increase toward the tropics. As Table 2.2 and Figure 2.18 show, tropical snails tend to be more strongly sculptured than are most temperate forms; their aperture is more likely to be elongate or to be bordered by strong teeth on the outer lip. For example, most cold-temperate thaidid snails are sculptured with weak axial growth lines or with evenly spaced spiral cords; the Northeast Pacific *Thais lamellosa* is highly exceptional in being adorned with well-developed axial ribs or even knobs in some morphs. A large proportion of tropical thaidids possess knobs or even spines at the shoulder, and some genera (*Drupa, Morula*) invariably have strong apertural teeth.

Many tropical gastropods, but few temperate ones, possess folds on the columellar side of the aperture where the retractor muscles are attached. For example, the mostly tropical Turbinellidae are usually characterized by three or more well-developed columellar folds; but *Vasum truncatum* from South Africa has these folds very poorly developed, and is one of only two warm-temperate members of the family.

Figure 2.17   Examples of repaired shell injuries. (Photograph by F. Dixon.)
a. *Terebra gouldi*, Kaneohe Bay, Oahu, Hawaii (inflicted by *Calappa hepatica*)
b. *Monodonta labio*, Aimeliik mangrove, Palau
c. *Nerita plicata*, Ngurukthapel, Palau
d. *Thais lamellosa*, Garrison Bay, San Juan Island, Washington

Table 2.1   Latitudinal gradients in the frequency of repaired shell injuries among rocky-shore thaidid snails of western America. Data are based on my own collections and those in the U.S. National Museum, Washington, D.C.

Species	Region	n	F
Thais emarginata	Washington, British Columbia	3	0 to 0.13
T. canaliculata	Washington, British Columbia	1	0.06
T. lamellosa	Washington, British Columbia	4	0 to 0.20
Acanthina spirata	California	2	0 to 0.22
A. brevidentata	Costa Rica to Ecuador	6	0.10 to 0.46
Thais melonis	Costa Rica to Ecuador	6	0.10 to 0.63
T. biserialis	Costa Rica to Ecuador	5	0.09 to 0.44
T. delessertiana	Peru	4	0.08 to 0.14
T. chocolata	Peru and Chile	4	0 to 0.16
Acanthina calcar	Chile	1	0
A. monodon	Tierra del Fuego	2	0 to 0.29

n = number of populations examined.
F = range of frequency of injuries per shell in a population.

Figure 2.18 Dorsal (*left*) and apertural (*right*) views of temperate (a and b) and tropical (c to f) species of *Nassarius*.

    a. *N. obsoletus*, Woodmont, Connecticut

    b. *N. mendicus*, near Friday Harbor, Washington

    c. *N. luteostoma*, Mata de Limón, Costa Rica

    d. *N. graniferus*, Ngemelis, Palau

    e. *N. vibex*, Amuay Bay, west coast of Paraguana Peninsula, Venezuela

    f. *N. versicolor*, Venado Beach, Panama

Tropical shallow-water species have more sculpture and often a more thickened outer lip than do temperate species. The ribs parallel to the outer lip (as in e and f) are regularly spaced and are highly developed in many tropical but few temperate species of *Nassarius*. (Photographs by F. Dixon and M. Montroll.)

Table 2.2 Latitudinal gradients in the incidence of some predation-related traits of low intertidal rocky-shore snails. Data are taken from my own collections.

Locality	Number of species	Percentage of species with—			
		Toothed apertures	Elongate apertures[a]	Inflexible operculum	Strong external sculpture
Cold temperate					
Vancouver Island, British Columbia	17	5.9	0	0	5.9
Boothbay Harbor, Maine	5	0	0	0	0
Warm temperate					
Plymouth, England	12	8.3	8.3	0	0
Isla San Lorenzo, Peru	11	9.1	0	9.1	0
Montemar, Chile	20	5	0	10	10
Tropical					
Playa de Panama, Costa Rica	15	40	47	20	20
Panama City, Panama	20	15	30	25	30
Fort Point, Jamaica	15	13	20	20	33
Playa Chikitu, Curaçao	10	20	10	30	0
Dakar, Senegal	13	15	23	7.7	7.7
Takoradi, Ghana	7	0	0	14	14

[a] Ratio of length to breadth greater than 2.5.

The subgenus *Charitodoron* of the genus *Mitra* is composed of three South African species that lack the columellar folds characteristic of all other members of the genus and family (Cernohorsky, 1976).

In snail families where spire height tends to increase as aperture size decreases (Trochidae and Turbinidae), tropical species tend to have higher spires (and therefore smaller apertures) than do their temperate relatives; but in groups where aperture size and spire height are independent, or are positively correlated (Conidae, peristerniine Fasciolariidae, Thaididae), spire height often decreases toward the tropics (Vermeij, 1977b). Temperate species of *Thais* and of related genera almost always have taller spires than do tropical species of *Thais*, *Morula*, or *Drupa*, but temperate trochids such as *Gibbula*, *Monodonta*, and *Tegula* often have lower spires and a much broader aperture than do such tropical genera as *Tectus*, *Trochus*, and *Thalotia*—or even tropical species of *Tegula* and *Monodonta*.

Some of these trends are evident even within a single species. The Australian *Dicathais aegrota* is characterized by nodose shell sculpture in the warm-temperate parts of its range (Western Australia and New

South Wales), but has continuous spiral cords in cold-temperate Victo-ria and Tasmania (Phillips and Campbell, 1973). Mediterranean popula-tions of *Thais haemastoma* never exhibit the large recurved knobs frequently found in West African populations (the so-called species *T. forbesi*).

Shallow-water gastropods from soft sediments exhibit a trend seem-ingly opposite to that of the rocky-shore snails discussed above, but again this trend may reflect an intensification of predation in general and shell destruction in particular toward the tropics. Tall-spired snails such as *Terebra, Turritella, Pyramidella, Turris*, and *Rhinoclavis* are unusual in temperate waters, but commonplace on tropical sand-flats and mud-flats. The turrids, cerithiids, and turritellids that do occur in cooler seas are often less slender than their warm-water counterparts. In addition, the neat, evenly spaced axial ribbing so characteristic of tropi-cal *Vexillum, Harpa, Nassarius, Rhinoclavis*, and *Epitonium* is less dis-tinct and less common among temperate soft-bottom snails (Figure 2.19). This trend is particularly well illustrated by the genus *Nassarius*.

Possibly, the increasing emphasis toward warmer waters on tall spires and axial sculpture among sand- and mud-dwelling snails is related to the concomitant increase in calappid crabs. The only calappids found in cold-temperate waters are forms such as *Hepatus* with subequal claws not specialized for lip peeling. Members of *Calappa, Acanthocarpus*, and related genera that do possess the characteristic peeling tooth on the right claw are found exclusively in tropical and warm-temperate waters.

A second possibility is that most sand-dwelling gastropods in the tem-perate zones are also commonly found in other habitats such as gravel, mud, or even solid rock. The British *Nassarius reticulatus* and the Chil-ean *N. gayi*, for example, are as much at home on sand as in stony habi-tats. The same seems to be true of many northern buccinids and of the northeast Pacific cerithiid *Bittium eschrichti*. By contrast, sand-dwelling snails in the tropics are almost never found on gravelly or rocky ground and therefore may exaggerate features useful in moving through soft substrates, since they need not be adapted to a wide range of substrate types.

Microgeographical Gradients in Predation

The latitudinal patterns in predation intensity and the associated architectural adaptations of gastropods may be seen on a smaller scale along at least two microgeographical gradients. Shell features that

Figure 2.19 Dorsal (*above*) and apertural (*below*) views of some tropical sand-dwelling gastropods. (Photographs by M. Montroll.)

a. *Conus arenatus*, Kikambala, Kenya
b. *Strombus gibberulus*, Ngemelis, Palau
c. *Polinices tumidus*, Ambatoloaka, Nosy-Be, Madagascar
d. *Oliva reticularis*, south coast of Aruba, Netherlands Antilles
e. *Rhinoclavis aspera*, Pago Bay, Guam
f. *Vexillum plicarium*, Ngemelis
g. *Terebra affinis*, Cocos Island, Guam

confer mechanical resistance to crushing become progressively more evident in a downshore direction on most rocky coasts (Vermeij, 1974a). Although littorines and certain other high intertidal snails are often beaded or even spiny on the external shell surface, the sculpture of lower-shore gastropods (when present) is usually coarser, and the individual knobs or spines are relatively larger. Note that this overall pattern of increased sculpture in a downshore direction runs counter to the sculptural trends within the family Littorinidae where, as discussed earlier, the irregular surface created by the nodules may function to radiate excess heat.

Connell (1961a, b, 1970, 1975) has been able to show experimentally that the diversity of predators and their impact on barnacles increase from high to low shore levels. While a similarly conclusive statement about predation on gastropods must await further work, the physiological or temporal limitation that periodic emergence imposes on potential predators makes it highly probable that upper-shore gastropods experience less intense predation than their lower-shore kin.

The deep sea supports a large diversity of predatory fishes, gastropods, crustaceans, and cephalopods (see Dayton and Hessler, 1972); but the morphology of these predators rules out crushing as a mechanism of predation on molluscs. Bright (1970) found that few of the deep-sea fishes he studied from the Gulf of Mexico had molluscan remains in their stomach; those that ingested hard-shelled echinoderms swallowed them whole. Moreover, the list of deep-sea gastropod genera published by Rex (1976) reveals a high proportion of species with loosely coiled or small shells, which would offer little resistance to being crushed. Rex has hypothesized that the impact of predation increases from the upper continental slope (at a depth of 500 m) to the abyssal rise, but then sharply decreases on the abyssal plane below a depth of 4,000 m.

It may be briefly mentioned here that fresh-water gastropods exhibit drastically reduced armor compared to their marine counterparts (Vermeij and Covich, 1978). Even the highly sculptured shells of Lake Tanganyika and the rivers of southeastern North America and Southeast Asia lack a number of antipredatory features common among marine forms. For example, no fresh-water gastropod exhibits teeth on the inner margin of the outer shell lip, and only one genus (the Southeast Asian neogastropod *Rivomarginella*) has an aperture whose length-to-width ratio exceeds 3. Correspondingly, such important molluscivores as lobsters, stomatopods, predatory gastropods, wrasses, asteroids, and octopods are entirely marine. While some fresh-water crabs do eat snails (Hamajima et al., 1976), most of them appear to be omnivorous and are not well adapted for crushing hard-shelled prey. Astacid and cambarid crayfishes, various birds and mammals, and especially fishes (such as Cichlidae and the African lungfish *Protopterus*) are known to crush molluscs and clearly have a selective impact on them; still, they may be less powerful than their marine counterparts (Vermeij and Covich, 1978).

Alternative Interpretations

Predation pressure is probably not the only biotic factor responsible for the latitudinal and microgeographical gradients evident in the architecture of gastropods. For example, boring sponges and encrusting calcareous algae may play a crucial role in the biology of epifaunal gastropods. Boring sponges (Clionidae) attack living gastropods in shallow waters throughout the world (see Figure 2.20); they may penetrate to the inner surface of the shell and cause extensive shell repairs to be made. Living tropical snails covered with calcareous lithothamnia seem generally less affected by boring sponges than are less encrusted temperate species. It is possible that boring sponges may be repelled by calcareous algae, or that they are smothered as the algal encrustation grows. MacGeachy and Stearn (1976) have found that sponges infesting bases of the coral *Montastrea* in the West Indies were not affected or killed by subsequent encrustation with calcareous algae, but it is not known whether sponges can settle on a shell or coral skeleton already occupied by encrusting lithothamnia.

Experiments are needed to ascertain whether encrusting calcareous algae benefit a living snail in any way; and, if so, whether an irregular outer shell surface created by strong sculpture promotes settlement by algae. Since shell rugosities have been shown to attract settling barnacles and other epizoans (Sacchi, 1970), the possibility arises that complex shell topography promotes settlement by encrusting algae. Such encrustation would also aid the gastropod in thickening the shell and perhaps contribute to its predator resistance. Wodinsky (1969) found that when shells of *Strombus raninus* were covered with various artificial materials through which an *Octopus* could drill, the *Octopus* attempted to extract the prey from the aperture without drilling. In any event, it is clear that natural encrustation is widespread among tropical hard-bottom gastropods, and that its function in preventing shell destruction remains in the realm of speculation.

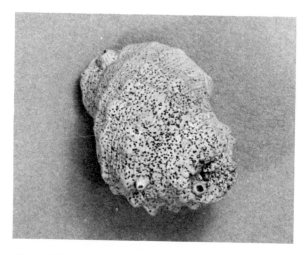

Figure 2.20 *Thais delessertiana* from Callao, Peru, severely bored at the apex
by clionid sponges. Shells with a thick covering of encrusting lithothamnion are
rarely attacked by borers, but temperate and tropical rock-dwelling snails
encrusted with barnacles or oysters are often riddled by boring sponges, polychaetes, algae, and fungi. (Photograph by M. Montroll.)

Conclusions

Many geographical and microgeographical patterns in gastropod shell
form reflect an increase in predation intensity toward the tropics and
toward conditions of decreasing physiological stress. As temperature
rises in an equatorward direction, there is an increasing emphasis on
shell features that promote resistance to crushing, and a corresponding
increase in the strength and selective impact of shell-destroying predators. These predation-related patterns exaggerate, modify, or occasionally are in conflict with patterns resulting from direct adaptation to
physical gradients or from nonadaptive responses to external conditions.

Are these patterns unique to gastropods? If so, they would be of little
interest to the biologist who is trying to extract broader principles about
the interrelationships of organism and environment from the architecture of plants and animals. In the next two chapters, I shall argue from
empirical evidence that the gastropods are not alone in reflecting geographical and smaller-scale gradients in predation.

CHAPTER 3

Bivalves

$\mathbf{I}$F THE latitudinal and microgeographical patterns in gastropod shell architecture are causally related to gradients in intensity of predation, then similar patterns should be evident in other skeletonized groups of organisms. Bivalve molluscs constitute an important group in which to evaluate this proposition. In contrast to the gastropods, they are an evolutionarily conservative class of molluscs in which evolution has generally been slow and architectural innovations rare (Simpson, 1953; Stanley, 1968, 1973). Particular modes of life are almost always associated with a specific shell architecture, no matter what the taxonomic position of a given species. Thus the oyster shape, in which one valve is cemented to the substratum with the other valve acting as a lid, has evolved independently in the subclasses Pteriomorpha (Ostreacea, Pectinacea), Heteroconchia (Chamacea), and Anomalodesmata *(Chamostrea)* (see Yonge, 1967). A demonstration of predation-related latitudinal and microgeographical patterns in a group as potentially unresponsive as the Bivalvia would furnish strong evidence that such patterns are widespread and characteristic of most benthic animal groups that have exoskeletons.

Shell Form and Physical Habitat

In spite of their conservatism, bivalves are numerically important animals in most hard-bottom and soft-bottom marine communities. They may be epifaunal (cemented, attached by a byssus, or free-living on the bottom), semi-infaunal or endobyssate (partially buried in soft bottoms), boring or nestling (burrowing or occupying cavities in rock or wood), or infaunal (buried in soft sediment).

Stanley (1968, 1969, 1970, 1972, 1975b) has carefully documented the architectural features characteristic of clams in various environments and has demonstrated that bivalve form and sculpture vary systematically along microgeographical gradients. This variation in shell architecture has been interpreted largely as differential adaptation to mobility and to the position occupied relative to the sediment surface.

Shell sculpture varies widely among clams, yet some microgeographical patterns in the type and expression of ribs and spines are evident (Stanley, 1970). Regularly spaced radial or concentric ribs, common on the shells of many cardiids, venerids, astartids, carditids, and other shallowly buried infaunal clams (Figure 3.1), probably stabilize the animal in the sediment by increasing the surface area of contact between the two and by creating resistance to being dislodged. This type of sculpture is not found in more rapid burrowers, whose shells are smooth and often compressed in order to minimize resistance to locomotion through the sediment. In a few genera (*Strigilla* and *Acila* are examples) fine oblique sculpture may also facilitate burrowing (Stanley, 1969; Seilacher, 1973). Clams that are deeply buried in sand or mud, such as *Ensis*, *Mya*, *Lutraria*, and *Saxidomus*, usually lack sculpture except for concentric growth lines (Figure 3.2); in these bivalves the sediment itself effects positional stability. Sedentary epifaunal clams vary from being smooth (many mytilids), to radially plicate (pectinid scallops), to scaly or spinose (*Anomia*, *Chlamys*, *Spondylus*, *Ostrea*, *Tridacna*); but the evenly spaced concentric ridges so commonly found in shallow-burrowing infaunal bivalves are virtually unknown among Recent epifaunal clams. The functional significance of sculpture in epifaunal bivalves has been little studied except in scallops, where the radial ribs are generally thought of as giving the thin, light shell resistance to crushing and other damage (Waller, 1972).

The extent to which the valves interlock with one another or cover the soft parts of the clam also varies microgeographically. Most epifaunal, semi-infaunal, and shallowly infaunal bivalves have complex heterodont or taxodont hinge dentition (Figures 3.1 and 3.3), which prevents the valves from shearing upon one another. In cardiids, trigoniids, and other

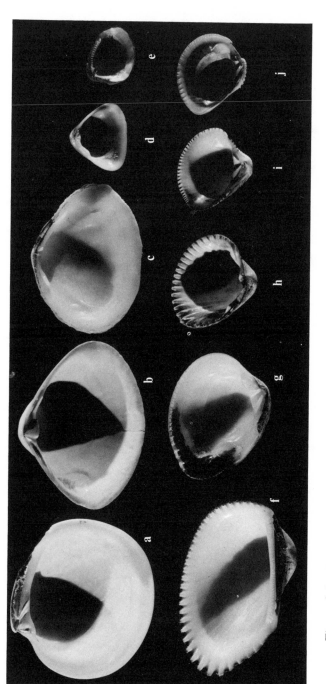

Figure 3.1   External (*above*) and internal (*below*) views of valves of some shallowly burrowing infaunal clams.

a. *Dosinia elegans*, Sanibel Island, west coast of Florida
b. *Spisula solidissima*, Sunset Beach, New Jersey
c. *Quidnapagus palatam*, Achang Bay, Guam
d. *Tivela bicolor*, Lumley Beach, Sierre Leone
e. *Acila castrensis*, Friday Harbor, Washington
f. *Anadara antiquata*, Nosy-Be, Madagascar

g. *Gafrarium tumidum*, Rendrag, between Koror and Babeldaob, Palau
h. *Cerastoderma edule*, Dawlish Warren, south coast of England
i. *Anomalocardia brasiliana*, Itamaraca, Pernambuco, Brazil
j. *Chione cancellata*, Sanibel Island

Crenulate inner valve margins are visible in f through j; the other five species have smooth margins. A taxodont hinge, with many small teeth of similar size, is exhibited by e and f; the other valves illustrated have fewer teeth and show variations of heterodont hinge dentition. (Photographs by F. Dixon.)

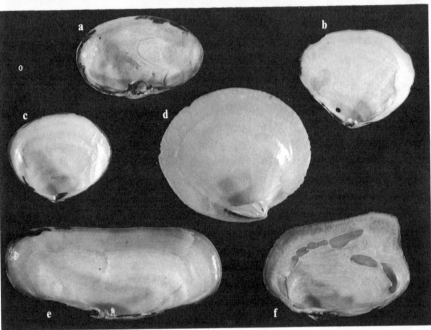

Figure 3.2 External *(above)* and internal *(below)* views of deeply infaunal clams. (Photographs by F. Dixon.)

    a. *Mya arenaria*, near Perry, Maine
    b. *Macoma nasuta*, Garrison Bay, San Juan Island, Washington
    c. *Scrobicularia plana*, Dawlish Warren, south coast of England
    d. *Codakia orbicularis*, Pear Tree Bottom, near Runaway Bay, Jamaica
    e. *Tagelus affinis*, Venado Beach, Panama
    f. *Panopea generosa*, Friday Harbor, Washington

Figure 3.3 External *(above)* and internal *(below)* views of epifaunal clams. (Photographs by F. Dixon.)

a. *Tridacna crocea*, a nestler, Ngemelis, Palau
b. *Isognomon radiatus*, Pear Tree Bottom, near Runaway Bay, Jamaica, on mangroves
c. *Argopecten irradians concentricus*, Turkey Point, Florida
d. *Arca zebra*, Discovery Bay, Jamaica
e. *Septifer bilocularis*, Rendrag, between Koror and Babeldaob, Palau

shallow burrowers capable of very rapid locomotion, the hinge can prevent dislocation even of widely gaping valves. Another method of reducing valve shear is interlocking of the valves along the ventral margin. In many epifaunal and shallowly infaunal clams the inner valve edges possess crenulations or plications, which may or may not be visible on the shell exterior as radial ribs (Figure 3.1).

Still another method of preventing valve shearing, one that is found in most attached epifaunal and many endobyssate clams, is the development of a very large posterior adductor muscle, which in oysters and forms resembling oysters is the only adductor present (Yonge, 1967; Stanley, 1972). Together with a strong hinge ligament, the adductor in these clams replaces or complements the hinge teeth in keeping valve shear to a minimum (see Figure 3.3).

Deep-burrowing clams (Figure 3.2) lack most of the adaptations used by epifaunal and shallowly infaunal clams to reduce valve shearing. They are characterized by reduced hinges (Runnegar, 1974), smooth inner valve margins, relatively small adductor muscles, and often a permanent posterior and sometimes anterior gape for the extrusion of siphons and foot respectively (Stanley, 1968, 1970). The reduction in shear-resistant features is often associated in deeply infaunal bivalves with extensive dorsal and ventral fusion of the mantle lobes which, together with the permanent gape, gives the soft parts of the clam the appearance of a U-shaped tube (Owen, 1958; Stanley, 1968). This morphology enables the clam to place the siphons near the sediment-water interface while keeping the bulk of the body in the physically stable deep layers of sediment.

It is clear, then, that many aspects of bivalve form can be understood by studying the relation of the living clam to the sediment. Life position and rate of locomotion in clams are closely correlated with overall body form and, conversely, body form is a good predictor of bivalve life habits.

Shell Form and Geography

Several geographical patterns in bivalve shell architecture suggest that physical factors associated with sediment depth, mode of life, and degree of mobility are not the only determinants of bivalve form. Carter (1968), for instance, has pointed out that radial ribbing and crenulate margins are very rare in fresh-water clams. This curious state of affairs is not merely an artifact of taxonomy, since families which in the sea contain many species with crenulate margins have a few tropical fresh-water representatives in which these shear-resistant structures are lacking (for example, in the Donacidae and Mytilidae). If stability in the sediment or (in some cases) rapid burrowing are the prime functions for this shell sculpture, then why should clams in the often unstable sediments of fast-flowing rivers and streams not be commonly endowed with these features?

There is also some evidence (Table 3.1) that the incidence of marginal crenulation among soft-bottom infaunal shallow-water clams increases from temperate to tropical latitudes. More data are needed to substantiate this point, but the pattern seems evident even within single families, including epifaunal ones. To take the epifaunal Mytilidae as an example, large temperate species such as *Mytilus edulis*, *M. californianus*, *Perna*, and *Semimytilus algosus* have smooth inner valve margins, while most species of the predominantly tropical to warm-temperate genera *Brachidontes*, *Hormomya*, and *Septifer* have interlocking shell margins with or without radial sculpture on the exterior of the valves. Many tropical oysters, spondylids, plicatulids, and other epifaunally cemented bivalves have markedly zigzag or crenulate margins, but temperate representatives are only weakly plicate or altogether smooth at the inner margin (for instance, *Crassostrea virginica* in the northeastern United States, *Ostrea edulis* in Western Europe, and *Pseudochama exogyra* in California).

In conjunction with this trend, the frequency of bivalves with a persistent posterior gape decreases quite sharply toward the tropics (Vermeij and Veil, 1978). About 30 percent of the shallow-water infaunal bivalves in cold-temperate New England and the Pacific coast of the United States are permanent gapers, while only 12 to 13 percent of the species in tropi-

Table 3.1 Incidence of crenulate or serrate inner valve margins in infaunal shallow-water bivalves. Data have been compiled from my own collections (GJV) and from species lists in the sources cited.

Locality	Species		Source
	Number	% with crenulated inner valve edge	
Cold temperate			
San Juan Island, Washington	20	25	GJV
Warm temperate			
Southern California	15	27	Peterson, 1975
Tropical			
Sanibel Island, Florida	19	53	GJV
Jamaica	41	37	Jackson, 1973
Venado Beach, Panama	19	58	GJV
Mahé, Seychelles	67	31	Taylor, 1968

cal America have a persistent posterior siphonal gape. As in the case of crenulate margins, this trend is also exemplified in single families (the Mactridae, for instance). The decline in gapers toward warmer waters does not mean that deep sediments remain unoccupied by tropical bivalves; rather, the deep-burrowing cold-water gapers are replaced in the tropics by equally deep-burrowing but tightly closing lucinaceans, which collect food by means of a more or less temporary mucus tube at the anterior end of the animal (Allen, 1958). Most tropical bivalves that retain a posterior gape are rapid burrowers such as *Ensis*, *Tagelus*, and *Solemya*.

Is there a biotic explanation for these geographical changes in form, one that is consistent with the microgeographical trends discussed in the preceding section? I believe that differences in the nature and intensity of predation can explain the overall increasing emphasis on tight valve closure toward the tropics, but supporting evidence for the most part is still anecdotal or circumstantial, or both. In the next section I review what is known about predators of clams and their methods of attack, and I shall suggest some interpretations of clam-shell architecture from the standpoint of predation.

Predation and Bivalve Shell Form

The principal mechanisms by which predators eat bivalves include crushing the shell (crustaceans, fishes, birds), prying apart the valves (asteroids, birds, octopod and gastropod molluscs, decapod crustaceans), drilling (octopods and gastropods), swallowing or enveloping whole (birds, mammals, fishes, asteroids, gastropods, sea anemones), and siphon cropping (fishes). Turbellarian flatworms such as *Stylochus* may invade the shell opening and be locally important as predators of oysters and other bivalves, but they are not considered further here.

Thickened valves and certain types of plicate sculpture may confer resistance to shell damage in bivalves. In southern California, for example, small (2 to 4 cm broad) *Cancer antennarius* in one day can eat five to seven individuals of the weakly attached, smooth, thin-shelled mussel *Mytilus edulis*, but only two to four individuals of the thicker, rougher, and more tightly adhering *M. californianus* of similar size (2 to 4 cm). Both *C. antennarius* and *Pachygrapsus crassipes* prefer these two species of *Mytilus* over *Septifer bifurcatus*, a thick-shelled, tightly attached mussel with crenate inner valve margins (Harger, 1972b).

Buttressing radial plications, seen as ridges on the outside and as furrows on the inside of the shell, are particularly characteristic of the thin shells of pectinid scallops and cardiid cockles, groups noted for their rapid swimming and jumping movements respectively. In scallops, the plications are formed of foliated calcite and provide sturdiness to a shell that must necessarily be thin so as to impede movement minimally.

Many tropical pteriomorph bivalves (such as *Pinna, Pteria, Malleus, Isognomon*) are able to retract the mantle edge a considerable distance from the shell edge (see Yonge, 1953). The whole shell, but particularly the distal portion, is rather thin and flexible and is therefore vulnerable to being frayed or pecked by fishes that encounter the exposed edges of these epifaunal and semi-infaunal bivalves. The retractibility of the mantle may well protect such clams against predation. Although *Pinna* and *Pteria* are known from the Mediterranean, the pliant and roomy shells of these pteriomorphs are not found in cold-temperate or polar waters.

As with gastropods, a good case can be made that crushing becomes more important as a source of bivalve mortality toward the tropics. Menzel and Hopkins (1955) lamented that the predatory activities of the large xanthid crab *Menippe mercenaria* and the drumfishes *Aplodinotus* and *Pogonius* make it impossible to set out oysters (*Crassostrea virginica*) in open trays for growth experiments in Louisiana, while further north such experiments can be carried out without special precautions. *M. mercenaria*, a warm-temperate species ranging north to Cape Hatteras, North Carolina, can crush adult oysters 8 cm long (see also Menzel and Nichy, 1958); while the xanthids *Neopanope sayi* and *Panopeus herbstii* and the blue portunid crab *Callinectes sapidus*, which range north to Cape Cod and beyond, are only important as predators of young oysters less than 5 cm long (Lunz, 1947; Carriker, 1951; Landers, 1954; Menzel and Hopkins, 1955; McDermott, 1960; Darnell, 1961; Tagatz, 1968).

Fishes that crush bivalves in their jaws (rays and puffers) or by means of pharyngeal teeth (wrasses) are primarily tropical or warm temperate in distribution; the few species that do inhabit cold-temperate waters, such as the eastern North American wrasse *Tautoga onitis*, usually eat only small bivalves (see, for example, Olla et al., 1974). Other bivalve crushers, including stomatopods, spiny lobsters, and turtles, are also largely tropical predators. Important temperate exceptions include nephropid lobsters of the genera *Homarus* and *Nephrops* (de Figueiredo and Thomas, 1967; Squires, 1970), some palinurid lobsters (Hickman, 1972), and the twenty-three species of the crab genus *Cancer* (Nations, 1975). These latter crabs have been shown to be capable of consuming large mussels (Mytilidae) in Europe and on the Pacific coast of North America (Ebling et al., 1964; Muntz et al., 1965; Harger, 1972a).

The development of spines, coarse radial and concentric ribs, and thick shells among many tropical and warm-temperate epifaunal and shallowly

infaunal bivalves may be an adaptive response to the increased importance of crushing predators, but this relationship has never been experimentally investigated. If these types of sculpture are adaptations against
predation, then their rarity in fresh-water clams would imply that crushing is a relatively unimportant factor in the selective environment of
fresh-water bivalves. Again, this needs confirmation.

Birds appear to be a major cause of mortality of many shallow-water
bivalves and other animals on temperate and polar shores. Drinnan
(1957) and Davidson (1967) estimate that the oystercatcher *Haematopus
ostralegus* accounts for 21 percent to perhaps as much as 75 percent of
the winter mortality of the cockle *Cerastoderma edule* in Britain. Hughes
(1970) concluded that in North Wales the oystercatcher accounts for
most of the 10 percent mortality of one-year and older *Scrobicularia
plana*, an infaunal deposit-feeder. Oystercatchers are also known to
attack mussels and other bivalves of rocky and sandy shores (Webster,
1941; Davidson, 1967; Heppleston, 1971; Dare and Mercer, 1973; Baker,
1974; Hartwick, 1976).

When attacking mussels and cockles, the oystercatcher may hammer
the prey by directing blows with the bill at the commissure between the
valves, or it may jab the bill between the opened valves of a clam under
water while the latter is feeding (Drinnan, 1957; Norton-Griffiths, 1967;
Heppleston, 1971; Baker, 1974). Clams taken by the latter method are
larger than those that can be successfully hammered. The critical size
above which bivalves become immune to hammering depends both on
prey characteristics (shell thickness, strength of byssal attachment) and
on the hardness of the sediment upon which the clam rests as it is being
hammered (Drinnan, 1957; Norton-Griffiths, 1967). The observations by
Norton-Griffiths suggest that interlocking, or at least tightly closed,
valves may offer greater resistance to hammering than do valves that
spring partly open. It should be noted, however, that the cockle, a
radially ribbed clam with strongly crenulated margins (see Figure 3.1), is a
favorite food of the oystercatcher, which can harvest specimens up to 35
mm in length (Drinnan, 1957; Hancock, 1972).

Gulls (genus *Larus*) are known to break bivalves and other shelled
invertebrates by dropping them upon rocks or other hard ground from
the air. In a collection of shells from a gull drop zone near San Juan
Island in Washington State, Spight (1976c) found that 34 percent of the
shells had been broken. Among the shells regularly fractured in this way
by *Larus glaucescens* in the Puget Sound area is the large intertidal cockle
Clinocardium nuttalli, which at a length of 10 cm can provide a gull with
a sizable meal.

Notwithstanding the various temperate crabs, lobsters, and birds, it seems that few subtidal clams at high latitudes are regularly subjected to crushing predation. Even in the intertidal zone the impact of oyster-catchers and crabs is seasonal, and on many coasts gulls and wading birds are important as predators in only a few localities.

From the work on oystercatchers it is evident that large prey may be attacked in a different way than smaller prey. If the prey cannot be crushed, predators attempt to pry the valves apart in order to obtain the soft parts. When a prey clam is near the maximum size of vulnerability to predation, the portunid crabs *Callinectes sapidus, Carcinus maenas*, and *Macropipus puber* will attempt to insert one or both claws between the shell valves of oysters and mussels (Carriker, 1951; Ebling et al., 1964; Seed, 1969b). The same technique has been observed by Smith (1975) as the ghost crab *Ocypode ceratophthalmus* attempted to eat large beach clams *(Donax faba)* in Kenya.

Valve prying is the rule for a number of other predators on clams. For example, several gastropods are known to use the outer shell lip in forc-ing apart the valves of their prey. Variations of this behavior have been described in the buccinid *Buccinum undatum* (Nielsen, 1975), the melongenid genus *Busycon* (Magalhaes, 1948; Carriker, 1951; Menzel and Nichy, 1958; Paine, 1962), some large Fasciolariidae (Wells, 1958b; Paine, 1963a, b), and some normally drilling muricids when attacking very young mussels (Seed, 1969b). In *Busycon*, the thick-shelled, strong-lipped species *(B. carica* and *B. contrarium)* specialize on a diet of tightly closing bivalves, while the thin-shelled species with weaker lips *(B. cana-liculatum* and *B. spiratum)* are restricted to thin-shelled or gaping clams (Paine, 1962).

Large muricids of the genus *Muricanthus* in the Gulf of Mexico and western tropical America feed on *Tagelus* without resorting to drilling, since this fast-burrowing bivalve has small but permanent anterior and posterior gapes through which the snail can insert the proboscis directly (Stump, 1975). Tightly closing infaunal clams are drilled at the shell commissure by *Muricanthus* (Wells, 1958a; Radwin and Wells, 1968) (see Figure 3.4), a habit also practiced by *Thais haysae* when drilling oysters in the northern Gulf of Mexico (Chapman, 1955), by *Acanthina spirata* in California (Hemingway, 1975), and by *Ceratostoma foliatum* when preying on the tightly closing venerid *Protothaca staminea* on the coast of Oregon (Kent, 1978). On the Pacific coast of Panama, E. Zipser and I found that *Muricanthus ambiguus* attacked virtually all species of bivalve it was offered except the thick-shelled, radially ribbed, and mar-ginally jagged *Cardita laticostata*. Other thick-shelled clams with finer marginal denticulation, including species of *Chione* and *Protothaca*, were successfully attacked on the anteroventral or posterioventral edge (Figure 3.4). It would be interesting to learn which architectural features, if any, confer resistance to edge-drilling muricids.

Certain tropical members of the mesogastropod genus *Cymatium* (*C. muricinum* and *C. pileare*, for example) attack bivalves by inserting the proboscis between the valve margins and anesthetizing the prey with a secretion from the salivary gland (Houbrick and Fretter, 1969). No damage is done to the prey shell.

Octopod molluscs and forcipulate sea stars apparently use suction to force apart the valves of a prey bivalve (Carter, 1968; Fotheringham, 1974). Octopods rasp out the soft flesh, but asteroids extrude their stomach between the valves to effect digestion. In most cases the prey shell is left intact by the sea star after feeding, but the large subtidal *Orthasterias koehleri* in the Puget Sound region apparently can rasp a hole at the ventral valve commissure of the large, thick-shelled, and tightly closing venerid *Humilaria kennerleyi* (Mauzey et al., 1968).

Several features of bivalve shells in principle could reduce the impact of prying or shell-invading predators. A plicate or wavy commissure might make it difficult for crabs, birds, and snails to wedge the bivalve open, particularly if the valves do not gape widely while the clam is feeding. Rudwick (1964) has shown that the linear distance between the opposing valve edges is smaller along most of the commissure in a bivalve with marginal plication than in one of similar size and with similar angular gape that has smooth shell margins. Only at the crests and troughs of the wavy edge would the linear distance between the two valves be greater, but the size of these troughs and crests can be minimized if the plications are strongly angular rather than smoothly rounded (see Figure 3.3). It is interesting in this respect that certain clams, such as tridacnids and cardiids with markedly plicate margins, gape widely while feeding. However, as already noted, oystercatchers prey heavily on the cockle *Cerastoderma edule*; and I have seen *Cymatium muricinum* in Guam feed successfully on another cockle, *Fragum fragum*, by inserting the proboscis through the saw-toothed posterior margin. Moreover, these cockles and other bivalves have jumping or swimming responses to predatory gastropods and asteroids (see Ansell, 1969; Bloom, 1975; Nielsen, 1975), and it is doubtful that these clams have adapted morphologically to avoid being eaten by these slow predators. Morphological adaptations may still be expected in sedentary or very slow-moving clams and should be the rule in all bivalves when the major predators are active crabs, fishes, or birds.

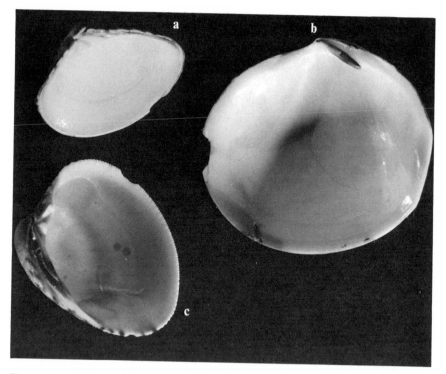

Figure 3.4 Valves drilled at the edge by *Muricanthus radix* on the Pacific coast of Panama. (Photograph by F. Dixon.)
a. *Chione subrugosa*   b. *Dosinia dunkeri*   c. *Protothaca grata*

Carter (1968) has suggested that shear-resistant crenulate valve margins may render certain clams immune to starfishes and other predators that pull apart the valves to feed. In order for this to be an effective means of avoiding predators, the forces upon the shell should be primarily parallel rather than perpendicular to the commissure. It is doubtful that sea stars resort to such shearing when attacking clams, though this point requires experimental confirmation. Moreover, the incidence of crenate margins increases toward the tropics, while the impact and number of species of bivalve-eating forcipulate sea stars sharply decrease. The only sea star known to attack clams extraorally in the tropics is the Indo-West-Pacific *Acanthaster brevispinus* (Lucas and Jones, 1976).

A more plausible, though equally untested, hypothesis concerning the occurrence of crenate margins is that they enable the bivalve to close more tightly in the presence of a potential predator, thereby perhaps depriving the enemy of important chemical cues. Carriker and Van Zandt (1972) have shown conclusively that the eastern North American muricid oyster drill *Urosalpinx cinerea* finds its oyster prey by chemoreception, and that it will not commence drilling if the oyster remains tightly closed. Only when the valves are slightly gaping and oyster metabolites escape into the surrounding water is the snail aware of the prey. The smaller gape associated with a crenate margin could in some cases minimize the dispersal of telltale chemical stimuli to attract prospective predators. Permanently gaping clams could, of course, always be chemically detected by a predator, and the equatorward reduction in the relative number of gaping species may be further testimony to the increased importance of tight valve closure (Vermeij and Veil, 1978).

Still another adaptation that reduces predation by shell invaders is the development of very large, powerful adductor muscles. Hancock (1965) has experimentally shown that mussels *(Mytilus edulis)* from Denmark with relatively large posterior adductors are eaten at a much slower rate by European sea stars *(Asterias rubens)* than are British mussels, whose adductors are relatively smaller. In the northeast Pacific, intertidal *Mytilus californianus* can attain sizes too large for successful predation by *Pisaster ochraceus* (Paine, 1976b), and it is likely that the size limit is imposed by the size and power of the posterior adductor muscle. Subtidally, however, Paine (1976a) has found very large (2,600 g) *Pisaster* capable of overpowering even the largest (255 mm long) *M. californianus*; but again, large mussels are consumed at a slower rate than are smaller individuals (see also Hancock, 1965; Paine, 1974; Seed, 1969b).

It must be pointed out that some bivalves have very few morphological adaptations to predation even in the absence of behavioral escape responses. The deposit-feeding Tellinidae for the most part are thin-shelled and highly vulnerable to all kinds of predators; like many small suspension-feeding mactrids, semelids, lyonsiids, pandorids, and cuspidariids, they have small, fragile shells with smooth margins and weak hinges and offer only token resistance to even the clumsiest and least specialized temperate predators. Such species are best regarded as weeds, which survive the depredations of their enemies by high reproductive rates.

Predation by drilling is a quantitatively important source of mortality for many bivalves, especially small, thin-shelled species (Ansell, 1960; Reyment, 1967; Jackson, 1972; Stump, 1975; Thomas, 1976). In thick-shelled fossil Neogene *Glycymeris* from the southeastern United States, for example, Thomas (1976) finds that from 0 to 60 percent of the valves are drilled, mostly by naticids. Among dead shells recovered from a lagoon in Nigeria, Reyment (1967) found the frequency of drilled valves to be 58 percent for *Ostrea* species, 36 percent for the cockle *Cardium papillosum*, and about 25 percent for *Nucula nitida* and *Iphigenia rostrata*. Drilling seems to be a minor source of mortality for other species in the sample: 7 percent for *Dosinia isocardia*, 5.5 percent for *Crassatella paeteli*, and 3 percent for *Venerupis decussata* and *Mactra nitida*. In Scotland, *Natica alderi* is responsible for about 45 percent of the mortality of first-year *Venus striatula* (= *V. gallina*), but cannot take shells greater than 15 mm in length. Ansell (1960) found only a 5 percent mortality from drilling in populations of the Australian beach clam *Notospisula*; some size classes, however, are much more affected than others.

No morphological features of bivalves have been definitely associated with resistance to or repulsion of drilling gastropods. For example, I can find no consistent patterns in shell sculpture relative to the frequency of drilled valves in the data given by Reyment (1967), Taylor (1970), and Stump (1975). The only characteristic known to inhibit drilling is a greatly thickened valve; either the predator cannot handle thick-shelled prey, as is the case with certain muricids from Florida (Radwin and Wells, 1968) with an unspecialized radula that can only drill through the relatively thin *Brachidontes exustus*, or thin-shelled prey are preferred over thicker-shelled clams, as in the naticid *Polinices duplicatus* in eastern North America (Carriker, 1951).

It is unclear to what extent octopods employ drilling when they are feeding on bivalves. In Guam I observed an *Octopus* (probably *O. cyaneus*) take the thin-shelled *Modiolus auriculatus* in the laboratory without drilling. Fotheringham (1974) assumed that all *Protothaca staminea* eaten by *Octopus bimaculatus* and *O. bimaculoides* were drilled, and from that assumption concluded that over 50 percent of the mortality of this clam near San Diego was attributable to predation by *Octopus*.

Features that might prevent a bivalve from being swallowed or enveloped whole might include a spiny or otherwise sharp external sculpture, and tight valve closure. Careful studies of the intraorally digesting sea star *Astropecten irregularis* in Denmark by Christensen (1970) reveal that tightly closing venerid and corbulid clams take much longer to be digested than do species with a less efficient seal and a lower tolerance for reduced oxygen (Mactridae, Cardiidae, Donacidae). For example, young (0.8 to 1.3 mm long) *Spisula subtruncata* are eaten at a rate of ten per sea star per day and are retained in the stomach for an average of nine hours. By contrast, the tightly closing, less oxygen-demanding young *Venus gallina* (0.8 to 1.2 mm long) are eaten at a rate of 0.2 per sea star per day and are retained for an average of sixty-one hours. The Mediterranean *Astropecten arantiacus* may retain *Corbula gibba* in its stomach for two to three weeks before this tightly closing bivalve is released, often without being digested (Massé, 1975).

Several volutid, melongenid, and olivid snails also feed by asphyxiating their prey; they store the prey in the large foot or swallow them whole (Olsson, 1956; Marcus and Marcus, 1959; Ponder, 1970, 1973). Aquarium observations on *Oliva sayana* in Florida showed that this sand-dwelling predator successfully attacks smooth-shelled bivalves such as *Donax variabilis* and *Laevicardium* spp., but tends to avoid such roughly sculptured clams as the common venerid *Chione cancellata* (Olsson and Crovo, 1968).

Many wading birds (other than oystercatchers) are known to feed in part on bivalves and usually swallow them whole. In general, the prey are small epifaunal or shallowly infaunal clams (see Reeder, 1951; Recher, 1966; Goss-Custard, 1969; Seed, 1969b; Davidson, 1971; Burton, 1974). Eiders and other ducks also take large quantities of mussels and other small prey clams at or near the sediment surface in the intertidal and shallow subtidal zone (Olney and Mills, 1963; Olney, 1965; Dunthorn, 1971; Stott and Olson, 1973; Canton et al., 1974).

Many fishes swallow bivalves whole, especially in temperate waters. Since most fishes obtain their prey by a suction or gulping of the mouth, and since the suction force that can be produced per unit area is inversely related to mouth size, most fishes are restricted to small bivalve prey (Alexander, 1967).

Certain flatfishes (Pleuronectiformes) and cod (Gadidae) crop the extruded siphons of bivalves (Nesis, 1965; Edwards and Steele, 1968; Edwards et al., 1970; Trevallion et al., 1970; Braber and de Groot, 1973; Deniel, 1974). These same fishes may swallow small individuals whole. As Trevallion and colleagues have conclusively shown, siphon cropping often leads to rapid siphon regeneration and does not necessarily kill the clam. Present inadequate evidence suggests that siphon cropping is largely a temperate phenomenon, but the abundance of fishes that nip and graze tropical surfaces probably implies that cropping could be equally important on warmer shores.

Evidence is accumulating in the parasitological literature that bivalves and other animals infected by parasites differ from uninfected individuals in behavior and that they may be more available to predators. This topic has been reviewed by Holmes and Bethel (1972). The deposit-feeding tellinid bivalve *Macoma balthica*, for example, is normally buried in the sediment; but individuals infected with a gymnophalid trematode often lie on the surface of the mud, where they are more easily spotted and captured by oystercatchers (Hulscher, 1973). Feare (1971a) observed that the proportion of the snail *Thais lapillus* infected by the trematode *Parorchis acanthus* on exposed rocky surfaces in Yorkshire is 13 percent, a figure identical to the incidence of parasitized *Thais* in the diet of oystercatchers. Individuals of *Thais* aggregated in crevices showed a much lower incidence of infection (1 percent) and were found to be less vulnerable to predation by oystercatchers. Parasites therefore may influence locomotion, habitat preference, and other characteristics relevant to the vulnerability of a prey individual to predation. This means that adaptations such as tight valve closure that would be effective defenses for uninfected prey might become unsatisfactory upon infection by a debilitating parasite. Studies evaluating the effectiveness of antipredatory adaptations therefore need to take into account the effects of parasitism.

Conclusions

If the foregoing data and interpretations about bivalve shell form are correct and representative, it can be concluded that tropical clams have generally more predator-resistant shells than do temperate species, although the contrast may not be as great as in the primarily epifaunal gastropods considered in Chapter 2. The most pervasive latitudinal trend among bivalves is an increasing equatorward emphasis on sturdy valves that completely enclose the soft parts. This trend, together with the strong sculpture often seen in tropical epifaunal clams, reflects greater resistance to predators that crush, drill, or forcibly pry open the valves. These trends affect not only those bivalves that can burrow away from certain of their slower enemies, but also sedentary forms incapable of behavioral escape when under attack. Correspondingly, the tropical predators are more fully endowed with specialized shell-destroying equipment than most of their temperate counterparts and may in general characterize themselves as, "We shell overcome."

Predation intensity seems to decline toward the deep sea and with increasing sediment depth. Deep-sea fishes in the Gulf of Mexico apparently ingest shelled invertebrates only rarely, and when doing so swallow them whole (Bright, 1970). Very few predators can exploit deep-burrowing bivalves; they include some drilling naticids and some large forcipulate sea stars (for instance, the Northeast Pacific *Pisaster brevispinus* and *Pycnopodia helianthoides*) (Mauzey et al., 1968). Even such long-billed birds as the curlew *(Numenius)* are incapable of extracting large, deeply buried prey (Edington et al., 1973; Burton, 1974; Woodlin, 1978).

Together with the downshore increase in predator resistance described earlier for gastropods, these patterns suggest a proliferation of morphological antipredatory adaptations toward habitats or regions of reduced physiological stress. If the arguments in Chapter 1 are correct, then this greater emphasis on predation-related adaptations is in part related to the easing of physiological limitations toward favorable environments.

CHAPTER 4

Sessile Organisms

I N THE preceding two chapters I have pointed out that one of the reasons for a decline in morphological antipredatory features among molluscs toward higher latitudes could be the increasing importance of behavioral escape responses to slower predators. Although such an explanation is difficult to refute for mobile gastropods and even for most infaunal bivalves, it is unacceptable for sedentary epifaunal bivalves and other sessile marine plants and animals. Nevertheless, these sedentary organisms exhibit striking gradients in form and chemistry that apparently reflect changes both in productivity (availability of nutrients in the water column) and in grazing or predation pressure. In the present chapter I shall describe and interpret these gradients in selected epifaunal sessile organisms, emphasizing the groups that have received worldwide ecological attention: algae, scleractinian corals, barnacles, brachiopods, sponges, and holothurians.

Benthic Algae

To the casual observer, one of the most striking geographical patterns in the low intertidal and shallow subtidal zones on rocky coasts is the change from a predominance of algae in temperate and polar regions to one of photosynthesizing animals in the tropics. This pattern results both from the dramatic increase in species of photosynthesizing animals toward the tropics and from a corresponding though less dramatic reduction in species of algae (Wells, 1957; Bakus, 1969; Stehli and Wells, 1971; Veron, 1974). For example, West Indian waters harbor about fifty reef-building (hermatypic) scleractinian coral species (Glynn, 1973b; Porter, 1974a) as well as many zooxanthellate gorgonians (Kinzie, 1973), sea anemones, zoanthids, and hydrozoan corals. (For a review of photosynthesizing animals see McLaughlin and Zahl, 1966.) Various compilations reveal that less than four hundred species of algae (not counting the blue greens) coexist regionally in Atlantic tropical America (see Vroman, 1968; Bakus, 1969). By contrast, two species of sea anemone (both of the genus *Anthopleura*) are the only conspicuous photosynthesizing animals on the cold-temperate coasts of Washington state and British Columbia, while more than five hundred species of algae inhabit this part of the Northeast Pacific (Bakus, 1969). The diversity of algae is even higher in central California, Great Britain, and New Zealand and reaches a dizzying 1,010 species in the remarkable marine flora of South Australia (review in Bakus, 1969). The trend of increasing diversity toward higher latitudes is most spectacular among the brown algae (Phaeophyta), less so among the red algae (Rhodophyta), and least among the green algae (Chlorophyta).

Superimposed on the diversity gradient is a marked decrease in algal body size toward the tropics. This decrease in size, which affects most groups of macroscopic algae, serves to accentuate the contrast between tropical and higher-latitude shores. The infralittoral fringe on most temperate and polar coasts is dominated by large brown algae that belong to the orders Laminariales (kelps) and Fucales. The fronds of these plants are generally long and strap-like, being usually fleshy in texture (Figure 4.1). Higher on most shores these luxuriant canopy plants are replaced by smaller fucoids. These fucoids are particularly well developed on the coasts of the Northern Hemisphere, but on the whole are rarer in the Southern Hemisphere (see, for example, Stephenson, 1944, 1948; Knox, 1960). In the tropics and subtropics the low-shore kelps are replaced by very much smaller fucoids of such genera as *Padina*, *Sargassum*, *Turbinaria*, and *Dictyota*, with fronds that are often tough and leathery rather than fleshy (Figure 4.1). Aside from a few encrusting forms such as *Ralfsia*, brown algae are not found in the middle intertidal zone on tropical coasts.

Red algae exhibit a similar though rather less conspicuous decrease in size toward the equator. Throughout the world the rhodophytes are important understory species, and encrusting calcareous corallines are characteristic of nearly all rocky shores. The greatest development of these wave-resistant encrusters, however, is found on the wave-exposed shores of tropical fringing and barrier reefs, where they form intertidal algal ridges. Small middle to high intertidal red algae are distributed the world over; like the fleshy green algae *(Enteromorpha, Ulva,* and *Chaeto-morpha)*, they change little in appearance with latitude. The larger green algae of the lower shore and shallow subtidal zone, including the cal-careous Codiaceae and Dasycladaceae as well as coenocytic forms, become increasingly important as the equator is approached.

The decrease in plant size toward the tropics and the increase in impor-tance of zooxanthellate animals may be related to patterns of productivity (Bakus, 1969). Nutrient levels in most tropical waters are low and con-stant, whereas in areas of upwelling and along most temperate and polar shores the surface waters are rich in nutrients for part or all of the year. Nutrient levels in these colder areas apparently are sufficient to maintain plants of large size and of high surface-to-volume ratio, but in the tropics such large plants may exceed the nutrient capacity of the water. More-over, high nutrient levels may permit very rapid growth that is not attain-able in tropical waters. Instead, photosynthesis may be supplemented to various degrees with other sources of food that are filtered out of the water column or actively ensnared by specialized structures. Reef-build-ing scleractinian corals, for example, entrap zooplankton or small demer-sile animals such as polychaetes at night by use of the tentacles, but the degree of reliance upon such animal food seems to vary widely. Porter (1976) has speculated that large-polyped corals such as the West Indian *Mussa, Favia,* and *Isophyllia* rely more heavily on animal food than do such small-polyped forms as *Porites* (see also Porter, 1974b; Johannes and Tepley, 1974).

It is tempting to hypothesize, as have Bakus (1969) and Earle (1972b), that the increasing emphasis toward the tropics on calcification, an encrusting growth habit, and perhaps chemical deterrents among photo-synthesizers reflects a more intense grazing pressure than in the cooler temperate and polar zones; yet the effect of productivity cannot be elim-inated. On cold shores most of the characteristics interpretable as adapta-tions against herbivores are typical of relatively small, low-growing spe-cies and tend to be absent among the fast-growing canopy algae (Paine and Vadas, 1969b). Thus, calcification is characteristic of encrusting or low-growing articulated coralline red algae; sulfuric acid is produced by the brown alga *Desmarestia;* and the encrusting habit has been developed among numerous understory red and even some brown algae (such as

a

b

c

d  e

f g

Figure 4.1 Impressions of temperate and tropical marine plants and plant-like animals. (Photographs by J. W. Porter.)
a. A kelp bed at 1 m depth, California
b. Low intertidal shore at Montemar, Chile, showing the kelp *Lessonia nigrescens*
c. A *Porolithon* algal ridge at Moro Tupo, San Blas Islands, Panama
d. Calcareous *Halimeda* at 4 m depth in the Cayman Islands, West Indies
e. A calcareous *Penicillus* at 3 m depth in a *Thalassia* bed, Cayman Islands
f. The coral *Acropora palmata* growing as parallel fingers on a fore-reef slope, north coast of Jamaica
g. The coral *A. palmata* growing in a low-energy environment at 2 m depth, San Blas Islands

Ralfsia). Presumably, the greater biological defense of these seaweeds compensates for, or is in any case associated with, a slower rate of growth and a higher probability of being devoured completely by grazing animals. By contrast, large kelps can continually replace the portions of their fronds that are consumed by herbivores and maintain themselves, often for many years, by high rates of biomass turnover (Mann, 1973). The greater incidence of calcification among tropical red and green algae, and the possibly common occurrence in the tropics of chemically noxious algae (*Caulerpa*, *Avrainvillea*, and others) and of nematocyst-bearing cnidarians, might reflect nothing more than the differential elimination of fast-growing large species that depend on high nutrient levels in the water column for their continued survival in the face of heavy grazing pressure. In fact, while fast-growing algae do occur on tropical shores, they tend to be weedy and ephemeral and cannot attain a size sufficient to withstand the onslaught of grazers.

A disturbing, though perhaps minor, complication of this interpretation is the existence of certain tropical cnidarians that lack stinging nematocysts and obtain their food exclusively by means of photosynthesis of their associated zooxanthellae. These forms include certain zoanthids and the octocorals *Xenia* and *Clavularia* (Goreau et al., 1971). Functionally, these animals are exactly like benthic algae in that they obtain inorganic nutrients from the water column. This is a peculiar adaptation, for two reasons, for animals in a low-productivity environment. First, animal food in the form of active zooplankton would seem to be a convenient if not necessary food source in an environment where nutrient levels are low and where many other photosynthesizers have acquired important roles in the community by exploiting animal foods. Second, stinging nematocysts, thought to contribute heavily to the remarkable immunity of cnidarians to predation, would be useful for slow-growing animals in an unproductive environment. It is possible that the cnidarian "plants" have some other chemical defense against potential grazers, since they appear to be exploited by very few herbivores (see Hiatt and Strassburg, 1960; Randall, 1967). In any case the role of nematocysts in defense, and the role of animal food in supplementing the diet of photosynthesizing animals in nutrient-deficient tropical waters, probably require clarification and reexamination.

Grazing

Even if the increasing incidence of antiherbivore adaptations toward the tropics is related to the decline in productivity, this trend is surely exaggerated by the latitudinal change in the nature and the sources of grazing. On colder shores, most herbivores are relatively sluggish animals — or else they are active only at certain times of the year. In the tropics and subtropics, on the other hand, these slow animals are joined, and to some extent replaced, by active herbivores grazing the year round. What follows is a brief and speculative review of grazing, intended to expose shortcomings in our data as well as formulate some preliminary generalizations.

The most important marine herbivores are found among teleost fishes, echinoids (sea urchins), molluscs (chitons and gastropods), decapod crustaceans (some crabs, hermit crabs, and spiny lobsters), reptiles (turtles and lizards), birds, and mammals. Certain tube-dwelling nereid polychaetes are known to hold onto, garden, and feed upon drifting green algae such as *Ulva* and *Enteromorpha*, and this herbivory is probably of considerable importance in some intertidal communities on muddy sand in the Puget Sound region (Woodin, 1974); how widespread this habit is cannot yet be properly evaluated.

Teleost fishes have an enormous impact on benthic algae and marine grasses, but this intensive grazing is largely limited to the tropics (Bakus, 1964, 1966, 1967, 1969; Earle, 1972a, b; Wonders, 1977). The dearth of algae on tropical coasts and the immediate, rapid growth of algal mats in fish-exclusion cages (Stephenson and Searles, 1960; Stephenson, 1961; Randall, 1961; Bakus, 1967; Earle, 1972b; John and Pople, 1973) suggest that teleosts exploit much of the algal biomass produced on tropical reefs. After a period of thirty days Vine (1974) found more than a tenfold difference in biomass between the algae on plates placed in cages at a depth of 2.5 m and those on control plates grazed by fishes in the Red Sea. Birkeland (1977) found that caged plexiglass plates set in coral reefs and coral communities on the two coasts of Panama grew from 1.5 to 2.3 times as much biomass after 133 days as plates available for grazing. Vine (1974) has shown that in the Red Sea, grazing intensity falls sharply with increasing water depth.

Territoriality of adult reef fishes creates considerable spatial variation in grazing intensity. This is well illustrated by Vine's (1974) work on the Red Sea damselfish *Pomacentrus lividus*. This small (10 cm long) territorial fish excludes other much larger fishes from its territory in shallow water, thereby protecting algae within the territory from being grazed as intensely as they are outside the defended area. The greater growth of filamentous and fleshy algae in the territory, in turn, provides the damselfish with a larger food supply, and inhibits settlement and growth of the encrusting calcareous red algae (lithothamnia) so conspicuous elsewhere on open surfaces of the reef (Vine, 1974). On the south coast of Curaçao in the West Indies, Wonders (1977) has also found that encrusting lithothamnia can settle and grow only in the presence of grazing fishes. When the latter are excluded from an area, filamentous and fleshy algae rapidly smother encrusting and endolithic forms, either by shading them or by acting as sediment traps.

There is some evidence that fishes are selective in their herbivory. Earle (1972b) reports that in the Virgin Islands the calcareous green algae *Avrainvillea* and *Penicillus* generally are not consumed by fishes, although *Holacanthus arcuatus* does include them in its diet. Among surgeonfishes (Acanthuridae) in Hawaii, only members of the genus *Naso* appear capable of eating such tough, leathery algae as *Sargassum*, *Dictyota*, and *Padina*; while *Turbinaria*, *Udotea*, *Chnoospora*, and *Dictyosphaeria* are rarely eaten by any surgeonfish (Jones, 1968). Similar observations have been made by Randall (1967), who pointed out that the only West Indian fishes capable of eating *Sargassum* are the two species of sea chub *(Kyphosus)*, two species of the angelfish genus *Pomacanthus* (Chaetodontidae), and the triggerfish *Melichthys niger* (Balistidae). Encrusting calcareous red algae are available to some parrotfishes (Scaridae), which rasp the substratum (Figure 4.2), but not to most other herbivorous fishes. Rabbitfishes (Siganidae) in the Indo-West-Pacific are limited to such tender algae as *Enteromorpha* and *Ulva* (Hiatt and Strassburg, 1960).

While temperate-zone fishes often take significant quantities of algae in their diets, few if any species appear to specialize on plant food. In the kelp communities of southern California, at least three fishes *(Girella nigricans, Medialuna californiensis,* and *Hermosilla azurea)* feed on algae, especially filamentous forms (Quast, 1968). The large clingfish *Sicyases sanguineus,* which ranges from southern Peru to southern Chile, has an extraordinarily catholic diet, and about half of the forty-eight individuals examined by Paine and Palmer (1978) contained large amounts of red, green, and brown algae in their stomachs. About 40 percent of the diet of the flounder *Pseudopleuronectes americanus* feeding intertidally and subtidally in Passamaquoddy Bay, New Brunswick, during summer months, consists of tender algae such as *Enteromorpha, Cladophora,* and *Acrosiphonia* (Wells et al., 1973). No species seems to feed to any significant degree on large kelps (Earle, 1972b). Moreover, I know of no temperate fishes that take sea grasses for food, despite the exploitation of these plants by surgeonfishes and other teleosts in the tropics (Randall, 1965).

Echinoids are important herbivores on nearly all coasts. While the number of sea-urchin species is nowhere very large (shallow-water hard

Figure 4.2 A parrotfish nipping algae at Grand Cayman, West Indies. (Photograph by J. W. Porter.)

bottoms in the West Indies support only six, for instance), echinoids form well-marked zones in the low intertidal and shallow subtidal zones of both temperate and tropical shores; they can scrape the rocks bare of algae even where production is very large. In St. Margaret's Bay, Nova Scotia, herbivores (mainly the urchin *Strongylocentrotus droebachiensis*) remove about 10 percent of algal production in the bay as a whole (Miller and Mann, 1973; Mann, 1973), but the biomass of algae in the presence of the urchins may be as much as an order of magnitude less than that in their absence. (See Kitching and Ebling, 1961, for similar results in Ireland.) In shallow tide pools on the open coast of Washington, Paine and Vadas (1969a) have shown that experimental removal of *S. purpuratus* results in an increase in the number of algal species because of the greater survival of newly settling plants. Later there is a diversity decrease owing to the competitive exclusion of many algae by the intertidal canopy species *Hedophyllum sessile* or the subtidal *Laminaria complanata*. A similar pattern occurs through the natural removal of urchins from a local area by the sea star, *Pycnopodia helianthoides*, which feeds on urchins and effects the mass escape of other urchins nearby (Paine and Vadas, 1969a; Dayton, 1975a). Working in deeper pools on the same coast, Dayton found no eventual competitive exclusion by certain canopy species when he removed *S. purpuratus*. In the Virgin Islands, Sammarco and associates (1974) showed that experimental removal of *Diadema antillarum* resulted in a tenfold increase in biomass, and in the relative increase in abundance of large brown algae *(Padina, Dictyotopteris,* and *Turbinaria)* and green algae (the highly calcified *Halimeda opuntia*).

Tropical echinoids appear to be important herbivores of marine grasses (Moore and McPherson, 1965; Randall, 1965; Earle, 1972b). I know of no case where temperate marine angiosperms are eaten by sea urchins.

Like echinoderms, grazing molluscs occur at all latitudes. Several groups of limpet-like molluscs, notably the chitons, abalones (Haliotidae), and acmaeid and patellid limpets, have diversity maxima in temperate latitudes (Knox, 1963; Vermeij, 1973b); but many other groups, including keyhole limpets (Fissurellidae), periwinkles (Littorinidae), nerites (Neritidae), Cerithiidae, pulmonate limpets (Siphonariidae), and Turbinidae, are most diversified in the tropics. Top shells (Trochidae) appear to be well represented on all shores where winter freezing is rare or absent.

Data on the impact of herbivorous molluscs are still scanty, but it appears that the greatest influence of grazing gastropods and chitons is felt in the intertidal zone. Removal of chitons, limpets, and littorines from experimental plots generally results in a marked increase in algal biomass, and often in an initial rise in abundance of fugitive species such as diatoms and the green algae *Ulva* and *Enteromorpha* (Jones, 1948; Stephenson and Searles, 1960; Castenholz, 1961; Vegas, 1963; May et al., 1970; Breen, 1972; Haven, 1973). Most studies suggest that molluscs that feed on microalgae are unselective in the species they exploit (Castenholz, 1961; Vegas, 1963; Dahl, 1964; Foster, 1964; Haven, 1971, 1973; Dayton, 1975a). In her study of middle intertidal limpets and high intertidal littorines in Washington, however, Nicotri (1977) found that canopy diatoms are taken relatively more often than are understory forms, and that blue-green algae are very little exploited. Taylor (1971), moreover, has suggested that tropical acmaeid limpets can crop algae to a lower height than can co-occurring siphonariid limpets, and that certain rock-boring algae are available to acmaeids but not to other grazing snails.

The large Californian acmaeid limpet *Lottia gigantea* is reminiscent of the damselfish *Pomacentrus lividus* in defending a territory of algal turf from intrusions by closer-cropping species of *Acmaea* (Stimson, 1970, 1973). Similar territoriality has been found in other large, low intertidal limpets in South Africa (Branch, 1976); and some territorial interactions have been seen among co-occurring middle intertidal chitons and limpets in central California (Conner, 1975).

Various more specialized molluscan herbivores may be found in the low intertidal zone, but their effect on standing crop or composition of the algae seems to be small. For example, Dayton (1975a) found that the large chiton *Katharina tunicata* in the Northeast Pacific feeds mostly on the competitively dominant brown alga *Hedophyllum sessile*, but does not prevent its prey from eventual predominance. A variety of temperate chitons and limpets feeds more or less exclusively on encrusting red algae (lithothamnia) (see Dayton, 1975a; Demopulos, 1975; Branch, 1975a; Simpson, 1976), yet these lithothamnia are a characteristic feature of low intertidal zones throughout the world wherever turbidity is not excessive (Hedgpeth, 1969b). In the tropical Pacific, large snails of the genus *Turbo* take mostly filamentous and encrusting forms, but apparently are unable to feed on large erect algae; their impact is likely to be small compared with that of fishes (Tsuda and Randall, 1971).

On most temperate shores (except the Northwest and Southwest Atlantic) a number of limpets have become highly specialized on one or a few closely related species of kelp (Branch, 1971; Vermeij, 1973b). Some of these limpets, such as the Chilean *Scurria scurra* or the Californian *Acmaea insessa*, may hasten the demise of their host alga (*Lessonia nigrescens* and *Egregia laevigata* respectively) by excavating deep cavities in the stipe and making the plant more vulnerable to being torn away by wave surge (Black, 1976). In the case of *A. insessa*, however, it is primarily the older portions of the plant that are eaten, and Black (1976) has suggested that grazing by the limpet may actually stimulate the growth and reproduction of *Egregia* (see Figure 4.3).

Other groups of animals are local or less important in their impact on plants. Some temperate palinurid lobsters of the genus *Jasus* are known to feed partly on benthic algae (Holthuis and Sibertsen, 1967; Beurois, 1975). Brachyuran crabs of the families Grapsidae, Xanthidae, and Majidae may be locally important consumers of algae, but they have been little studied (Crane, 1947; Hiatt, 1948; Kramer, 1967; Hedgpeth, 1969a; Griffin, 1971). In contrast to the pointed claws of carnivorous crabs, the tips of the chelae of herbivorous species are spoon-shaped or otherwise excavated (Crane, 1947; Griffin, 1971). Hermit crabs are usually thought of as scavengers or predators, but Ameyaw-Akumfi (1975) has shown that the West African intertidal *Clibanarius chapini*

Figure 4.3   The limpet *Scurria scurra* excavating a deep pit in the stipe of the chilean kelp *Lessonia nigrescens* at Montemar, Chile. (Photograph by J. W. Porter.)

and *C. senegalensis* take in large quantities of *Enteromorpha, Cladophora, Rhizoclonium*, and other filamentous or delicate algae. If this habit is more widespread, hermit crabs could be locally important herbivores, especially on tropical shores.

Some sea turtles graze on marine sea grasses and algae (Stephenson and Searles, 1960; Randall, 1965) and may have been quantitatively important before man seriously depleted their populations. In the Galapagos the endemic marine iguanid lizard *Amblyrhynchus cristatus* (Figure 4.4) grazes sublittoral algae (Houvenagel and Houvenagel, 1974). Among mammals the order Sirenia contains a number of herbivores, including the tropical Atlantic manatees (*Trichechus* spp.) and the Indo-West-Pacific dugong *(Dugong dugon)*, which are known to feed primarily on sea grasses, particularly the tender species (Randall, 1965;

Lipkin, 1975). The extinct gigantic Pleistocene and Recent Steller's sea cow *(Hydrodamalis gigas)* from the Northeast Pacific, apparently fed on kelps (see Dayton, 1975b). The principal marine herbivorous birds are geese (Anseridae) and ducks (Anatidae), which locally consume large quantities of *Zostera* sea grass in temperate regions. (For a review see Morse, 1975.)

Predators on photosynthesizing animals often show considerable prey specificity owing to the various defenses that many of their prey possess. The plant-like animals whose predators are best known are the reef-building scleractinian corals (Robertson, 1970; Glynn, 1973b; Randall, 1974). Their principal predators include fishes, decapod crustaceans, gastropods, asteroids, and echinoids. Most notable among the coralli-vorous fishes are certain parrotfishes (Scaridae), butterflyfishes (Chaeto-dontidae), blennies (Blenniidae), wrasses (Labridae), puffers (Tetraodontidae, Canthigasteridae), damselfishes (Pomacentridae), trig-gerfishes (Balistidae), and filefishes (Monacanthidae). Parrotfishes scrape off living coral tissues (Figure 4.5), while most of the other groups bite off bits of skeleton or crop the polyps themselves (Figure 4.6) (see also

Figure 4.4   The Galapagos marine iguana *(Amblyrhynchus cristatus)* grazing algae on the shores of Academy Bay. (Photograph by J. W. Porter.)

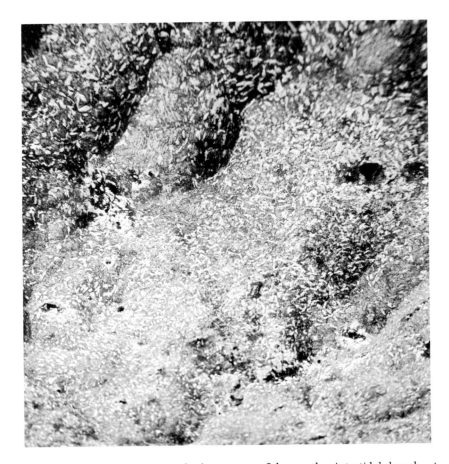

Figure 4.5 Scrape marks made by a parrotfish on the intertidal bench at Eniwetok, Marshall Islands. In the Indo-West-Pacific, parrotfishes also scrape living corals. (Photograph by J. W. Porter.)

Hiatt and Strassburg, 1960; Randall, 1967, 1974; Glynn et al., 1972; Hobson, 1974). Glynn (1973b) has suggested that the sharp spines on the skeletons of mussids and other large-polyped corals (Figure 4.7) may serve to discourage fishes from scraping or biting off pieces of the skeleton; but chemical defenses must also play a deterrent role, especially in members of the large genus *Acropora*, whose skeletons are generally fragile.

Neudecker (1977) seems to be the only investigator to have measured quantitatively the potential impact of fishes on corals. He transplanted the branching coral *Pocillopora damaecornis* from a grazing-free environment in Apra Harbor, Guam, to sites at various depths on a more exposed reef at the northern end of the island. At depths of 15.2 m and 30.5 m, corals lost about 35 g weight after one week because of removal of branch tips. No grazing of branch tips occurred at a depth of 2.4 m on the surf-swept reef margin. Neudecker was also able to show that *P. damaecornis* actually grow faster at depths where grazing is intense than at the reef margin (6.5 mm of growth at 15.2 m, 3.8 mm at reef margin).

Kaufman (1977) has shown that fishes in the West Indies may seriously damage and kill corals even when the latter are not used for food. The territorial damselfish *Eupomacentrus planifrons* maintains algal lawns in the midst of living coral at a depth of 10 to 30 m. The lawns, used by the fish for grazing, are enlarged when the fish bites neighboring corals; these are then quickly colonized by algae. Kaufman estimates that 10 to 40 percent of the surface area on the fore-reef terrace is occupied by algal lawns of *E. planifrons*.

A number of crustaceans are known to eat corals. Majid crabs of the genus *Mithrax* pinch off corallites with their chelae (Glynn, 1973b), and certain hermit crabs (*Aniculus* and *Trizopagurus*) break away pieces of the skeleton (Glynn et al., 1972). Various coral-associated xanthid crabs, such as *Trapezia*, *Tetralia*, and *Domecia*, apparently feed on coral mucus and may occasionally ingest living coral tissue. (For a review see Castro, 1976.)

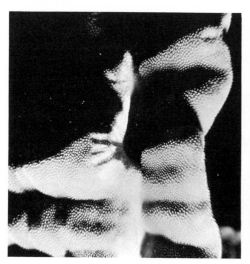

Figure 4.6 The coral *Porites* (*left*) is scarred by the bites of the smooth puffer *Arothron meleagris* (*right*) at Isla Clarion, Mexico. (Photographs by J. W. Porter.)

a

b

Figure 4.7   Corals with complex skeletons. (Photographs by J. W. Porter.)
a. *Meandrina meandrites*, 10 m depth, Grand Cayman, West Indies
b. *Isophyllia sinuosa*, 12 m depth, Grand Cayman. This species has a spiny
   skeleton that might discourage fishes from attacking.

The best known corallivorous molluscs are gastropods of the family Coralliophilidae, but others are known in the Ovilidae *(Jenneria)*, Calliostomatidae *(Calliostoma javanicum)*, Epitoniidae, Muricidae *(Drupella)*, and among aeolid nudibranchs (Robertson, 1963, 1970; Glynn et al., 1972; Glynn, 1973b; Harris, 1975; Hadfield, 1976; Taylor, 1976). On the Pacific coast of Panama, the cowry-like *Jenneria pustulata* accounts for about 80 percent of the biologically induced mortality of corals (Glynn et al., 1972), but in the West Indies Ott and Lewis (1972) found that *Coralliophila abbreviata* and the polychaete *Hermodice carunculata* have only a small effect on *Montastrea* and other reef corals.

Various echinoids, including the pencil urchin *Eucidaris* and the needle-spined black *Diadema* (Figure 4.8), occasionally rasp coral colonies in much the same way that they normally scrape surfaces covered with algae (Bak and van Eys, 1975). The best known corallivorous echinoderms, however, are the sea stars of the genera *Culcita, Nidorellia*, and especially *Acanthaster* (Chesher, 1969, 1972; Goreau et al., 1972; Glynn, 1973a, 1976; Laxton, 1974). Contradictory claims have been made concerning the activities of *Acanthaster* when feeding on corals (Figure 4.9): Laxton (1974) and others have argued that *Acanthaster* prefers branching and plate-like corals to encrusting forms, while Glynn (1976) has found experimentally that prey preference depends on the presence of commensal crustaceans. In the presence of the crab *Trapezia ferruginea* or the pistol shrimp *Alpheus lottini*, which live commensally in pocilloporid corals, *Acanthaster* avoids species of *Pocillopora* and will feed on encrusting *Porites* or *Pavona* species; in the absence of these crustaceans, the sea star will ascend and feed on *Pocillopora* colonies. Goreau and colleagues (1972) suggest that *Culcita* prefers encrusting corals as prey, and all investigators seem to agree that corals which inhabit depressions are more or less immune from asteroid predation.

It thus seems reasonable to conclude that metabolically active fishes become relatively more important as grazers toward lower latitudes, while the more slow-moving molluscs and echinoids are the main consumers of benthic photosynthesizers in temperate and polar waters. The only temperate fish that has been demonstrated to have a potentially significant effect on sedentary intertidal organisms is the Chilean clingfish *Sicyases sanguineus* (Paine and Palmer, 1978), but even this fish spends much of its time clinging to rocks and may be less active than such tropical reef species as parrotfishes, surgeonfishes, and sea chubs. Although temperate urchins are, and Steller's sea cow may have been, locally effective in cropping algal populations, most temperate grazers (chitons, gastropods, fishes including *Sicyases*, and crustaceans) apparently are

Figure 4.8 The urchin *Diadema antillarum* grazing *Acropora cervicornis* at a depth of 4 m, Discovery Bay, Jamaica. (Photograph by J. W. Porter.)

limited to smaller plants, and do little to depress algal standing crops significantly. Thus, even though the standing crop of algae has been reduced in areas of the Northeast Pacific where the predatory sea otter (*Enhydra lutris*) was locally exterminated, as along the coast of central California, the increase in abundance of herbivorous echinoids, chitons, and gastropods (especially *Haliotis*) did not lead to the local extinction of kelp (Lowry and Pearse, 1973). The effect of the sea otter in waters of the Aleutian Islands may be more substantial; Estes and Palmisano (1974) report that the densities of urchins (8 per square meter) and chitons (less than 1 per square meter) in the Rat Islands, where

Figure 4.9 The coral-eating sea star *Acanthaster planci* at 2 m depth on rock, Isla Clarion, Mexico. (Photograph by J. W. Porter.)

otters are present, are substantially lower than in the Near Islands (densities 78 and 38 per square meter respectively), where the otters are absent. Kelps are abundant to a depth of 25 m in the Rat Islands, but virtually absent except in the intertidal zone in the Near Islands.

In the tropics fishes seem to interfere with all but the chemically noxious and highly calcified photosynthesizers. They and other herbivores leave a much lower standing crop of algae, and the size refuge that larger temperate algae enjoy from their attackers may be less attainable in tropical areas. This, in turn, may be the result both of less effective herbivory on colder shores and of more rapid growth caused by seasonally high nutrient levels.

Other Sessile Animals and Their Predators

Sessile filter-feeding animals also are affected by predators that graze or browse the rock surface. When a parrotfish rasps the surface with its plate-like beak, it removes not only low-growing algae, but also young individuals and newly settled spat of barnacles, serpulid worms, snails, bivalves, and other animals. The geographical patterns discernible in photosynthesizing organisms thus should appear as well among epifaunal filter feeders.

Among sponges, many tropical species are mildly to acutely toxic to fishes (Bakus, 1974; Bakus and Green, 1974). In the Indo-West-Pacific and Eastern Pacific most sponges live in depressions or under ledges where fishes and other predators cannot easily reach them; but in the West Indies, where the proportion of poisonous or unpleasantly spiculed sponges is also high, sponges form a dominant element on the outer face of reefs as well as in cryptic environments (Reiswig, 1973). When cryptic sponges on the undersurfaces of stones are exposed to surface predators in the Indo-West-Pacific, they are removed in a matter of hours (Bakus, 1967). Sponges in the cold waters around San Juan Island in Washington and in Antarctica for the most part are not toxic and experience little fish predation (Bakus, 1964, 1974; Bakus and Green, 1974; Dayton et al., 1974). Instead, asteroids and dorid nudibranchs are the principal consumers of cold-water sponges (Mauzey et al., 1968; Bloom, 1976). Although it is likely that spicular characteristics and chemical defenses discourage, or reduce the efficiency of, sponge predation by dorids (Bloom, 1976), few sponges seem to be completely immune from attack. Among the gigantic sponges of Antarctica, Dayton and co-workers (1974) have not been able to correlate sponge spicular or other morphology with relative vulnerability to predatory asteroids and dorids.

Holothurians (sea cucumbers) show a similar increase in toxicity to fishes toward the tropics (Bakus and Green, 1974; Bakus, 1974). In the Indo-West-Pacific and Caribbean regions holothurians are common epifaunally and seem to have few predators. Sea cucumbers in the Puget Sound region have a number of asteroid predators (Mauzey et al., 1968), yet they occur in vast numbers in exposed places.

Articulate brachiopods are today primarily a cold-water group whose members usually live attached by the peduncle to hard substrata. In the tropics most species are found under overhangs or in other dark, cryptic situations (Jackson et al., 1971). The cold waters of the Northeast Pacific and New Zealand, however, support a number of brachiopods that live on the open bottom (Rudwick, 1962). In spite of their thin shells, Northeast Pacific brachiopods possess very little edible flesh and, according to C. W. Thayer, are rarely eaten by crabs, fishes, drilling gastropods, and other predators that could easily penetrate the shell. Thayer has found that since brachiopods cannot reattach once they are dislodged from the substratum, the grazing activities of echinoids restrict the distribution of brachiopods in the Puget Sound region. If these conclusions are valid for brachiopods in general, then the increased grazing pressure from fishes and echinoids on tropical coasts would restrict the habitat range of brachiopods there more than grazing has at higher latitudes. Although grazing is certain to affect byssally attached bivalves as well (Newman, 1960), this process is likely to be less calamitous than for brachiopods, since most bivalves can reattach when dislodged.

Like brachiopods, but in contrast to corals and other animals with symbiotic algae in their tissues, the balanomorph barnacles are in abundance, if not in diversity, a predominantly temperate group. The midlittoral zone in the intertidal is, in fact, often referred to as a barnacle or balanoid zone, especially in northwest Europe, the Mediterranean, the east and west coasts of North America, warm-temperate South Africa, southern South America, and the warm-temperate parts of Australia, New Zealand, and Japan (Stephenson, 1944; Stephenson and Stephenson, 1952, 1954a, b; Lewis, 1955, 1964; Pérès and Picard, 1955; Picard, 1957; Habe, 1958; Knox, 1960; Bennett and Pope, 1960; Hodgkin and Michel, 1961; Morton and Miller, 1968; Ricketts and Calvin, 1968; Luckens, 1970; Lipkin and Safriel, 1971). Barnacles are reduced in numbers or entirely absent on many cold-temperate shores in the Southern Hemisphere, as well as on polar coasts, and locally on all temperate shores where they may be accompanied or replaced by bivalves or algae (Southward, 1958a; Bennett and Pope, 1960; Knox, 1960; Kenny and Hayson, 1962; Holthuis and Sibertsen, 1967; Morton and Miller, 1968; Hedgepeth, 1969a; Penrith and Kensley, 1970a, b; Arnaud, 1974; Simpson, 1976). In the tropics, barnacles are abundant in Brazil and on the west coasts of America and Africa (da Costa, 1962; Lawson, 1966; Vermeij and Porter, 1971; Glynn, 1972; Reimer, 1976). Barnacles are rare on wave-exposed rocky shores in the Caribbean and most of the Indo-West-Pacific, but they may be locally abundant on certain volcanic shores and in bays or lagoons sheltered from wave action (Purchon and Enoch, 1954; Endean et al., 1956a, b; Lewis, 1960; Newman, 1960; Plante, 1964; Southward, 1975). Where barnacles do occur on open tropical shores (usually as large species of *Tetraclita* and *Balanus*), they occupy the summits of the local topography or they are restricted to steep faces.

Latitudinal patterns are difficult to discern among barnacles. Many temperate species of *Balanus*, for example (*B. balanoides* in the North Atlantic, *B. glandula*, *B. cariosus*, and *B. nubilis* in the North Pacific), are cylindrical in shape and have a wide apical aperture. Many tropical species of *Balanus* and *Tetraclita* are more conical in appearance and have smaller apertures (Figure 4.10). The genus *Tetraclita*, whose members inhabit most warm-temperate and tropical coasts (except tropical West Africa), is characterized by low shells whose thickened walls are invested with a network of tubules. These features of warm-water barnacles, and the distinct trend for low-shore barnacle species to be thicker-shelled and larger than species higher on the shore (Connell, 1961a, b, 1970; Dayton, 1971; Achituv, 1972; Luckens, 1975), may be related to gradients in predation by muricid gastropods. Most muricids drill their victim through the apical plates — and more rarely, through the lateral shell wall (Connell, 1961a, b, 1970; Radwin and Wells, 1968; Dayton, 1971; Luckens, 1975). A. R. Palmer has suggested to me that the wall structure in *Tetraclita* and the strong ribbing on the shell of *Balanus cariosus* (Figure 4.10) and certain other barnacles may inhibit predation by drilling gastropods.

Data to support a possible latitudinal trend in predation intensity by gastropods are unfortunately lacking, but Connell (1961a, b, 1970) has presented evidence that snail predation on barnacles decreases upshore in Scotland and in the Puget Sound region. He found that nearly all of the 27 percent monthly mortality of *Balanus glandula* in Washington at low shore levels is caused by *Thais lamellosa*, while at higher levels *T. emarginata* accounts for a much smaller fraction of the 20 percent monthly mortality of this barnacle. Near the top of its vertical range *B. glandula* is immune from predation by *Thais*, since there is insufficient time during high tides for an individual *Thais* to drill and consume a barnacle (Connell, 1970). Similar high-shore refuges from predation exist in Scotland (Connell, 1961a), in New Zealand (Luckens, 1975), and doubtless on other temperate as well as tropical shores.

Additional barnacle predators include asteroids such as *Pisaster* and *Leptasterias* in Washington (Paine, 1966a; Mauzey et al., 1968; Menge, 1972a, b). Blennies may remove barnacles in tide pools (Gibson, 1969, 1972), and the Chilean fish *Sicyases* is also known to scrape and eat barnacles from vertical faces in the intertidal zone (Paine and Palmer, 1978). R. T. Paine informs me that in Chile the crab *Acanthocyclus* specializes on a diet of barnacles. We have seen the xanthid *Eriphia squamata* feeding on barnacles, both in the field and in the laboratory, on the Pacific coast of Panama (Zipser and Vermeij, 1978; see also Crane,

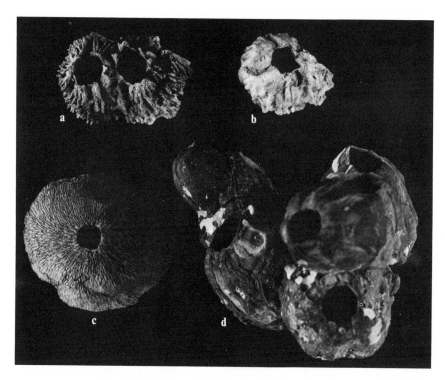

Figure 4.10   Some temperate and tropical barnacles. (Specimens collected by
A. R. Palmer; photograph by F. Dixon.)
   a. *Balanus cariosus*, Friday Harbor, Washington
   b. *B. glandula*, Friday Harbor
   c. *Tetraclita stalactifera panamensis*, Naos Island, Panama
   d. *Balanus* sp., Melones Island, Panama

1947). Stephenson and Searles (1960) demonstrated experimentally that
barnacles on the Great Barrier Reef flourish only where fishes do not
graze constantly over the rocks; indeed, the comparative rarity of barnacles on tropical hard surfaces is attributed to intense grazing by fishes
and other animals (Newman, 1960). The microgeographical distribution
of tropical barnacles is consistent with this interpretation; the greatest
abundance of these animals is found in sheltered lagoons and on heavily
wave-washed cliffs, where fishes typically do not forage.
   Limpets and chitons kill settling barnacles in the course of their algal
grazing (Dayton, 1971), but can do little against most adult barnacles. In
fact, Lewis and Bowman (1975) showed that growth rate and locomotion
of the British limpet *Patella vulgata* both are greatly restricted when it
lives on rocks primarily occupied by barnacles. I have commonly seen

Patella virtually hemmed in by barnacles; the rim of the shell then is often deformed and thickened.

In this chapter and the preceding one I have presented evidence which, though incomplete and by no means always conclusive, has led me to concur with Bakus (1969) that antipredatory defenses are developed to a greater degree and are found in an increasing proportion of species along a gradient from high to low latitudes. With qualifications, such gradients have been shown to exist in all the groups for which data are available: gastropods, bivalves, barnacles, sponges, holothurians, and marine sessile photosynthesizers. A parallel microgeographical trend of increasing armor seems to exist on most shores from high to low intertidal levels, and in soft sediments from deep to shallow layers. In conjunction with these trends in prey defense, the armament and probably the effectiveness of predators also appear to increase.

It is tempting to conclude from considerations in Chapter 1 that the geographical patterns in prey defense and predator armament result, at least in part, from increased muscular capabilities of tropical predators and grazers as compared to those in cooler waters. Calcification and other processes that lead to the production of defensive mechanisms may correspondingly be easier or metabolically less "costly" for tropical prey organisms than for their temperate and polar counterparts. Moreover, increased seasonality away from the equator may limit the activity and foraging effectiveness of both ectothermic and endothermic predators. A similar temporal restriction may apply to high intertidal as compared to lower-shore predators. In deep sediments reduced oxygen levels and high resistance to locomotion may reduce foraging efficiency of predators. Empirical data therefore seem to point to the conclusion that the most powerful and specialized predators, and the most heavily armed prey, are found in areas where physical conditions interfere minimally with physiological processes.

Connell (1972) has reached similar conclusions from considerations of population dynamics. If populations fluctuate less (or at least more predictably) in physiologically favorable environments than in stressful ones, then the predators in favorable environments can track populations of individual prey species without having to tolerate the severe oscillations typical in more rigorous environments. As a result, the predator can specialize on a smaller number of prey species and become better designed to exploit those species (see also MacArthur, 1972). If resources are unpredictable, these specializations are not feasible and the diet should be less specialized.

It must be emphasized, however, that no definitive data are yet available on latitudinal gradients in overall predation or grazing pressure. Until such data are obtained, the modifying effects of productivity and other factors on the geographical patterns in predation-related adaptations cannot be fully evaluated.

PART TWO

Interoceanic Patterns
of Adaptation

CHAPTER 5

Predation and Grazing

I F LATITUDINAL patterns in organic form were determined exclusively by the direct and indirect effects of temperature and correlated factors on physiological processes, then we should expect organisms living under identical conditions in unconnected biogeographical areas to be essentially alike in armament and in other adaptations. Such striking convergence is well known for leaf shape and leaf size in tropical rain-forest trees, and for plant growth forms in geographically isolated pockets of Mediterranean-type climate (Richards, 1952; Mooney and Dunn, 1970; Parsons and Moldenke, 1975). In the sea, too, community convergence has been widely recognized. Thorson (1957), for example, described "*Macoma*" communities from shallow, subtidal sandy muds on various Arctic, temperate, and even tropical coasts; and he found these assemblages of molluscs and polychaetes to have strikingly similar species composition, trophic structure, and life-history patterns. Stephenson and Stephenson (1949, 1972) were struck by the apparently universal biological division of the intertidal zone into three subzones, each of which is characterized by the predominance of a particular group of organisms: the littorinid zone (littoral fringe at the top of the shore), balanoid zone (often occupied also by algae, bivalves, and limpets), and algal zone (sublittoral fringe) (see also Knox, 1960; Lewis, 1964).

Yet closer inspection and comparative studies have revealed that this convergence is incomplete and that important differences among communities exist in such properties as trophic structure, diversity, and the fineness with which the available resources have been partitioned among the species. For example, the littoral fringe extends to a higher level, and is divisible into a greater number of subzones, in the West Indies than in tidally comparable portions of the tropical Pacific (Stephenson and Stephenson, 1950, 1972; Lewis, 1960; Salvat, 1970). These differences have been found to transcend the limits of taxonomy.

In this chapter I shall focus attention on interoceanic (east-west) differences in the degree to which predation-related adaptations are developed. By using criteria set forth in Chapters 2, 3, and 4, I shall try to show that communities in the Pacific and Indian oceans are often characterized by greater intensities of predation and grazing than are ecologically similar communities in the Atlantic; and that physiologically favorable open-surface habitats exhibit a much greater regionality in this respect than do stressed or cryptic habitats.

Tropical Molluscs and Their Predators

Open-surface gastropods living in the low intertidal zone of tropical rocky shores exhibit some striking regional differences in the degree to which certain predation-related shell characteristics are expressed (Vermeij, 1974b, 1977b). Shelled gastropods of the Indo-West-Pacific, and to a lesser extent of the Eastern Pacific, exhibit a higher incidence and better development of strong sculpture, narrowly elongated apertures, and outer-lip dentition than do the snails on comparable shores in the tropical Western and Eastern Atlantic (see Table 5.1 and Figures 5.1 through 5.4). There is a slight but insignificant trend for inflexible opercula to be better represented in Indo-West-Pacific faunas than among snails of other tropical biotas. Spire height among Indo-West-Pacific species tends to be lower than in the other tropical oceans when whole gastropod faunas are compared, but regional variation in spire height may follow either of two trends depending on the specific family (Vermeij, 1977b). In groups where the aperture becomes relatively smaller or narrower as spire height increases (Trochidae, Turbinidae, Planaxidae, *Thais*), Indo-West-Pacific and to an insignificantly lesser extent Eastern Pacific gastropods have higher spires than their Atlantic relatives; but where no relation exists between spire height and aperture size, or where higher spires are associated with broader apertures (*Conus*, peristerniine Fasciolariidae, *Vasum*, *Morula*-like Muricidae), species in the Indian and Pacific oceans have lower spires than their kin in West Africa, Brazil, and the West Indies.

The interoceanic patterns in shell architecture are evident not only in comparisons of whole gastropod faunas, but also within narrowly defined taxonomic categories. Thus, Indo-West-Pacific rock-associated species of *Conus* have thicker shells, lower spires, and narrower apertures than do ecologically similar species of *Conus* in the West Indies and West Africa (Vermeij, 1976); Eastern Pacific species are intermediate in these respects. In the *Thais* subgenus *Mancinella*, the Western Atlantic species (*T. deltoidea*) is higher-spired and less heavily knobbed than are species in the Eastern Pacific (*T. triangularis* and *T. speciosa*) and the Indo-West-Pacific (*T. mancinella* and *T. tuberosa*). The large *Drupa-Morula* group of muricids, characterized by well-developed apertural dentition, nodose or spinose sculpture, and low spires, is confined to the Indo-West-Pacific.

The proportion of species with columellar folds (and therefore with a large area of attachment for retractor muscles) also varies among tropical regions. On rocky shores of the tropical Indian and Western Pacific oceans, 20 to 37 percent of the open-surface snail species have columellar folds. Values elsewhere are significantly lower: 0 to 13 percent in the Red Sea, 10 to 17 percent in the Eastern Pacific, less than 10 percent in the Western Atlantic, and less than 10 percent in the Eastern Atlantic. These differences are accounted for in part by the distribution of mitrids and cypraeids, which are relatively most diversified in the Indo-West-Pacific. Peristerniine fasciolariids, which have small folds on the anterior part of the columellar lip, are found on most tropical shores, but are rare or absent in West Africa.

Although these predation-related differences in gastropod shell architecture were originally recognized from purely morphological criteria (Vermeij, 1974b), they have since been confirmed experimentally. Working in Guam, I tested the relative vulnerability of West Indian and Western Pacific shells to crushing predation by using a number of xanthid and parthenopid crabs (*Carpilius convexus, C. maculatus, Eriphia sebana,* and *Daldorfia horrida*) as predators (Vermeij, 1976). Crabs were offered living Guamanian snails as well as related Jamaican shells into which Guamanian hermit crabs were introduced. The Jamaican shells had been collected alive, dried in air, and then rewetted before being "reincarnated," a process that apparently does not significantly alter the mechanical properties of molluscan shells (Currey, 1975). Using West Indian and Western Pacific open-surface trochids, neritids, and cones, it was found that the West Indian species in every case were vulnerable to crab crushing at much larger sizes (both absolute and relative) than were their

Table 5.1   Incidence of predation-related shell traits among gastropods in tropical rocky-shore assemblages. Data are taken from my own collections.

Locality	Number of species	Percentage of species with—			
		Toothed apertures	Elongate apertures[a]	Inflexible operculum	Strong external sculpture
Low intertidal, open surfaces:					
Indo-West-Pacific					
Aimeliik Reef, Palau	24	50	25	21	25
Ngemelis, Palau	14	64	57	14	50
Nosy Komba, Madagascar	13	46	46	46	31
Nosy-Be, Madagascar	11	55	45	36	36
Kikambala, Kenya	14	50	36	7.1	36
Pulau Salu, Singapore	12	8.3	17	25	25
Pago Bay, Guam	46	37	41	13	38
Kahuku, Hawaii	12	67	42	0	25
The Fjord, Sinai	15	33	27	13	40
Marset el Et, Sinai	14	33	47	6.7	29
Ras Muhamad, Sinai	15	27	40	13	20
Wadi Taba, Sinai	8	38	50	25	13
Eastern Pacific					
Punta Carnero, Ecuador	8	37	25	0	25
Panama City, Panama	20	15	30	25	30
Morro de Taboga, Panama	14	23	54	23	29
Playa de Panama, Costa Rica	15	40	47	27	20
Playas del Coco, Costa Rica	7	29	14	14	14
Western Atlantic					
Fort Point, Jamaica	15	13	20	20	33
Discovery Bay, Jamaica	18	17	11	11	33
Boca Playa Canoa, Curaçao	6	0	17	17	17
Vieux Habitants, Guadeloupe	7	0	0	0	14
Pointe des Chateaux, Guadeloupe	5	0	0	20	0
Boca Grandi, Curaçao	6	17	17	17	0
Playa Chikitu, Curaçao	10	20	10	30	0
Cahuita, Costa Rica	14	20	14	14	14
Recife, Brazil	8	25	13	13	25
Amuay Bay, Venezuela	10	10	0	30	10
Eastern Atlantic					
Takoradi, Ghana	7	0	0	14	14
Abadzi, Ghana	7	0	0	29	14
Tema, Ghana	6	17	0	17	33
Dakar, Senegal	13	15	23	7.7	7.7
Low intertidal, cryptic:					
Indo-West-Pacific					
Ngemelis, Palau	16	56	56	17	31

Locality	Number of species	Percentage of species with—			
		Toothed apertures	Elongate apertures [a]	Inflexible operculum	Strong external sculpture
Nosy-Be, Madagascar	14	43	7.1	0	14
Pago Bay, Guam	23	48	22	8.7	35
The Fjord, Sinai	6	0	17	17	0
Marset el Et, Sinai	5	0	0	0	20
Eastern Pacific					
Morro de Taboga, Panama	21	43	29	0	24
Venado Beach, Panama	12	33	25	0	33
Western Atlantic					
Discovery Bay, Jamaica	11	45	18	0	18
South coast, Curaçao	12	42	42	11	17
Fernando de Noronha, Brazil	7	29	29	29	29
Recife, Brazil	6	33	17	0	17
Eastern Atlantic					
Takoradi, Ghana	12	25	17	17	17
Dixcove, Ghana	10	20	10	20	20
Dakar, Senegal	15	6.7	27	13	13
High intertidal:					
Indo-West-Pacific					
Arakabesan, Palau	12	13	0	25	8.3
Ngemelis, Palau	12	67	0	42	17
Nyali, Kenya	13	54	0	38	31
Nosy-Be, Madagascar	12	50	0	33	42
Majunga, Madagascar	6	33	0	17	17
Pago Bay, Guam	9	22	0	33	22
Pulau Subar Darat, Singapore	8	37	0	13	13
Puerto Galera, Philippines	9	56	0	33	22
Nasugbu, Philippines	7	29	0	0	14
The Fjord, Sinai	8	25	0	25	0
Marset el Et, Sinai	6	33	0	33	0
Eastern Pacific					
Academy Bay, Galapagos	6	33	0	17	17
Naos Island, Panama	6	33	0	17	17
Playa de Panama, Costa Rica	8	13	0	13	13
Western Atlantic					
Fort Point, Jamaica	13	23	0	23	23
Boca Playa Canoa, Curaçao	10	20	0	20	20
Vieux Habitants, Guadeloupe	9	22	0	22	11
Saba	8	25	0	25	13
Cahuita, Costa Rica	11	27	0	27	18
Fernando de Noronha, Brazil	4	25	0	25	0
Eastern Atlantic					
Takoradi, Ghana	5	20	0	20	0
Dakar, Senegal	4	50	0	25	0

[a] Ratio of length to breadth greater than 2.5.

Figure 5.1   Dorsal *(above)* and apertural *(below)* views of Indo-West-Pacific snails from open rocky surfaces. (Photographs by M. Montroll.)

    a. *Conus miliaris*, Kikambala, Kenya
    b. *Cypraea caputserpentis*, Kahuku, Oahu, Hawaii
    c. *Vasum turbinellus*, Umatac, Guam
    d. *Morula granulata*, Kikambala
    e. *Drupa ricinus*, Kikambala

f. *Thais armigera*, Aimeliik Reef, Palau
g. *Turbo setosus*, Pago Bay, Guam
h. *Trochus ochroleucus*, Pago Bay
i. *Cerithium nodulosum*, reef off Malakal, Palau
j. *Latirolagena smaragdula*, Aimeliik Reef

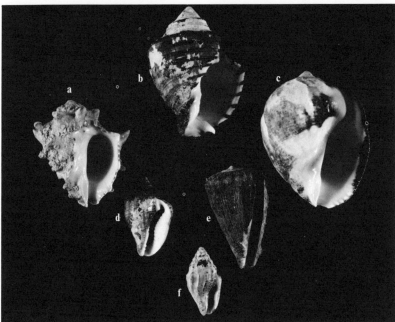

Figure 5.2   Dorsal *(above)* and apertural *(below)* views of Eastern Pacific open-surface snails. (Photographs by F. Dixon.)

    a. *Thais speciosa*, Playa de Panama, Costa Rica
    b. *Opeatostoma pseudodon*, Perico Island, Panama
    c. *Thais melonis*, Ayangue, Ecuador
    d. *Columbella major*, Morro de Taboga, Panama
    e. *Conus princeps*, Naos Island, Panama
    f. *Anachis fluctuata*, Morro de Taboga

Figure 5.3   Dorsal *(above)* and apertural *(below)* views of Western Atlantic open-surface snails. (Photographs by F. Dixon.)

      a. *Fissurella nodosa*, Fort Point, Jamaica
      b. *Cittarium pica*, Boca Playa Canoa, Curaçao
      c. *Thais deltoidea*, Discovery Bay, Jamaica
      d. *Columbella mercatoria*, near Moule, Guadeloupe
      e. *Leucozonia nassa*, Pointe des Chateaux, Guadeloupe
      f. *Conus mus*, Rio Bueno, Jamaica

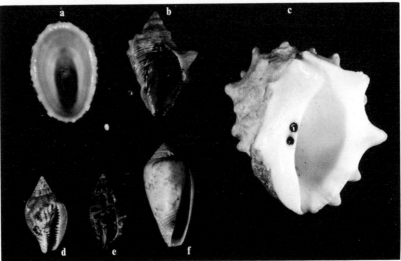

Figure 5.4   Dorsal *(above)* and apertural *(below)* views of Eastern Atlantic open-surface snails. (Photographs by F. Dixon.)

       a. *Fissurella nubecula*, Takoradi, Ghana
       b. *Thais haemastoma*, Abadze, Ghana
       c. *T. nodosa*, Aberdeen, Sierre Leone
       d. *Columbella rustica*, Takoradi
       e. *Trachypollia nodulosa*, Aberdeen
       f. *Conus mercator*, Anse Bernard, Dakar, Senegal

ecological and taxonomic counterparts in Guam. For example, *Daldorfia horrida* broke the West Indian *Conus mus* to a shell length of 39 mm (the largest seen on the north coast of Jamaica), while native epifaunal cones from the reef flats of Guam, which often exceeded 45 mm in maximum length, were immune from lethal attack beyond a length of 31 mm. This difference stems not only from the taller spire of *C. mus* (apical half-angle 42.9° as compared to 46.3 to 55.3° for Guamanian species), but also from its much thinner shell (no more than 1.1 mm thick for *C. mus*, 1.4 to 2.8 mm for the Guamanian species studied). The Guamanian *Trochus niloticus* could be broken to a diameter of 40 mm by adult (110 to 138 mm wide) *Carpilius maculatus*, but the West Indian *Cittarium pica* had not yet reached immunity at a diameter of 57 mm. As suggested in Chapter 2, this difference results from the tighter coiling of *Trochus* rather than from a difference in shell thickness.

I also conducted studies on the vulnerability of Caribbean and Indo-West-Pacific species of *Nerita* to predation by the common Guamanian intertidal xanthid crab, *Eriphia sebana* (Vermeij, 1976). A crab 56.1 mm wide was unable to crush the Indo-West-Pacific *N. plicata* and *N. albicilla* beyond a shell diameter of 23 mm and a shell thickness of about 1.9 mm; West Indian neritids (*N. peloronta*, *N. versicolor*, and *N. tessellata*) are much thinner (0.8 to 1.5 mm) and can be killed up to sizes characteristic of large adults. Reynolds and Reynolds (1977) have since found that species of *Nerita* in the northern Gulf of California (*N. funiculata* and *N. scabricosta*) have a shell thickness (about 1.0 mm for shells 17.0 mm in diameter) intermediate between those of West Indian and Indo-West-Pacific species, and that *Eriphia squamata* is also of intermediate strength compared to the species in the other two biotas.

In experiments with the spiny puffer *Diodon hystrix* in Panama, A. R. Palmer (1978 and personal communication) has given further support to the hypothesis that Eastern Pacific gastropods are more resistant to crushing than are related Caribbean species. A puffer 45 cm long can eat the Eastern Pacific *Leucozonia cerata* up to 39.5 mm in length, whereas it can take Atlantic *L. nassa* up to about 48 mm in shell length. A 29 cm puffer can eat the Eastern Pacific *Thais melonis* up to a shell length of 20.5 mm, and Atlantic *T. deltoidea* up to 22.2 mm. Similar differences in vulnerability were found in *Conus* and *Nerita*; a small (21 cm) puffer was only able to crush Eastern Pacific *Nerita scabricosta* less than 15.4 mm in diameter, while the same fish crushed all Atlantic *N. versicolor* offered (up to 23.5 mm).

I obtained contradictory results in some trials with a large (37 cm long) specimen of the Eastern Pacific spiny lobster *Panulirus gracilis*; this lobster was able to peel the lip and kill specimens of the Eastern Pacific thick-shelled, smooth *Thais melonis* up to a size of 35 mm, while the Atlantic form *(T. deltoidea)*, with large knobs at the shoulder, was able to survive the lobster's attacks above a length of 30 mm.

In curious harmony with the interoceanic patterns in gastropod shell form, hermit crabs specialized for living in shells with long, narrow apertures are found only in the Indo-West-Pacific. *Dardanus guttatus*, for example, is a large hairy hermit whose strongly flattened body is well suited to fit in the shells of *Conus*, *Cypraea*, and narrow-apertured species of *Strombus* and *Lambis*. Elsewhere in the tropics, such shells are relatively less abundant and are inhabited by small individuals of hermit-crab species that can also live in shells with more rounded apertures. Holthuis (1959) noted that *Paguristes depressus* off the coast of Surinam can live in snail shells of various geometries, but that its body is peculiarly flattened only when inhabiting *Conus* shells. On the Pacific side of Panama, I found it difficult to introduce local hermit crabs into shells of *Conus*, *Drupa*, and *Morula* (which I had brought from Guam).

Within each of the four tropical biotas, regional variations in shell architecture are surprisingly limited. The Red Sea, for example, contains a large endemic molluscan element, but the incidence and expression of antipredatory armor is no less than in other parts of the Indo-West-Pacific. *Drupa ricinus hadari*, an endemic Red Sea subspecies, has the teeth on the outer lip less well developed than does the widely distributed Indo-West-Pacific *D. r. ricinus*; but the Red Sea form is distinctly knobbier than is the typical East African or Western Pacific form. A similar situation exists in *Vasum turbinellus* (Figure 5.5). Several groups endemic to the Central Pacific have noticeably higher spires than do their Western Pacific and Indian Ocean relatives. Good examples of this trend are the Easter Island population of the widely distributed *Conus miliaris*, and the Central Pacific *Drupa speciosa* (related to the Indo-West-Pacific *D. rubusidaeus*), *Vasum armatum* (close to *V. turbinellus*), and *Thais affinis* (related to *T. armigera*) (see Abbott, 1959; Emerson and Cernohorsky, 1973; Kohn and Riggs, 1975). The last three examples, however, are strongly sculptured shells, and there is a large representation of other heavily armored gastropods in the faunas of Hawaii, the Line Islands, and French Polynesia (Kay, 1967, 1971; Salvat, 1970, 1971).

The principal regional differences within the Western Atlantic and Eastern Pacific biotas are related to sand scouring (Vermeij and Porter, 1971; Vermeij, 1974b). The expression of armor (especially that of strong

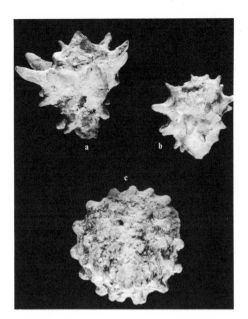

Figure 5.5   Dorsal *(left)* and apertural *(right)* views of Red Sea open-surface snails from The Fjord, Sinai. (Photographs by F. Dixon.)
  a. *Vasum turbinellus*
  b. *Drupa ricinus hadari*
  c. *Tectus dentatus*

sculpture) seems to be reduced among gastropods typically found on sand-scoured rocky surfaces; indeed, the interoceanic differences in gastropod shell architecture seen between assemblages of these two areas disappear on sand-abraded rocks. In the Western Atlantic, sand scouring is particularly characteristic of mainland Brazilian shores. Eastern Pacific rocky shores characterized by sand abrasion seem to be widespread along the coast of mainland tropical West America and, like the coasts of Brazil, are inhabited with an impoverished intertidal biota of barnacles, small mussels, and limpets (see, for instance, Paine, 1966a, for Costa Rica; and Vermeij and Porter, 1971, for southwest Ecuador).

Gastropods known or presumed to be annuals do not exhibit any clear interoceanic patterns in armament. Small, apparently annual cerithiids, for example, seem morphologically much the same in all the tropical seas (see Houbrick, 1974a) and, like many shelled opisthobranchs, are best regarded as opportunistic weeds. In unshelled opisthobranchs and certain prosobranchs, there may well exist east-west differences in chemistry or in the extent of warning coloration, but these have been neither recognized nor investigated.

It is difficult to evaluate the relative antipredatory shell morphology of rocky-shore bivalves in the various tropical oceans, and to date no relevant experiments have been performed. One example of a possible east-west difference may be seen among mussels of the family Mytilidae: the common Indo-West-Pacific *Septifer bilocularis* has a larger, thicker shell—and probably a relatively larger adductor muscle—than such ecological analogues as *Brachidontes exustus* and *B. dominguensis* in the Western Atlantic, and *B. niger* and *B. puniceus* in West Africa. I cannot detect interoceanic differences among oysters (*Ostrea* and *Crassostrea*), *Chama, Spondylus, Arca, Barbatia*, or *Isognomon*. Some of these bivalves may turn out to be weedy opportunists (*Isognomon?*), but this is clearly not the case with others (chamids, ostreids, spondylids, arcids).

Rocky-shore bivalves are not the only open-surface animals to show regional uniformity from east to west on tropical shores. Among regular echinoids (sea urchins), differences among species in the circumtropical genera *Diadema, Echinometra, Eucidaris*, and *Tripneustes* pertain only to obscure details of plate number and cannot be detected upon casual inspection in the field. Sea urchins with blunt stubby spines (*Eucidaris, Colobocentrotus, Heterocentrotus*) are able to wedge themselves more tightly into crevices than are other sea urchins and they are significantly more diversified in the Indo-West-Pacific than elsewhere in the shallow-water marine tropics. A similarly high diversity occurs among needle-spined or poison-spined echinoids in the Indo-West-Pacific (*Diadema, Echinothrix, Astropyga, Toxopneustes*). Hendler (1977) has shown that most urchins occurring intertidally on a reef flat on the Atlantic coast of Panama die at times of very low tide from desiccation or heat stress rather than from predation. Indeed, he argues that thin-walled epifaunal species, such as *Diadema antillarum* and *Lytechinus variegatus*, show the characteristics of weeds, at least in the intertidal zone: they grow rapidly, mature early (within a year), and exhibit wide fluctuations in population size. The somewhat more cryptic and thicker-walled species of *Eucidaris* and *Echinometra*, and the infaunal heart urchins *Brissus, Echinoneus*, and *Paraster*, are less weedy in character.

Open-surface balanomorph barnacles also exhibit a disconcerting regional homogeneity of form across the tropics. Although the abundance of barnacles is known to depend on the intensity of predation by fishes and other animals, there are no parallel trends in morphology; many species are either circumtropical or else differ among regions only by obscure characters. It is plausible that these barnacles, as well as the sea urchins and bivalves considered earlier, compete more intensely for space (or, in the case of urchins, for food) than do the more mobile shelled gastropods, and that selective pressures related to competition weigh more heavily than do those connected with predation. These speculations deserve further study through field experimentation, and much needs to be learned about which morphological and behavioral features of sedentary rocky-shore urchins, barnacles, and bivalves confer a competitive advantage.

High in the intertidal zone (littoral fringe and upper balanoid zone) rocky-shore snails are less well armored than their lower-shore counterparts, and the expression of armor is uniform throughout the tropics except in two separate regions of the Indo-West-Pacific (Vermeij, 1974a). These areas lie in the western Indian Ocean and in the Indo-Malaysian region (Indonesia, Philippines, New Guinea, and Solomons). Here the development of apertural teeth, strong sculpture, and other antipredatory armor is impressively high (Table 5.1), and many species (particularly some species of *Nerita* and of the littorinid genus *Tectarius*) attain large dimensions (Figure 5.6). Elsewhere in the tropics, assemblages of high intertidal gastropods principally differ from each other in adult body size; species of *Nerita* and of *Siphonaria* in the Eastern Pacific, for example, are giants compared to those in the tropical Atlantic and in the peripheral parts of the Indo-West-Pacific.

Mangrove-associated gastropods exhibit generally more fragile shells than do species on rocky shores, and the incidence of antipredatory features is low and uniform throughout the tropics (Table 5.1). Interesting regional differences do exist, however, with respect to body size, and are parallel to the size differences seen among the high intertidal snails of rocky shores. Mangrove-associated snails restricted to the Indo-Malaysian area are larger than are species from other parts of the tropics; Eastern Pacific mangrove snails are also apt to be larger than their ecological counterparts in the West Indies, Brazil, and West Africa. In all these cases large adult body size seems to be associated with high local species diversity (Vermeij, 1973c).

I have not been able to detect consistent interoceanic differences in the predation-related morphology of gastropods living in deep crevices or under large stones (Vermeij, 1974b). Species in this environment by and large are small in size, highly diversified in form, and relatively less armored than are sympatric open-surface species (Table 5.1). This impression is reinforced by the fact that the juveniles of many open-surface species are restricted to the undersurfaces of stones, where they are less exposed to such open-surface predators as crabs, fishes, birds, sea stars, and drilling molluscs (Kitching et al., 1966; Shepherd, 1973; Palmer and Frank, 1974). In the tropics many snails have determinate growth.

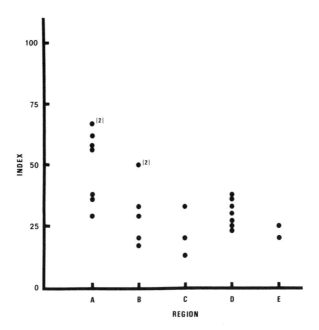

Figure 5.6 Distribution of predation-related shell traits among high intertidal gastropods on tropical shores. Each point represents one assemblage at one locality. The "index" refers to the percentage of species bearing either a toothed outer lip or an inflexible operculum. Further data may be found in Table 5.1. (Graph by E. Dudley; photograph by L. Reed.)

            A. Indo-Malaysian region and western Indian Ocean
            B. Peripheral Indo-Pacific: Hawaii, Guam, Red Sea
            C. Eastern Pacific
            D. Western Atlantic
            E. Eastern Atlantic

Adult size is achieved rapidly after a period of fast growth, during which the immature shell is often thin and apertural armature is scarcely, if at all, evident. Birkeland and Gregory (1975) found that juvenile *Cyphoma gibbosum* in Panama hide at the base or in furrows of gorgonaceans by day and come out to feed only at night; adult specimens, on the other hand, which have thick shells and an external mantle conspicuously colored to the human eye, graze on the gorgonaceans even by day.

My experience with molluscs from shallow sand and turtlegrass flats has been more limited than that with molluscs on rocky or mangrove shores, and the extent to which soft-bottom biotas in the tropical oceans differ from one another in expression of antipredatory features is still difficult to ascertain satisfactorily. The scanty numbers in Table 5.2 suggest, in a preliminary way, that the incidence of very tall spires and of narrowly

Table 5.2 Incidence of some possibly antipredatory shell traits in sand-dwelling gastropods. Data are taken from my own collections. Differences between the Indo-West-Pacific and New World tropics are significant at the 0.05 level.

Locality	Number of species	Percentage of species with—	
		Elongate apertures[a]	Tall spires[b]
Indo-West-Pacific			
Ngemelis, Palau	20	50	20
Nosy-Be, Madagascar	33	45	9.1
Kikambala, Kenya	19	58	11
Pago Bay, Guam	23	52	17
Eastern Pacific			
Venado Beach, Panama	35	31	14
Western Atlantic			
Cahuita, Costa Rica	8	38	0
Aruba	13	46	0
Amuay Bay, Venezuela	15	33	6.7
Sanibel Island, Florida	24	50	0
Turkey Point, Florida	21	48	10

[a] Ratio of length to breadth greater than 2.5.
[b] Apical half-angle less than 10°.

elongate apertures (believed to be effective against calappid crabs) is higher among sand-dwelling gastropods in the Indo-West-Pacific than in the New World tropics. Although more data are needed to establish this trend, particularly in the New World, differences among members of particular families seem to strengthen this conclusion. Shallow-water terebrids are mostly small in the Western Atlantic and are low in number of species compared to the Indo-West-Pacific and Eastern Pacific; in the latter two areas, moreover, the small species are joined by many large, thick-shelled forms. Among chank shells of the genus *Turbinella*, the Caribbean *T. angulata* and Brazilian *T. laevigata* have taller spires and a broader aperture than does the massive-shelled *T. pyrum* of India and Ceylon. In the Indo-West-Pacific, many species of *Nassarius* have a strongly thickened and toothed outer lip, and several species (*N. graniferus*, for example) have a strongly beaded or even spiny sculpture. Most of the common shallow-water species in the Eastern Pacific and Western Atlantic, however, have thinner lips. Sand-dwelling cones in the Eastern Pacific (such as *Conus ximenes*, *C. patricius*, and *C. orion*), Western Atlantic (*C. floridanus*, *C. puncticulatus*), and West Africa (*C. papilionaceus*) have taller spires and broader apertures than do such Indo-West-

Pacific species as *C. eburneus*, *C. quercinus*, *C. leopardus*, and *C. litteratus*. The only shallow-water sand-dwelling cones in the Western Pacific and Indian oceans with a conspicuously high spire and thin shell are those specialized to feed on molluscs and small fishes (for example, *C. textile*, *C. aulicus*, and *C. striatus*). They may be protected against predators by their spear-like radular tooth, which can inflict painful—sometimes fatal—wounds to careless human collectors. Fish-eating cones are known also in the other tropical seas (*C. purpurascens* in the Eastern Pacific, *C. ermineus* or *C. testudinarius* in the tropical Atlantic), but molluscivorous cones are restricted to the Pacific and Indian oceans (see Percharde, 1974).

As on rocky shores, the proportion of species with columellar folds is higher among Indo-West-Pacific sand-dwelling gastropods (17 to 37 percent) than in the New World tropics (14 to 23 percent). In the Indo-West-Pacific it is primarily the Mitridae and Vexillidae that contribute to the high proportion of species with columellar folds. These families are much less diversified elsewhere in the tropics. Other families of soft-bottom gastropods with columellar folds (Volutidae, Turbinellidae, Fasciolariidae, Marginellidae, Cancellariidae) are more equably distributed.

If little can be said about gastropods, even less is known about interoceanic differences in the shell architecture of tropical soft-bottom bivalves. The proportion of shallow-water Jamaican grassbed bivalves with crenulate margins is 36 percent (17 of 47 species), almost identical to the 37 percent (42 of 113 species) among soft-bottom bivalves from the Seychelles (calculated from species lists given by Taylor, 1968; Taylor and Lewis, 1970; and Jackson, 1973). The proportion of infaunal shallow-water bivalves with a persistent posterior gape is nearly identical on the two sides of tropical America (12 percent in the West Indies and 13 percent in mainland tropical western America), but seems to be higher than the 2.3 percent calculated for the seventy-six infaunal bivalve species at Mahé, Seychelles, in the Indian Ocean (see Taylor, 1968; Vermeij and Veil, 1978). However, persistent posterior gapes occur in about 15 percent of the Singapore infaunal bivalves studied by Vohra (1971). Thus, present inadequate evidence suggests that interoceanic differences in bivalve shell architecture either do not exist at all or are very slight.

Sandy open-ocean beaches not only show a striking regional uniformity in molluscan architecture from east to west, but also display few latitudinal gradients. There is a distinct latitudinal increase in diversity toward the tropics; in fact, virtually no large invertebrates seem to occur on polar beaches (Christiansen, 1965). Tropical beaches are especially rich in rapid brachyuran and anomuran crabs and in molluscs, but their diversity varies greatly from place to place: continental shores are richer in species than are the shores of islands, and some oceanic islands and archipelagos (Hawaii, southeastern Polynesia, Aldabra, Fernando de Noronha) seem to lack sandy-beach molluscs altogether (see Dahl, 1952; Kay, 1967; Taylor, 1971; Ansell et al., 1972). Dexter (1972, 1974, 1976) has shown that tropical Western Atlantic beaches have a less diversified fauna of molluscs, crustaceans, echinoderms, and polychaetes than do the wider, more finely grained beaches of western tropical America.

Despite these differences in diversity, sandy-beach molluscs display regional uniformity in architecture. The only latitudinal trend I have been able to detect is that temperate species of the beach-clam genus *Donax* have a more slender shell than do most tropical species (see Figure 1.1). Within the tropics most gastropods (*Bullia, Olivella, Terebra*) and bivalves (*Donax, Atactodea, Tivela*) are streamlined and smooth or finely ridged so as to facilitate burrowing and to present minimal surface area to the waves. The portunid and ocypodid crabs, small shore birds, naticid gastropods, and flatfishes known to prey on beach molluscs probably differ little in their relative capabilities or impact (see Hughes, 1966; Wade, 1967; Edwards, 1969; Jones, 1972; Smith, 1975).

Many species of *Donax*, and also some *Olivella*, display striking individual color variations, which have been experimentally linked by Smith (1975) to predation by ocypodid ghost crabs and other predators. The crabs appear to form search images, for either the most common or the most conspicuous prey, depending on prey density, so that different color forms are favored at different densities.

Among gastropods, then, the most profound interoceanic variations in architecture occur on open rocky surfaces; greater uniformity is found in such physiologically more stressful habitats as the high intertidal, mangrove swamps, and open-ocean sandy beaches. East-west differences also seem to be undetectable among the snails of cryptic environments, where predator mobility is limited. Although evidence is scanty, bivalves as a group seem to exhibit few east-west differences in shell form.

The interoceanic differences in predation-related gastropod shell features in open-surface environments led me to hypothesize that Pacific Ocean and Indian Ocean crushing predators are more powerful or are able to take larger, sturdier prey items than can corresponding predators in the tropical Atlantic. Thus far this hypothesis has been tested in only two groups, the brachyuran crabs and palinurid spiny lobsters. Inspection of the dimensions and shapes of crab crusher claws has revealed that molluscivorous crabs of the Indo-West-Pacific, and to a lesser extent of the Panama region, have larger, and presumably more powerful, claws than Atlantic congeners of the same carapace width (Vermeij, 1976, 1977a). The Indo-West-Pacific *Carpilius maculatus* has a crusher claw with mean height and thickness each 1.09 times those of equal-sized *C. corallinus* from the Western Atlantic. The crusher claw of the Indo-West-Pacific *Eriphia smithii* is about 1.20 times as high and thick as that of the Eastern Pacific *E. squamata* of equal size, and about 1.25 times as large as that of the Western Atlantic *E. gonagra*. *E. sebana*, another Indo-West-Pacific species, also has higher and thicker claws (height-to-carapace-breadth ratio, 0.527; thickness-to-carapace-breadth ratio, 0.332) than does either tropical American species. The difference between the Eastern Pacific *E. squamata* (ratios 0.458 and 0.312) and the Western Atlantic *E. gonagra* (ratios 0.429 and 0.295) is small—6 to 7 percent—but significant at the 0.05 probability level. Moreover, these differences refer to linear dimensions, whereas muscle power is proportional to cross-sectional area of muscle and claw; therefore, a claw that is 5 percent higher and 5 percent thicker than another actually has a cross-sectional area that is 10 percent greater.

Similar patterns hold within the genera *Ozius* and *Daldorfia*. In the soft-bottom Portunidae, the Indo-West-Pacific genus *Scylla* (with the single species *S. serrata*) has much larger and more specialized crusher claws (ratios 0.394 and 0.235) than does *Callinectes latimanus* from West Africa, the most robust member of a related Atlantic and Eastern Pacific genus (ratios 0.229 and 0.156). There are no striking interoceanic differences in claw size among species of *Callinectes*.

In addition to these patterns in relative claw size, there is a general trend for Pacific and Indian Ocean crabs to be larger as adults than their Atlantic congeners or other close relatives. The Indo-West-Pacific *Scylla serrata* attains a width of more than 20 cm; within the genus *Callinectes*, the Eastern Pacific *C. bellicosus* and *C. toxotes* commonly exceed 15 cm in width, while no tropical Atlantic species is so large. *Eriphia sebana* and *E. smithii* in the Pacific are as much as 6.5 cm in width, while the Eastern Pacific *E. squamata* is no larger than 5.5 cm. The largest Western Atlantic *E. gonagra* I have seen is only 4.8 cm wide.

An exception to all these rules occurs in the xanthid genus *Menippe*, in which the largest claws are found in the two Western Atlantic species, *M. nodifrons* (ratios 0.488 and 0.309) and *M. mercenaria* (ratios 0.489 and 0.265). Proportionally the smallest claws are found in *M. rumphii* (ratios 0.408 and 0.279), an Indo-West-Pacific species that is the smallest in the genus. The case of *M. mercenaria* is particularly interesting and

Table 5.3 Frequency of repaired shell injuries in some tropical species of *Nerita*. Data have been compiled by E. Dudley from my collections and from specimens in the U.S. National Museum, Washington, D.C., and the Academy of Natural Sciences, Philadelphia.

Species	Region	n	F
High intertidal:			
N. *plicata*	Indo-West-Pacific	19	0.15
N. *undata*	Indo-West-Pacific	9	0.36
N. *scabricosta*	Eastern Pacific	12	0.32
N. *versicolor*	Western Atlantic	15	0.11
N. *peloronta*	Western Atlantic	9	0
N. *ascensionis deturpensis*	Fernando de Noronha, Brazil	5	0
Middle and lower intertidal:			
N. *exuvia*	Western Pacific	5	0
N. *textilis*	Indian Ocean	6	0
N. *doreyana*	Indo-West-Pacific	5	0.18
N. *picea*	Hawaii	6	0.31
N. *sanguinolenta*	Red Sea	5	0
N. *albicilla*	Indo-West-Pacific	10	0.05
N. *polita*	Indo-West-Pacific	8	0.05
N. *funiculata*	Eastern Pacific	10	0.05
N. *tessellata*	Western Atlantic	9	0
N. *senegalensis*	West Africa	8	0.06

n = number of populations examined.
F = frequency of repaired injuries per shell in modal population.

puzzling, since this stone crab is a warm-temperate species common on the coasts of the southeastern United States and Cuba, but is absent from the tropical Atlantic proper. In fact, the coasts from North Carolina to Florida and Texas are even more remarkable in being occupied by several very large molluscivorous gastropods (*Fasciolaria, Pleuroploca, Busycon*, and *Muricanthus*), which for the most part do not extend to strictly tropical waters.

Whether other molluscivores exhibit regional variations in crushing ability remains largely an unresolved question at present. Kent (1978) has shown that the width and length of the crushing molars of spiny lobsters (*Panulirus*) change very little relative to body size among tropical species, except that the small West Indian *P. guttatus* seems to have relatively larger molars than do other species. There is also evidence that *P. gracilis*, a Panamic species, has smaller mandibles than does the widely distributed *P. penicillatus* from the Indian and Pacific oceans. No information is yet available concerning the relative size and power of the muscles that operate the mandibles, since these are hidden within the skeleton and cannot be measured without destroying the specimen.

One indirect method of measuring the intensity of crushing predation is to count the number of repaired injuries in the average shell of a given size in a population of snails. The greater the number of injuries per shell of a given size, the greater the cost of shell repair, and the greater the intensity of predation on smaller individuals that can be successfully overpowered by the predator. Work now in progress indicates that the frequency of repaired injuries is generally higher in the Indo-West-Pacific than in the Atlantic, with Eastern Pacific populations as usual falling between the two. Some sample values are given in Table 5.3.

Application of the Mann-Whitney U-test demonstrates that the frequency of repaired injuries is quite significantly higher in the Indo-West-Pacific *N. plicata*, *N. undata*, *N. picea*, and *N. doreyana*, and the Eastern Pacific *N. scabricosta*, than in comparably zoned species in the West Indies (*N. versicolor*, *N. peloronta*), Fernando de Noronha (*N. ascensionis deturpensis*), and West Africa (*N. senegalensis*). *N. textilis* in the Indian Ocean and *N. exuvia* in the Western Pacific, two very thick-shelled species from the high middle intertidal zone, are rarely attacked. The lower intertidal *N. albicilla* from the Indo-West-Pacific shows a higher frequency of repairs than does the comparably zoned Eastern Pacific *N. funiculata*, which in turn has significantly more injuries than the Western Atlantic *N. tessellata* and the Red Sea *N. sanguinolenta*.

Temperate Molluscs

From rather limited experience on temperate shores, it is my distinct impression that gastropods, bivalves, and chitons are generally larger and more heavily armored in the Pacific than in climatically comparable parts of the Atlantic. For example, the Chilean and Peruvian muricid *Crassilabrum crassilabrum* has a larger, thicker shell with an internally more reinforced dentate outer lip than do species of the closely related genus *Ocenebra* in Japan, western North America, and Europe (see Figure 5.7). Intertidal trochids of the genus *Tegula* on the Pacific coast of North America (*T. funebralis* and *T. brunnea*) and South America (*T. atra* and *T. tridentata*) have sturdier and often larger shells than their European counterparts in the genera *Monodonta* and *Gibbula*; intertidal trochids are entirely absent in the northwest Atlantic. Members of the archaeogastropod family Turbinidae, which are characterized by a calcareous operculum, occur on warm-temperate shores in Australia, New

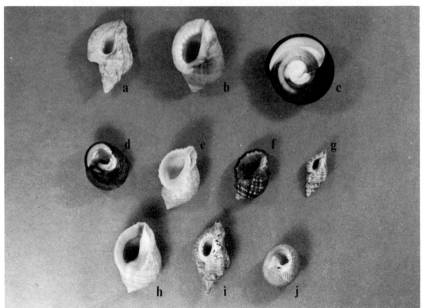

Figure 5.7   Dorsal *(above)* and apertural *(below)* views of some temperate rocky-shore snails. (Photographs by M. Montroll.)

    a. *Crassilabrum crassilabrum*, Montemar, Chile
    b. *Acanthina calcar*, Montemar
    c. *Tegula atra*, Montemar
    d. *T. funebralis*, Botanical Beach, Vancouver Island, British Columbia
    e. *Thais canaliculata*, Botanical Beach
    f. *T. emarginata*, Botanical Beach
    g. *Ocenebra interfossa*, Cliff Island, near Friday Harbor, Washington
    h. *Thais lapillus*, near Plymouth, south coast of England
    i. *Ocenebra erinacea*, near Plymouth
    j. *Gibbula cineraria*, near Plymouth

Figure 5.8   Variation in *Thais lamellosa* at San Juan Island, Washington. (Photograph by F. Dixon.)
- a. Garrison Bay; large form in shelter, strongly sculptured
- b. Cliff Island; high-spired form, well-developed spiral cords
- c. Collins Cove; small high-spired form, reduced sculpture
- d. Deadman's Bay; strongly sculptured form from exposed west side

Zealand, Japan, South Africa, western North America, and western South America, but are unknown in the Northwest Atlantic and in Argentina.

The thick-shelled Northeast Pacific *Thais lamellosa* (Figure 5.8) and western South American *T. chocolata* are remarkable among low-intertidal, cold-temperate muricids in being large (up to 10 cm long) and in possessing distinct knobs at the shoulder of the whorls in some populations. In other temperate muricids such knobs or frills are more delicate and are restricted to subtidal species (*Boreotrophon, Austrotrophon*) or to sheltered populations of species, which on more wave-exposed shores are smooth or weakly ridged (*T. emarginata* in western North America, *T. lapillus* in the North Atlantic). Other cold-temperate intertidal *Thais*, or members of related genera, are usually smaller than *T. lamellosa* and *T. chocolata*, and their sculpture consists of weak or strong spiral cords crossed by growth lines (*Neothais* in Australia and New Zealand, *Lepsiella* in New Zealand, *Thais cingulata* and *T. squamosa* in southern Africa, *T. lima* and *T. canaliculata* in the North Pacific).

Muricids with a strong spine on the edge of the outer lip are found in the temperate Eastern Pacific *(Acanthina* in North and South America, *Ceratostoma* in North America) and in Japan *(Ceratostoma)*. Elsewhere such muricids or forms convergent with them are absent.

Chitons and limpets are important constituents of the intertidal fauna on virtually all temperate shores. In the North Atlantic, however, chitons are limited to a few small species living on shells or under stones; no form equivalent to such large open-surface chitons as the Northeast Pacific *Katharina* and *Cryptochiton*, the Chilean *Acanthopleura* and *Enoplochiton*, or the Australian and New Zealand *Sypharochiton* occur either in the North Atlantic or along the coast of Argentina (Lewis, 1964; Olivier et al., 1968). Only in the Mediterranean are some chitons found on open intertidal surfaces (Starmühlner, 1969).

Large intertidal limpets, more successful in penetrating the Atlantic, are conspicuous in Western Europe and the Mediterranean *(Patella)*, South Africa *(Patella* and *Helcion)*, and the Southwest Atlantic *(Nacella)*. In the Northwest Atlantic, however, a single small limpet *(Acmaea testudinalis)* is found in the intertidal zone, where it occurs at middle and low shore levels (Stephenson and Stephenson, 1954a, b; Wallace, 1972). Limpets abound in great diversity on all Pacific shores.

The common mussels (Mytilidae) of cold-temperate North Atlantic rocky shores *(Mytilus edulis* and *Modiolus modiolus)* have smooth shells and smooth inner valve margins. Species of this morphology are also found in virtually all other temperate regions; indeed, M. *edulis* is known (as various subspecies) from the entire temperate Northern Hemisphere, as well as from southern South America, Australia, New Zealand, and Kerguelen Island (Soot-Ryen, 1955). However, cold-temperate mussels with crenate margins and radiating ridges are found only in the Pacific Ocean and in the south temperate zone: *Perumytilus purpuratus* and *Brachidontes granulatus* in Chile, *Choromytilus crenatus* on the west coast of southern Africa, and *Aulacomya maoriana* in New Zealand.

Epifaunal and shallowly infaunal bivalves in general are larger in the North Pacific than in the North Atlantic. To cite but a few examples: the European intertidal cockle *Cerastoderma edule* attains a shell length of not more than 5 cm, while the Northeast Pacific *Clinocardium nuttalli* may exceed 10 cm in length; North Atlantic *Mytilus edulis* may attain a length of 11 cm, but M. *californianus* may be more than twice as long (25 cm); the North Atlantic *Macoma balthica* (3.5 cm) remains considerably smaller than the Northeast Pacific M. *nasuta* (6 cm) and especially M. *secta* (10 cm). Such size differences are much less conspicuous among deep-burrowing mactrids (*Tresus* in the Northeast Pacific and *Lutraria* in the Northeast Atlantic, both up to 25 cm long) and solenid razor clams.

With a few exceptions, cold-temperate Pacific predators seem to be either larger or stronger than their Atlantic counterparts. Within the genus *Cancer*, proportionally the largest claws are found in the small (5 cm broad) *C. oregonensis* (claw-height-to-carapace-width ratio, 0.432; claw-thickness-to-carapace-width ratio, 0.254), and in the larger (13 cm broad), more southerly distributed Californian *C. antennarius* (ratios 0.319 and 0.189). *C. pagurus*, a European crab, is the largest species in the genus (27 cm broad), but its claws are similar in relative size to those of *C. productus* in the Northeast Pacific, *C. borealis* in eastern North America, *C. novaezelandiae* in New Zealand, and *C. polyodon* in Chile (ratios for males about 0.24 and 0.14 respectively) (Vermeij, 1977a). Species of *Cancer* are not found in southern Africa or on the temperate coasts of Argentina and Uruguay (Nations, 1975), nor is the genus replaced there by functionally equivalent forms; indeed, none of the approximately twenty-five species of brachyuran crabs known in the vicinity of Buenos Aires has large crushing claws (see Boschi, 1964). South Australia is the home of the largest known brachyuran crab, the xanthid *Pseudocarcinus gigas* (ratios 0.385 and 0.240), a species said to reach 36 cm across the carapace and possessing large, though fairly slender, claws with massive crushing molars.

Large predatory sea stars seem to be characteristic of most temperate Pacific shores: *Pisaster*, *Pycnopodia*, and *Orthasterias* in the Northeast Pacific; *Stichaster* in New Zealand and western South America; *Meyenaster* and *Heliaster* in Chile (see Dayton et al., 1977); and *Coscinasterias* in South Australia. Smaller sea stars such as *Leptasterias* accompany these forms in the North Pacific, and the large *Solaster*, which feeds on holothurians or other sea stars (Mauzey et al., 1968), is found in both the Pacific and the Atlantic. The only molluscivorous sea stars in the Atlantic tend to be small forms less than 10 cm or so in diameter: *Asterias* and *Leptasterias* in North America and Europe, *Marthasterias* in Europe and southern Africa (see Stephenson, 1944).

The huge North Pacific *Octopus dofleini* (up to 9 m across) is the largest member of its genus and ranges much farther north (to Alaska) than does the smaller Atlantic *O. vulgaris* (up to 3 m across), which is found as far north as Connecticut in southern New England and the North Sea in Europe. Neither the urchin- and snail-eating North Pacific sea otter (*Enhydra lutris*) nor the omnivorous scraping Chilean clingfish (*Sicyases sanguineus*) has any equivalent in the North Atlantic, or for that matter in any other temperate region. On the other hand, the lobster *Homarus* is an Atlantic genus with no Pacific equivalent.

Differences in predation intensity on the two coasts of North America may also be inferred from the distribution pattern of the circumboreal mussel *Mytilus edulis*. In New England and in Great Britain, this species occurs both in sheltered bays and on rocks exposed to heavy wave action; but in the Northeast Pacific (at least in California), *M. edulis* is restricted to sheltered bays or locally protected places on the open coast; it is replaced in areas of wave surge by the larger, stronger *M. californianus* (Harger, 1972a, b). At any given shell size *M. californianus* is more immune to predation by crabs than is *M. edulis*, and the same may be true with respect to predation by asteroids, although this requires confirmation (see Harger, 1972b).

Grazing

Bakus (1966, 1969) has suggested that the grazing intensity on algae and corals is greater in the Indo-West-Pacific than in physically comparable habitats in the Western Atlantic. Earle (1972a) and Glynn (1972) have further argued that the scarcity and low profile of benthic algae in the Eastern Pacific as compared to the Caribbean may also result from the more intense grazing by fishes, echinoids, and other herbivores. This intense grazing affects not only algae, but also such sessile animals as barnacles, bivalves, sponges, tunicates, and bryozoans (Bakus, 1964, 1966, 1967, 1969, 1974; Bakus and Green, 1974).

In the West Indies, for instance, demosponges are a conspicuous component of seaward reef faces in shallow water (Reiswig, 1973) and show little evidence of predation. The only fishes in the West Indies that regularly feed on sponges are certain specialized angelfishes of the genera *Pomacanthus* and *Holacanthus*, and the triggerfish *Cantherines* (Randall and Hartman, 1968). The situation is quite different in the Indo-West-Pacific and Eastern Pacific. Here fishes with generalized diets, as well as more specialized forms (such as the batfish *Platax*), prey heavily on sponges, and the latter are often rare on exposed surfaces (Bakus, 1964, 1966, 1969; C. E. Birkeland, personal communication). If interoceanic patterns in toxicity or spicule form of sponges exist, these have not yet been recognized.

Birkeland (1977) appears to be the first and the only investigator to have measured grazing intensity directly and by the same experimental method in more than one ocean. He set out plexiglass plates in various environments on the two coasts of Panama and weighed the biomass that had accumulated on the upper and lower surfaces of the plates after these had been left for 133 days. In a coral-reef environment on the Caribbean side, plates available for grazing by fishes had accumulated only 44 percent of the biomass present on protected plates; in a setting with small corals, grazed plates acquired 72 percent of the biomass of the protected plates. On the Pacific coast, grazed plates set in a small coral reef had 53 percent as much biomass as did plates protected from grazing. The biomass on grazed plates in a nearby area of strong upwelling and with almost no corals present was actually 1.4 times greater than that of protected plates. These results indicate that grazing intensity is highly variable from place to place, even within small areas of a single biogeographical province; expressed as ratios of biomass on grazed and ungrazed plates, they give no support to the contention that grazing intensity is greater on the Pacific coast than in the Caribbean. However, if the difference rather than the ratio of grazed and ungrazed biomasses is considered, a different picture emerges. In the Pacific coral community, there was a difference of 3.0 g between grazed and ungrazed plates; in the Atlantic coral-reef and coral-community plates, the differences were only 1.0 and 1.5 g respectively. Given that all the plates used by Birkeland were the same size (14 cm by 5 cm), it seems that Eastern Pacific grazers removed a greater biomass than did Atlantic grazers.

Unfortunately, Birkeland's experimental results cannot be compared with those of Vine (1974) in the Red Sea. Vine used different-sized plates and ran his experiments for 30 rather than for 133 days; he found grazed plates to have more than an order of magnitude less biomass than ungrazed controls. In future experimental work on interoceanic differences in grazing pressure, it will be important to keep plate area, depth, habitat, and duration of the experiments constant.

If grazing or browsing intensity does vary in comparable habitats among oceans, we might expect to see architectural differences among the sessile inhabitants, much as we see such differences among molluscs. Almost no work has been done on this subject, and I have already commented on the apparent regional uniformity of epifaunal bivalves and balanomorph barnacles. I have, however, been impressed that the brown algae *Sargassum cristaefolium* and *Turbinaria ornata*, two abundant species on wave-exposed intertidal benches and algal ridges in Guam and other high islands in the Pacific, are much coarser and spinier than are the corresponding Caribbean species, *S. vulgare* and *T. turbinata*. Possibly this greater spininess reflects a greater commitment to antiherbivore armor among the Pacific as compared to the Atlantic species.

Reef-building corals differ strikingly in diversity from one tropical region to another, but they exhibit an equally remarkable convergence of form. Wells (1957) has noted that specific architectural types can be recognized in the coral faunas of the Western Atlantic and Indo-West-Pacific, despite the fact that these two regions have very few coral genera in common. Thus, the "brain" coral form is represented in the Caribbean by *Diploria* and in the Western Pacific and Indian oceans by *Platygyra*. Thin platy corals in the Caribbean belong to the genus *Agaricia*, while morphologically identical types in the Indo-West-Pacific belong to *Leptoseris* and related genera.

Nevertheless, some architectural types do seem to be better represented in the Indo-West-Pacific than in the other tropical biotas. In the richest Western Atlantic reefs, off Jamaica and Panama, some fifty species of hermatypic coral have been recorded (Glynn, 1973b; Porter, 1974a). In the Western Pacific, on the other hand, R. H. Randall has found approximately three hundred species of colonial coral in Guam, and similar indications of riotous diversity come from other parts of the Western Pacific and central Indian oceans (Wells, 1957; Pichon, 1971; Rosen, 1971; Stehli and Wells, 1971). Only about twenty hermatypic coral species are recognized from Eastern Pacific reefs; most of these belong, or are very closely related, to Indo-West-Pacific species (Porter, 1972a, 1974a). West Africa supports only about nine species of coral (Laborel, 1974). C. E. Birkeland has pointed out to me that the great richness of Indo-West-Pacific as compared to Caribbean and West African coral faunas is particularly evident among fast-growing, branching forms (many *Pocillopora*, *Stylophora*, *Montipora*, and especially *Acropora*). In the Caribbean, only three species of *Acropora* occur, while in West Africa and West America this genus is entirely lacking; dozens of species are found throughout the Indo-West-Pacific. The West American fauna may be unusual in that branching *Pocillopora* species are the predominant corals despite the low overall diversity of species (Porter, 1974a; Glynn, 1976).

It remains an open question whether the disproportionate abundance and diversity of branching corals in the Eastern Pacific and Indo-West-Pacific is related to more intense predation there. Laxton (1974) argued that, since *Acanthaster planci* on the Australian Great Barrier Reef seems to prefer branching over massive corals, the latter should be relatively more prominent on Indo-West-Pacific reefs where this coral-eating, crown-of-thorns sea star is common. In a series of elegant experiments, however, Glynn (1976) has been able to show that *Acanthaster* in Pacific Panama eats the branched *Pocillopora* only when colonies of this coral are devoid of snapping shrimps (*Alpheus lottini*) and crabs (*Trapezia ferruginea*). These two coral-associated crustaceans interfere with the feeding of *Acanthaster* by snapping or nipping at the asteroid's tube feet. Since *Trapezia* and *Alpheus* live with pocilloporid corals throughout the Pacific and Indian oceans, it is tempting to agree with Glynn (1976) that it is the branching rather than the more massive forms which should predominate in *Acanthaster*-occupied reefs. In the Red Sea, however, Ormond and colleagues (1976) have shown that, while *Pocillopora* tends to be shunned by *Acanthaster*, branching *Acropora* is always preferred over such large-polyped corals as fungiids, *Favia*, and *Cyphastrea*; the large-polyped corals may have more powerful nematocysts than do species of *Acropora*. Still another complicating factor is that branching corals may be superior to massive forms in competing for light (Lang, 1973; Porter, 1974a), and it is possible that the greater diversity of coral species with overtopping-growth forms in areas with high overall coral diversity reflects favorable conditions for the evolution and retention of competitively superior species.

Asteroid predators such as *Acanthaster*, *Nidorellia*, and *Culcita* affect corals only in the Pacific and Indian oceans. This statement holds with equal force for many other coral predators as well. Parrotfishes (Scaridae), triggerfishes (Balistidae), butterflyfishes (Chaetodontidae), blennies (Blenniidae), wrasses (Labridae), puffers (Canthigasteridae, Tetraodontidae), and damselfishes (Pomacentridae) include in their ranks many species for which coral constitutes an important part of the diet in the Pacific and Indian oceans; but few Atlantic species eat coral (Hiatt and Strassburg, 1960; Randall, 1967, 1974; Glynn et al., 1972; Glynn, 1973b; Hobson, 1974). The hermit crab genus *Aniculus*, reported by Glynn and associates (1972) to pinch off pieces of coral with their chelae on Pacific reefs off Panama, is restricted to the Indo-West-Pacific and Eastern Pacific. Similar provincial restrictions hold for the common corallivorous gastropods *Drupella* (Indo-West-Pacific), *Jenneria* (Eastern Pacific), and *Quoyula* (Indo-West-Pacific and Eastern Pacific).

Evidence from coral-associated gastropods further supports the premise that grazing may be more intense in the Indian and Pacific oceans than in the Atlantic. West Indian coral-eating snails, belonging mostly to the Coralliophilidae and Calliostomatidae, are thin shelled relative to open-surface molluscs that do not feed or live on corals; none of the corallivores is known to excavate pits or cavities in its host. Indo-West-Pacific coralli-vores in the Coralliophilidae, Muricidae, and Epitoniidae are not noticeably thin or otherwise poorly armored relative to other open-surface snails. Among the Coralliophilidae, the limpet-like *Quoyula madreporarum* excavates deep cavities near the base of pocilloporid corals (Robertson, 1970), and the related *Magilus antiquus* has become a worm-like borer in corals.

Like these excavating coralliophilids, bivalves attached to or boring into living corals are rare outside the Pacific and Indian oceans. (For a review see Hadfield, 1976.) Although rock-boring clams in such genera as *Lithophaga*, *Gastrochaena*, *Pholas*, and *Botula* are known throughout the tropics as well as in the temperate zones, borers partial to living corals are known only from the Indo-West-Pacific—and even there seem to have very patchy distributions. The fragile mytilid *Fungiacava eilatensis*, for example, bores into living fungiid corals in the Gulf of Aqaba, the Maldive Islands, and the Marshall Islands, but thus far is unknown from the Philippines, Indonesia, and other large regions of the Indo-West-Pacific (Goreau et al., 1969). *Lithophaga lessepsiana*, found in the small solitary coral *Heteropsammia*, is known only from the Red Sea and western Indian Ocean (Arnaud and Thomassin, 1976). The byssally attached bivalve *Pedum spondyloideum* often is found in clefts of living coral in the Indo-West-Pacific. In these cases the partial escape from predation achieved by penetrating hard rock may be made more effective by association with nematocyst-bearing corals, since snails and other potential predators usually will not crawl on living coral tissue.

Moreover, snails living on the shells of other gastropods show a geographical pattern in their ability to excavate. *Sabia conica*, an abundant scatophagous hipponicid widely distributed in the Indo-West-Pacific, excavates a deep pit on host shells and is difficult to dislodge from its site of attachment. By contrast, epizoic snails elsewhere in the tropics (*Crepidula incurva* in the Eastern Pacific, *C. porcellana* in West Africa, *Acmaea pustulata* and *A. leucopleura* in the West Indies) make a very shallow pit or no excavation at all and are more easily sheared off. On the temperate coasts of Peru and Chile, the limpet *Scurria parasitica* grinds deep scars on the shells of other limpets and chitons (Figure 5.9), but other temperate epizoic limpets (*Crepidula* sp. in Chile, *C. adunca* and *Acmaea asmi* in the Northeast Pacific, *C. convexa* in the Northwest Atlantic) do not scar their host shells.

It is tempting to infer a connection between the peculiar habits of *S. parasitica* and the grazing activities of the clingfish *Sicyases sanguineus*. This remarkable intertidal fish not only scrapes vertical rock faces with its teeth while the body is attached to the cliff by a ventral sucker, but it probably can also insert its teeth under, and thus dislodge, the shells of fissurellid and acmaeid limpets, which it then swallows or scoops out. Possibly the excavating habit of *S. parasitica* protects this 2-cm-long limpet from predation by the clingfish; one fish was found to have a specimen of the limpet in its stomach (Paine and Palmer, 1978). The clingfish appears to be unique among intertidal animals anywhere in the world, both for its extraordinarily broad diet (barnacles, molluscs, crabs, echinoids, algae) and for its

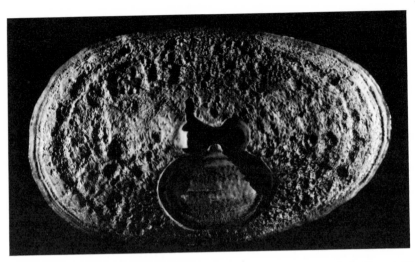

Figure 5.9   Scar left by the limpet *Scurria parasitica* on the keyhole limpet *Fissurella crassa* at Montemar, Chile. (Photograph by F. Dixon.)

potentially significant impact on the community in which it lives (Paine and Palmer, 1978).

   Some interoceanic patterns in the morphology of comatulid crinoids also may result from differences in grazing by fishes. Meyer and Macurda (1977) remarked that the proximal pinnules of many Indo-West-Pacific crinoids are stiff, large, and often spinose, and shield the vulnerable oral disc from attack. Such pinnules are unknown in Caribbean comatulids.

   Differences in grazing intensity may also account for the higher proportion of chitons and limpet-like gastropods on open rocky surfaces in the Atlantic and to a lesser extent Panamic regions than in the tropical Indo-West-Pacific (see also Vermeij, 1973c). For example, keyhole limpets (Fissurellidae) occur on open intertidal surfaces in the West Indies (six species on the north coast of Jamaica), West Africa (two species in Senegal), the Eastern Pacific (two species in Panama), and Brazil (one species), but are wholly absent from large portions of the Indo-West-Pacific. The only common occurrences of open-surface fissurellids in the Indo-West-Pacific are reported from the mainland coast of Queensland (Endean et al., 1956b) and from Ceylon (Atapattu, 1972); and in both cases, the limpets involved (*Montfortula* and *Clypidina*) lack the apical keyhold characteristic of the Atlantic–East Pacific *Fissurella*. As previously mentioned, limpets and chitons are relatively more prominent in number of species and in abundance on temperate than on tropical shores, and may be sheared off easily as juveniles by reef fishes and other rock scrapers.

   Relatively little has been published about possible differences in grazing pressure among temperate or polar regions. Lubchenco and Menge (1978) have demonstrated experimentally that overall grazing pressure seems to be substantially less in New England than in Washington state. In the Northwest Atlantic, sea urchins do not generally penetrate the intertidal zone except in pools, and snails (*Littorina* and *Acmaea*) are the principal herbivores. These molluscs are less effective, eat fewer species of algae, and graze more slowly than do the intertidal urchins (*Strongylocentrotus purpuratus* and *S. droebachiensis*) in the Northeast Pacific. Urchins also seem to be rare on intertidal surfaces throughout much of Great Britain (Lewis, 1964), and grazing pressures there may be comparable to those in New England. Temperate Pacific grazers (chitons, urchins, and the North Pacific sea cow *Hydrodamalis*) also tend to be larger in size than their North and South Atlantic counterparts.

Whether these apparent differences in grazing pressure from region to region are reflected in architectural differences among the sedentary biotas remains an unresolved question. Paine and Palmer (1978) noted that low intertidal plants and animals in Chile were either very small (encrusting and small articulated red algae) or very large (the barnacle *Megabalanus psittacus*, the brown algae *Durvillea* and *Lessonia*, fissurellid keyhole limpets, the chitons *Enoplochiton* and *Acanthopleura*), with few species being of intermediate size. They suggested that this bimodality in size might be related to the grazing of the clingfish *Sicyases*, and that such a bimodality is less conspicuous on other temperate shores. There seem to be few differences in the predation-related architecture among temperate kelps, at least to the untrained observer; nor do barnacles or other sessile epifaunal animals show any regional patterns except that many species in the Pacific are larger than Atlantic forms. It would be interesting to compare the proportion of sponge species in different regions according to spicule form or chemical repulsion.

Trophic Structure and Symbiosis

Several additional lines of evidence suggest that biotic interactions as a rule have proliferated and have been perfected more in the Indo-West-Pacific than elsewhere in the tropics. On the coasts of West Africa, Brazil, and the West Indies, comparatively few gastropod species are carnivorous; but in the Eastern Pacific and especially the Indo-West-Pacific, a majority of gastropod species prey on other animals (Figure 5.10). As mentioned earlier, the Florida region is somewhat exceptional in the Western Atlantic for having a large diversity of fasciolariid, melongenid, and muricid predators (see also Paine, 1963a; Radwin and Wells, 1968).

Wells (1957) suggested that symbiotic relationships are developed to a

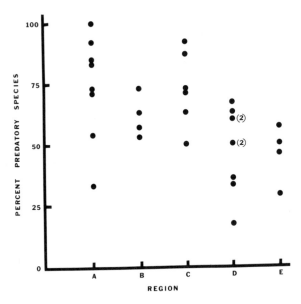

Figure 5.10 Percentage of predatory snail species in various tropical, low intertidal, rocky-shore assemblages of shelled gastropods. A predatory snail is defined
as any species feeding primarily on other animals. Except as noted, each point
refers to one assemblage at one locality. (Graph by E. Dudley; photograph by L.
Reed.)   A. Indo-West-Pacific   D. Western Atlantic
          B. Red Sea   E. Eastern Atlantic
          C. Eastern Pacific

higher degree on Pacific and Indian Ocean reefs than in the Atlantic tropics. Unfortunately, this interesting proposition has never been seriously
investigated, but some preliminary findings do predispose me to believe it.
For example, zooxanthellate bivalves occur in the Indo-West-Pacific (Tridacnidae and the cardiid genera *Corculum* and *Fragum*), but not in the
Caribbean (see Kawaguti, 1950; Stasek, 1961). D. L. Meyer tells me that
Indo-West-Pacific crinoids harbor a larger number of commensal species
than do ecologically similar Caribbean forms (see also Fishelson, 1974, for
data on Red Sea crinoids). Of the xanthid and hapalocarcinid crab genera
associated with living branching corals, the great majority are limited to
the Pacific and Indian oceans (Garth, 1974; Castro, 1976). Acroporid corals
in the Indo-West-Pacific may have commensal crabs of the genera *Tetralia*
and *Domecia*, while in the Western Atlantic only *Domecia* is commensal
with *Acropora*. The incidence of commensalism at the generic level in
caridean shrimps is 59 percent in the Indo-West-Pacific, 50 percent in the
Eastern Pacific, and only 41 percent in the tropical Atlantic (Bruce,
1976).
    Should further work prove that symbiotic relationships are indeed more

common and more specialized in the Indo-West-Pacific than in the Atlantic, a good case could be made that biotic interactions as a whole have proliferated and have been perfected to a greater extent in the former region.

Nonmarine Environments

Although much of the literature on aquatic and terrestrial communities has stressed the extent to which communities under similar climatic regimes are convergent (Richards, 1952; Cody, 1968, 1970; Keast, 1969), marine communities are not alone in exhibiting marked regional differences. Janzen (1976) has amassed anecdotal and circumstantial evidence which suggests that reptiles in tropical Africa have evolved more elaborate antipredatory devices than have their tropical American counterparts, and that predation upon them by both specialized and incidental carnivores is more intense. Most island communities are characterized by reduced predation and grazing pressure and differ from continental areas with similar vegetation in the greater biomass of reptiles, spiders, snails, fearless animals, heliophile animals, and so on; correspondingly, birds, large mammals, and insects as a rule are strongly reduced on islands (MacArthur and Wilson, 1967; MacArthur et al., 1972; Janzen, 1973a; Allan et al., 1973; Hamilton, 1973; Diamond, 1975). My qualitative impression is that the flora of the eastern United States contains a larger proportion of aromatic, and therefore presumably more distasteful, understory forest shrubs than does the flora of Europe. The African tropical rain forest supports relatively far fewer species of liane and vine than do the forests of tropical America or Southeast Asia (Richards, 1952, 1973). These and other differences may all be related to differences in predation or grazing pressure, or to the varying degrees to which biotic dependence has developed.

Similar regional differences probably occur in fresh-water fish and molluscan faunas (Roberts, 1972; Vermeij and Covich, 1978). Elaborate shell forms and specialized predator-prey interactions have been most dramatically evolved in such ancient basins as Lake Tanganyika, the Mekong and Mississippi river systems, and the cold-water lakes Titicaca, Ohrid, and Baikal (see also Brooks, 1950; Fryer and Iles, 1969, 1972).

Conclusions

The degree to which morphological antipredatory features have been developed by gastropods on open rocky surfaces is generally greater in the Pacific and Indian oceans than in the Atlantic, both in the tropics and in the temperate zones. Interoceanic differences also appear to be present in soft-bottom molluscs, but are not detectable among short-lived weedy molluscs or among those inhabiting open-ocean sandy beaches or the undersurfaces of rocks; there is also a puzzling circumtropical uniformity in the architecture of barnacles, epifaunal bivalves, and regular echinoids. More intense grazing pressure in the tropical Indo-West-Pacific and Eastern Pacific than in the tropical Atlantic is inferred from the distributional patterns of sponges, and from the degree to which certain animal-associated gastropod grazers excavate their host's skeleton. Indirect preliminary data suggest that symbioses or specialized biotic relationships may be more common and more widespread in the Indo-West-Pacific than elsewhere in the tropics.

From these patterns it appears that the greatest interoceanic differences occur in habitats where predation may be very intense, while the greatest uniformity is found in communities where the impact of predators seems to be diminished. The next chapter explores how productivity and diversity influence these architectural patterns.

CHAPTER 6

Relation to Present-Day Conditions

NOW that we have considered some of the interoceanic patterns in predation-related traits, let us inquire how and when the regional differences in the expression of these adaptations arose. First, in connection with present-day physical and biological conditions, I shall try to show that east-west variations in architecture for the most part are independent of wave regime, tide, or temperature. When the effects of productivity are considered, these too are found to be of little help in understanding patterns of gastropod form. Finally, the complex relation between armor and species diversity is examined. There is a trend for molluscan armor to be best expressed where present-day diversity is high, but the east-west pattern of diversity in the tropical oceans does not strictly conform to the interoceanic trend in antipredatory armor. The reasons for this partial correlation are explored toward the end of the chapter.

Waves, Tide, and Temperature

Regional variations in the shell architecture of molluscs are for the most part not correlated with purported differences in wave exposure. As noted in Chapter 2, several characteristics of gastropods are typically associated with severe wave stress. For example, it has occasionally been argued that the coarse sculpture of many open-coast gastropods diminishes the impact of waves on the shell by scattering the water streams and redistributing the forces over the shell as a whole (Vermeij, 1969). To my knowledge, this suggestion has never been experimentally tested; nonetheless, evenly spaced, rounded, or pointed knobs are very common in species restricted to wave-exposed rocks: the Indo-West-Pacific *Drupa morum*, *Morula granulata*, and *M. uva*; the West Indian *Fissurella nodosa*; the West African *Thais nodosa nodosa*; juveniles of the Californian limpet *Lottia gigantea*; and tropical American species of *Purpura*. In related species from more sheltered sites, sculpture becomes either more delicate (the long spines of many muricids, for instance), or more continuous (as with the radial ridges of the West Indian keyhole limpets *Fissurella barbadensis* and *Diodora listeri*).

If certain types of shell sculpture are effective in dissipating wave shock, then the higher incidence of such sculpture in the Indo-West-Pacific as compared to the tropical Atlantic may reflect nothing more than a generally more intense wave action in the Pacific and Indian oceans. Two lines of reasoning lead me to believe that interoceanic differences in wave action are not responsible for the observed patterns in external sculpture. First, while the windward shores of many Pacific islands indeed receive heavy wave exposure, so do many eastern Caribbean shores and the windward sides of Barbados and Fernando de Noronha. Algal ridges associated with strong turbulence, though relatively uncommon in the Atlantic, nevertheless are well developed in parts of the eastern Caribbean and off the Atlantic coasts of Panama and Brazil (Kempf and Laborel, 1968; Laborel, 1969; Glynn, 1973b). In the second place, rocky-shore gastropods from the northern Red Sea, where turbulence is greatly reduced relative to the open Pacific Ocean, have a high incidence of wave-baffling sculpture (as in *Drupa morum*, *D. ricinus hadari*, *Morula uva*, *M. granulata*, *Tectus dentatus*, and *Vasum turbinellus*). Moreover, a large proportion of Indo-West-Pacific gastropods is characterized by long, narrow apertures and by other geometrical configurations not optimally designed for withstanding severe turbulence. The West Indian trochid *Cittarium pica*, with its large foot and low-spired shell, for instance, seems much better adapted to wave stress than the Indo-West-Pacific species of *Trochus* and *Tectus*, whose feet are smaller and whose spires are considerably higher. In other words, present evidence does not reveal consistent differences in turbulence among the several tropical oceans, and the general correlation between shell architecture and turbulence is beset with numerous exceptions and complications.

In the temperate zones, shores facing the open Pacific seem to be exposed to generally heavier swells and waves than are the shores bordering the open Atlantic, and a fauna specialized to an open-coast existence has evolved in Pacific North and South America but not in the North Atlantic (Guiler, 1959a, b; Ricketts and Calvin, 1968; Marincovich, 1973). Typical molluscs of the wave-exposed coasts of western North America include the mussel *Mytilus californianus*, the limpets *Lottia gigantea* and *Diodora aspera*, and the snail *Thais canaliculata*; Chilean examples include the chitons *Acanthopleura echinata* and *Enoplochiton niger*, the limpets *Fissurella crassa* and *Scurria viridula*, and the limpet-like muricid *Concholepas concholepas*. Many of these open-coast species are large, with either thick shells or strong sculpture or both. Nevertheless, many of the most heavily armed molluscs in the temperate Eastern Pacific are not restricted to turbulent conditions. The Chilean muricids *Acanthina calcar*, *Thais chocolata*, and *Crassilabrum crassilabrum* are found at both exposed and sheltered sites, and the Northeast Pacific *Thais lamellosa* is typically found on relatively sheltered shores. In the North Atlantic, populations of *Thais lapillus* from the open coast have smaller, thinner shells with a wider aperture than do populations from less exposed localities (Kitching et al., 1966). A similar polymorphism has been recognized among New Zealand species of *Lepsiella* (Kitching and Lockwood, 1974). Thus, there is no consistent correlation between shell armor and wave exposure among temperate snails, just as there is none among tropical species.

Tidal factors are of paramount importance in imparting a vertical zonation to the shore community, but probably do not significantly affect the expression of molluscan architectural defense. Periodic aerial exposure represents a physiological stress for most intertidal organisms, since most inhabitants of the shore are of marine origin. During low tides feeding and growth generally cease, and highly mobile predators are usually confined to pools. The effects of this tidal stress are most severe when ebb tides come in the middle of a hot summer day or in the middle of a cold winter night, or if variations in sea level or wave intensity are greater than tidal amplitude. Gislen (1943), for example, has argued that the tidal regime at Pacific Grove, California, imposes less physiological stress on intertidal organisms than do the tides in central Japan. In California, summer low tides occur early in the morning, while in winter the lowest tides fall in the middle of the day. In Japan, on the other hand, exposure to air is longest on summer afternoons and on winter nights.

In the West Indies, Guam, and certain other parts of the tropics, unexplained variations in sea level and periods of reduced wave action often result in aerial exposure of habitats that ordinarily are submerged for most or all of an average tidal cycle. Such perturbations have resulted in mass mortality of the intertidal biota and seem to be most frequent on shores with a small tidal amplitude (see Glynn, 1968; Yamaguchi, 1975b).

These regional variations in tidal regime and stress are reflected in the vertical distribution of sessile plants and animals, but are not correlated with patterns in the expression of antipredatory traits. For example, tidal amplitudes in the Red Sea (2 m) and oceanic West Africa (1.5 to 2 m) are of the same magnitude (Lawson, 1966; Fishelson, 1971), yet West African open-surface gastropods appear to be structurally weaker than those in the Red Sea. Antipredatory characteristics of snails in the West Indies and Brazil are expressed to about the same degree, yet tides in Brazil range from 1.5 m to as much as 3.2 m in amplitude, whereas West Indian tides have amplitudes of less than 60 cm (Lewis, 1960; da Costa, 1962; Glynn, 1968, 1972; Furtado-Ogawa, 1970; Vermeij and Porter, 1971). Tides in the Indo-West-Pacific vary from being irregular and of small amplitude (many Pacific atolls) to being semidiurnal and with amplitude as high as 3 or even 4 m (northern Australia, Singapore, East Africa, Madagascar) (Purchon and Enoch, 1954; Plante, 1964; Salvat, 1970, 1971; Hartnoll, 1976). Yet the predation-related architecture of molluscs is quite similar throughout this large region.

One puzzling pattern, which might be governed by regional differences in tidal regime, is the restriction of many groups in the Caribbean to the subtidal zone, while the same groups often occur intertidally in the Indo-West-Pacific. The cemented thorny bivalve *Spondylus*, the crab *Carpilius*, the gastropod *Bursa*, and the coral *Acropora* all occur intertidally on rocky shores in much of the Indo-West-Pacific, but are restricted to permanently submerged surfaces in the West Indies and Brazil. Cone shells (*Conus*), cowries (*Cypraea*), mitrids, and large hermit crabs of the genus *Dardanus* are plentiful on open intertidal surfaces of the Pacific and Indian oceans, but are rare or little diversified in the Western Atlantic. Even within the Indo-West-Pacific, groups typical of the intertidal in most of the region may be subtidal in some areas. In Hawaii, such common Western Pacific intertidal species as *Cypraea tigris*, *C. schilderorum*, *Bursa bufonia*, and *Terebra* spp. are found only below the low-tide line (Kay, 1967). In the northern Red Sea, cones, mitrids, and *Bursa* occur intertidally, but cypraeids seem to be very rare in the intertidal zone. Even in Hawaii, some cowries (*Cypraea isabella*, *C. caputserpentis*), cones, and mitrids retain the intertidal distribution typical of them in the rest of their geographical range; they are, in fact, characteristic elements of the Hawaiian intertidal biota.

It is plausible that the small tidal amplitudes of the West Indies and per-haps even the Red Sea render intertidal life for certain corals and mol-luscs (especially cowries, which have shells covered with the mantle) untenable because of frequent exposure to heat and desiccation. Cool winter air temperatures may keep certain cowries and corals out of the intertidal zone in Hawaii and the northern Red Sea, but a similar expla-nation seems unlikely for the West Indies or northeast Brazil, where air temperatures almost always are greater than 15° C. Moreover, it is sur-prising that echinoids such as *Echinometra* and *Diadema*, which are notoriously sensitive to heat and other physiological stress (Glynn, 1968), never exhibit tidal submergence and are faithful to the intertidal zone in all the areas mentioned. Clearly, the reasons for such differences in inter-tidal distribution are not well understood and deserve further study; but it must be stressed that the absence of certain groups in the West Indies or Hawaii intertidal does not explain the reduced expression of armor in the Atlantic as compared to the Pacific and Indian oceans. The largely sub-tidal West Indian cones, cowries, corals, and *Carpilius* are less well adapted to predation than are many of their Indo-West-Pacific relatives in the intertidal zone.

Tides along temperate coasts have, on the average, greater amplitudes than on tropical shores, although they vary considerably from place to place. The Mediterranean and Baltic seas have virtually no tides, while the coasts of Argentina, the Bay of Fundy, and the English Channel may have tides exceeding 10 m in amplitude (Stephenson and Stephenson, 1954a, b; Lewis, 1964; Colman and Stephenson, 1966; Kühnemann, 1972). Again, the expression of armor does not seem to be related to tidal amplitude. Many heavily armored Chilean gastropods range from north-ern Chile to beyond the Chiloe Islands in the south, yet the tidal ampli-tude along that coast varies from less than 2 m to more than 6 m.

Regional differences in temperature regime also do little to explain away the interoceanic architectural patterns. One could argue that armor should be more limited in areas where temperatures are occasion-ally low (Eastern Pacific and West Africa) than in areas where tempera-tures are consistently high (Indo-West-Pacific and Western Atlantic); but the data contradict such reasoning. It could further be argued that, if the West Indies were generally cooler than most of the Indo-West-Pacific, the reduced expression of armor among Western Atlantic gastropods might result from the more stringent thermal limitations on calcification and on muscular function. Reference to Table 1.2 will quickly reveal, however, that temperatures in the southern Caribbean and on the south coast of Puerto Rico are as constant and as high as those in Singapore, the Marshall Islands, and the central Indian Ocean.

Moreover, thermal variations within biogeographical regions are not reflected by the predation-related architecture of the biota. Molluscan species in Sierre Leone and Cameroon, where temperatures are relatively high and upwelling does not occur, are very similar in architecture to those on the coasts of such other West African countries as Ghana and the Ivory Coast, where upwelling in summer depresses sea-surface temperatures to below 20° C. In fact, intertidal species composition of molluscs and algae is much the same throughout tropical West Africa (Lawson, 1955, 1956, 1957, 1966; Longhurst, 1958). Hawaii, by and large, has lower sea-surface temperatures than do such other Indo-West-Pacific areas as the Marshalls, Guam, Singapore, and Kenya (Table 1.2); yet the Hawaiian intertidal gastropods differ from members of the same species in warmer parts of the Indo-West-Pacific only in being larger as adults *(Drupa morum, Conus ebraeus, Cypraea tigris)* or else not at all *(Drupa ricinus, Morula granulata, Mitra litterata, Conus chaldaeus)* (see Kohn, 1959b; Kay, 1967; Foin, 1972). Many molluscs in Bermuda (32° N) are also large compared to West Indian members of the same species, but in other ways do not differ from them (Abbott and Jensen, 1967). It is interesting that the larger crabs of Easter Island, Hawaii, Bermuda, and other marginally tropical areas not only are morphologically indistinguishable from individuals in more strictly tropical localities in the same biogeographical region, but have much the same adult size (see Verrill, 1908; Edmondson, 1954, 1959, 1962; Garth, 1973).

In the temperate zones, as we see in Table 1.2, temperature variations are less drastic in the Pacific, South Africa, and Europe than in eastern North or South America. Nevertheless, European intertidal snails are not consistently better protected against crushing predators than are the gastropods of New England or Argentina.

I conclude that, while calcification, muscular contraction, and other processes connected with predation and the adaptations to it are temperature dependent, the east-west variations in temperature within climatic zones still cannot explain the differences in expression of predation-related traits. Factors other than temperature, tidal regime, and turbulence must be responsible for these differences.

Productivity

Variations in shallow-water productivity (quantity of nutrients in the water column) among tropical regions apparently affect the distribution of corals and other photosynthesizing organisms, but do not explain many other aspects of the distribution of sessile plants and animals, nor do they shed light on east-west variations in gastropod architecture. Seasonal upwelling of cold water on the west coasts of tropical America and Africa, and locally near the equator in the central Pacific and northwestern Indian Ocean, replenishes the surface water with nutrients. Plankton blooms that result from upwelling make the water more turbid and create a substantial food supply for suspension-feeding benthic animals. Nevertheless, coral reefs are poorly developed in areas of upwelling. This fact, already recognized by Ranson (1952), is most clearly evident on a small scale in the Eastern Pacific. Here coral reefs thrive to a depth of about 10 m in the Gulf of Chiriqui (northwest Panama) and off the coast of Colombia, where cold, nutrient-rich water never reaches the shallow zone inhabited by corals; but reefs are absent or much reduced in the upwelling waters of the Bay of Panama (Glynn et al., 1972; Dana, 1975). In the Pearl Islands off Panama, coral reefs are best developed on the north and northeast sides of the islands, where upwelling is relatively limited, and least on the south and southwest sides where upwelling is intense during the windy dry season (Glynn and Stewart, 1973).

The poor development of coral reefs in areas of upwelling appears to result, not from the competitive exclusion of adult corals by other sessile forms, but from the slower growth of juvenile corals (Birkeland, 1977). Newly settled corals in the Bay of Panama, where upwelling is intense and light penetration is reduced during the dry season, grow more slowly than do juvenile algae and barnacles. The reasons for this slower growth are not altogether clear, but perhaps turbidity is unfavorable for the growth of young corals, or perhaps newly settled corals are less efficient than other suspension feeders at taking up the abundant planktonic food particles. Consequently, the juvenile corals are at a competitive disadvantage to fast-growing barnacles, hydroids, and even some algae. As adults, however, corals are able to hold their own in the presence of these competitors in spite of the low winter temperatures that depress rates of calcification and growth (Glynn and Stewart, 1973; Birkeland, 1977).

Productivity, however, is not always the critical factor in determining the taxonomic or architectural composition of sessile communities. This is particularly evident in comparisons between Indo-West-Pacific and Caribbean reefs, which flourish under similar conditions of clear, warm, nutrient-poor water, yet differ in the relative abundance of certain groups. For example, sea whips (Gorgonacea) and soft corals (Alcyonacea) occur in shallow waters in both oceans, but the former predominate in the West Indies and are comparatively rare in the Indo-West-

Pacific, whereas the soft corals have precisely the opposite pattern of abundance (Wells, 1957; Goreau, 1959; Laborel, 1969; Kinzie, 1973). Sponges are common on open reef faces in the Atlantic, but not in the Pacific and Indian oceans.

Equally striking differences in the composition of sessile communities occur among geographically separated regions of upwelling. Algae are small and scarce in various parts of the Eastern Pacific (Bay of Panama, Cocos Island, and the mainland coast of Costa Rica), but are luxuriant in the Galapagos, at Malpelo Island (between Panama and the Galapagos), and throughout West Africa (Sourie, 1954; Lawson, 1955, 1956, 1966; Hedgpeth, 1969a; Earle, 1972a; John and Lawson, 1974; Bakus, 1975; Birkeland et al., 1975). Hydroids are not particularly abundant in West Africa, but they are so abundant in the Bay of Panama that their stinging nematocysts constitute a significant annoyance—to me and to other observers of the intertidal zone.

The reasons for these differences among tropical regions remain for the most part obscure and demand further inquiry. Differences in algal abundance probably reflect variations in grazing intensity (Bakus, 1975), but it is not always easy to see why grazing pressures should be so different from place to place within a single biogeographical region. One factor known to be important is topography. Reef fishes, including parrotfishes and surgeonfishes, require crevices and hollows to hide in at night when these herbivores are not actively foraging over the reef (Randall, 1965; Earle, 1972b; Hobson, 1974). In fact, Randall has demonstrated experimentally that grazing pressure caused by fishes in the immediate vicinity of a reef is so heavy that a band of bare sand lies between the reef proper and the grassbeds beyond. The grassbeds, at considerable distance from the hiding places of the fishes, are less heavily grazed than the surfaces immediately adjacent to the reef. On open coasts with limited topography, such as the steep shores of Malpelo Island or the oceanic coasts of West Africa, algae may persist and become luxuriant because of the virtual absence of hiding places for large herbivorous fishes (see also John and Pople, 1973). Where branching corals create extensive three-dimensional structures, numerous refuges are available for fishes and other potential consumers. However, factors other than topography are probably also involved. Much of the Costa Rican coast, for instance, seems topographically quite simple, yet algae are scarce in most places. If grazing rates in Eastern Pacific and Indo-West-Pacific reefs are higher than those in the Caribbean (see Chapter 5), it strikes me as unlikely that this is caused by differences in topography or productivity.

Interoceanic differences in gastropod shell architecture also do not correspond to the east-west variations in productivity (or, for that matter, topography), despite the plausible argument that high nutrient levels might support larger gastropod populations that are sufficiently stable for exploitation by highly specialized predators. Armor is well developed in western tropical America, where waters are seasonally enriched with nutrients, but is poorly expressed in West Africa, where comparable oceanographical conditions exist. West African and West Indian gastropods display similar antipredatory characteristics despite differences in productivity on the two coasts of the tropical Atlantic.

The only effect that high productivity seems to have on gastropods and many other mobile animals is on growth rate and adult size. I have noticed that snails on the west coast of the Paraguana Peninsula of Venezuela, where upwelling occurs, are on the average 1.5 times as long as the same species on the adjacent islands of Aruba and Curaçao (see Chapter 8). Eastern Pacific snails, crabs, and bivalves are commonly 1.2 times as long or as wide as their Atlantic counterparts. Although the appropriate measurements have not been made, it seems reasonable to assume that larger adult size in productive waters results from faster growth rates during the active growth phase.

Regionality and Diversity

Thus far I have discussed patterns in prey and predator armament with only occasional reference to diversity, one of the most widely studied assemblage characteristics in ecology and biogeography. Yet many authors have pointed out a general positive correlation between diversity and the inferred importance of biological interactions (competition, predation, symbiosis) in regulating species populations (see, for example, Paine, 1966a; Sanders, 1968; Slobodkin and Sanders, 1969; MacArthur, 1972; Rex, 1973, 1976). Since armor is most apparent in assemblages where biological interactions (notably predation) are judged to be important sources of selection, correlations between diversity and expression of armor might be expected. Although such correlations do, in fact, appear along various microgeographical, latitudinal, and east-west gradients, the relationship is a complex one and there are a number of important exceptions and qualifications.

Diversity is known to increase from high to low shore levels (Glynn, 1965; Taylor, 1968; Moore, 1972), as does the expression of gastropod armor. A similar concomitant increase in diversity and armor occurs among infaunal bivalves along a gradient of decreasing sediment depth (see Jackson, 1972, for data on diversity). Rex (1973, 1976) has shown that gastropod diversity along a gradient of increasing water depth from the temperate Western Atlantic continental shelf to the abyss first rises to a peak at a depth of about 4,000 m, then decreases as biomass declines. This diversity pattern parallels the presumed impact of predators, but is not associated with obvious changes in shell armor. This must be advanced as a tentative conclusion, since virtually nothing is known about the methods used by deep-sea predators to exploit their prey, or about the adaptations that prey species have acquired to avoid being eaten.

Latitudinally, species numbers in most phyla, classes, and orders of plants and animals increase dramatically toward the equator, along with skeletal armor (for data on diversity see Fischer, 1960; MacArthur, 1965; Stehli et al., 1969). Such a trend is also detectable in north-south transects of specific habitats, including salt marshes and the high intertidal zone (see Macdonald, 1969; Spight, 1976a). Table 2.3 provides still further evidence of increased diversity and armor toward the tropics in the lower intertidal zone.

In east-west comparisons the correlation between diversity and armor is still positive, but the architectural differences among gastropods of the various tropical oceans are not simply the result of differences in diversity. Armor among molluscs decreases in a west-to-east direction in the tropics, peaking in the Indo-West-Pacific and being least developed in the Atlantic (especially West Africa). Correspondingly, the overall diversity of gastropods, bivalves, crabs, and echinoids decreases from the Indo-West-Pacific through the Eastern Pacific and Western Atlantic to a low in the Eastern Atlantic (Stehli et al., 1967; Chesher, 1972). This pattern is repeated among many families and genera of animals (Table 6.1), but in many other taxa diversity in the Western Atlantic is often higher than in the Eastern Pacific. The anomaly is particularly large in animals such as fishes, crinoids, stomatopods, and corals, which are usually found in or near coral reefs and sea-grass beds; these environments are widespread in the Western Atlantic, but rare and of limited extent in the Eastern Pacific (Glynn, 1972).

Further discrepancies in the correlation between diversity and armor are revealed in regional comparisons of gastropods in specific habitats. Both at high shore levels and in the lower intertidal zone, the highest incidence and greatest expression of antipredatory armor occur in parts of the Indo-West-Pacific where diversity is correspondingly high; but high values of diversity are not invariably associated with greater immunity to predation, nor is it always the case that areas with low diversity

Table 6.1   Diversity of some animal groups in the tropical biotas.

Group	Number of species in—			
	Indo-West-Pacific	Eastern Pacific	Western Atlantic	Eastern Atlantic
Hermatypic Scleractinia	300+	20	50	10
Mollusca	6,000+	2,100	1,200	500
Gastropoda				
Littorinidae	28	11	13	5
Strombus	38	4	7	1
Cassidae	15	5	7	4
Oliva	44	9	3	1
Turbinellidae	15	1	8	0
Harpidae	8	1	0	1
Mitra	87	12	5	5
Crustacea				
Stomatopoda	150+	40	60	10
Brachyura	700+	390	385	200
Uca	12	29	12	1
Daldorfia	6	1	0	1
Ozius	6	3	1	0
Eriphia	3	2	1	0
Carpilius	2	1	1	0
Callinectes	0	3	9	3
Fishes	1,500+	650	900	280

Source: These data have been compiled from works by Abbott (1959, 1960, 1968), Cernohorsky (1976), Forest and Guinot (1962), Crane (1975), Briggs (1974a), Rehder (1973), Rosewater (1970, 1972), Kilburn (1975), Vokes (1964, 1966), Vermeij (1977a), Williams (1974), and Zeigler and Porreca (1969). R. B. Manning has made the estimates for stomatopods. The data for Cassidae do not include *Sconsia*, *Morum*, and allied genera. Figures given for fishes, brachyurans, molluscs, and Indo-West-Pacific corals are rough estimates.

are characterized by a de-emphasis of armament. West Indian rocky shores, for example, support a rich gastropod fauna of up to thirteen species, in which armor is generally less well expressed than in comparably diverse assemblages in East Africa or the Southwest Pacific (Vermeij, 1973c, 1974a). Armor among low intertidal Hawaiian gastropods is developed to about the same high degree as in more diversified assemblages in other parts of the Indo-West-Pacific (Table 5.1). In this case, however, most of the Hawaiian species are abundant throughout the Indo-West-Pacific and thus are commonly found in areas supporting a local diversity higher than that in Hawaii. This situation recalls the intense predation on Eastern Pacific corals; there are few species of these corals, nearly all of Indo-West-Pacific extraction.

The most convincing demonstration that snails from a species-poor habitat may be more resistant to predation than are snails from another region where diversity is higher comes from the work of Palmer (1978). He showed that the puffer *Diodon hystrix* can crush all individuals of the four species of Caribbean *Nerita* and of *N. funiculata* in the Eastern Pacific, but that large *N. scabricosta*, the only other Eastern Pacific species in the genus, is immune from predation by puffers. None of the seven Western Atlantic species of *Strombus* has a thicker, more heavily armored shell than any of the four Eastern Pacific species in the genus. *Vasum caestum*, the single Eastern Pacific member of its genus, has a shell as heavy and as impregnable as *V. muricatum*, the largest and probably the sturdiest of the five Western Atlantic species.

Among brachyuran crabs the correlation between relative claw size and species diversity is less flawed (Vermeij, 1977b). For example, two of the three Indo-West-Pacific species of *Eriphia* (*E. sebana* and *E. smithii*) have relatively and absolutely larger and stronger crusher claws than does the single Western Atlantic species (*E. gonagra*); in the Eastern Pacific, where two species occur, *E. squamata* has claws larger than those of *E. gonagra* but smaller than those of *E. smithii*. Some of the nine species of *Callinectes* in the Western Atlantic have proportionally larger crusher claws than do any of the three Eastern Pacific species, but this difference may be partially or completely offset by the larger overall size of the Pacific species.

I conclude from the evidence just presented that armor and diversity in shallow water both tend to increase latitudinally from the poles to the equator, and microgeographically from stressed to physiologically more favorable environments; but that the extent to which predation-related adaptations are developed in species-rich areas depends on factors not necessarily correlated with diversity. This last qualification notwithstanding, it is useful to view the increase in diversity along a shallow-water gradient not as a random accumulation of species, but as a nonrandom addition in which two types of species are strongly emphasized: specialized or powerful predators (Paine, 1966a; Connell, 1972), and prey species with a strong morphological commitment to defense. Organisms typically found in species-rich communities may also be superior competitors for limited resources. (See also Diamond, 1975, for a thorough analysis of the characteristics of birds found in species-rich communities as compared to birds restricted to species-poor communities.)

The Maintenance of Diversity

Two questions now arise. First, is the relation between diversity and armor along gradients in shallow water merely fortuitous—a case where the same circumstances that promote defense are also responsible for higher diversity—or is there a causal connection between diversity and the intensity of predation? Second, what factors other than diversity determine the expression of predation-related adaptations in organisms living in species-rich communities? To answer these questions, it is important to specify more precisely the ecological conditions that permit large numbers of species to coexist, and to inquire whether present-day differences in these factors can account for the observed interoceanic patterns in predation-related architecture.

At the scale of local species diversity, it has been experimentally shown that predators and grazers specialized to feed on a single species or a narrowly defined category of organisms cannot persist if their principal source of food is scarce or unavailable at certain times of the year or tidal cycle. Predatory snails of the genus *Thais*, for instance, cannot exploit the highest mussels or barnacles on the shore because the period during which these snails can feed (while the tide is in) is too short for completion of the drilling and ingestion processes (Seed, 1969b; Connell, 1970). Sea stars such as *Pisaster* and *Stichaster* have similarly imposed upper limits above which they cannot effectively exploit sessile prey (Paine, 1971, 1974). Severe wave stress impedes the activity and feeding effectiveness of tropical parrotfishes, temperate crabs, sea stars, and intertidal snails (Kitching et al., 1959, 1966; Stephenson and Searles, 1960; Ebling et al., 1964; Dayton, 1971; Menge, 1976). In short, animals that depend for their food on other benthic organisms are least deterred in feeding, and are therefore most likely to persist, in physically stable and unstressed environments where food supplies are constant and predictable, if not opulent. Stability of resources and a lack of physiological stress should thus promote the maintenance of high local diversity, at least among herbivores and predators (Sanders, 1968).

A second factor influencing local species numbers is habitat or geographical area. MacArthur and Wilson (1967) have suggested that species number S is related to area A of a habitat or a region by the empirical equation $S = CA^z$, where C and Z are constants that depend on dispersibility and other characteristics of the organisms in question. Species whose population density is low cannot maintain themselves in small isolated patches of otherwise favorable habitat, since the inevitable fluctuation in population size caused by climatic and other vagaries will eventually drive the population to extinction. For plants and animals with high powers of dispersal, these effects of area are much less important than for poor dispersers; hence the diversity of the former is much more area dependent than that of the latter. Therefore, two regions differing in area, but not in food density or quality, do not normally support the same abundance or species number of large or mobile consumers. For example, large mammalian carnivores and certain insectivorous birds are typical of large land masses, but are underrepresented or absent on small oceanic islands, where ecologically they are partially replaced by smaller, less mobile lizards and spiders (Keast, 1969, 1970; Allan et al., 1973; Janzen, 1973a). Until man introduced pigs and goats, poorly dispersing large mammalian herbivores were rare on oceanic islands and ecologically were usually replaced there by smaller and slower birds, lizards, and tortoises (Janzen, 1973a; Bakker, 1975; Morse, 1975).

In the sea the effects of area on diversity, although relatively little studied, are probably most strongly felt in such semiterrestrial environments as mangrove swamps, open-ocean sandy beaches, and the rocky high intertidal zone (Taylor, 1971; Ansell et al., 1972; Dexter, 1972, 1974, 1976; Vermeij, 1973c, 1974a). In all these environments diversity is greatest on continental shores or in large archipelagos where long, continuous stretches of suitable habitat abound. For mangrove crabs, molluscs, and trees, diversity is highest in and around Malaysia (see also Macnae, 1968; Sasekumar, 1974). High intertidal snails attain their highest diversity in the western Indian Ocean, the Indo-Malaysian island arcs, and in the West Indies. The richest sandy-beach biotas are found in western tropical America, India, and perhaps West Africa (see also Lawson, 1966).

On the scale of biogeographical provinces or regions, area is an inconsistent indicator of diversity (see Table 6.1). Within the tropics, the Indo-West-Pacific has by far the largest area as well as the highest total and local species diversity in most groups. Next in area is the tropical Western Atlantic; and the diversity of fishes, stomatopods, gorgonians, crinoids, scleractinian corals, ophiuroids, and spiny lobsters is correspondingly higher than in the Eastern Pacific (see Briggs, 1967a, b, 1974a; Chesher, 1972; Porter, 1972a, 1974a; George, 1974). On the other hand, the species richness of molluscs, echinoids, asteroids, and brachyuran crabs is somewhat greater in the Eastern Pacific than in the Western Atlantic (Abele, 1972, 1974; Chesher, 1972; Olsson, 1972) despite the smaller size of the Eastern Pacific. West Africa has the smallest area and lowest diversity of the tropical marine regions.

In regions where physical conditions and area are favorable to the maintenance of predator populations, the predators themselves often have a positive influence on local diversity. Certain predators, including intertidal sea stars, gastropods, and grazing sea urchins, are known to prefer or to feed exclusively on competitively dominant sessile prey species; these predators therefore reduce competition for space among the prey and create open spaces suitable for colonization by other, competitively less successful species (Paine, 1966a, 1971, 1974; Paine and Vadas, 1969a; Dayton et al., 1970, 1974; Connell, 1970; Sammarco et al., 1974; Menge and Sutherland, 1976). Local diversity will decline or remain the same if the predators become highly efficient at finding and destroying their prey, or if they take a large number of prey species more or less indiscriminately (Addicott, 1974; Menge and Sutherland, 1976). This seems often to be the case when adult suspension feeders and deposit feeders remove most or all settling larvae and metamorphosed juveniles, often including members of their own species (Dayton et al., 1974; Sutherland, 1974; Woodin, 1976). Diversity of prey also decreases when highly mobile grazing fishes scour all available surfaces at frequent intervals (Stephenson and Searles, 1960; Randall, 1961; John and Pople, 1973). A striking case of reduction in prey diversity occurred when the rapacious fish *Cichla ocellaris* was introduced from South America into Gatun Lake, Panama Canal Zone (Zaret and Paine, 1973). Nicotri (1977) has found that preferential grazing by mid and high intertidal gastropods in the Puget Sound area on canopy diatoms with a chain-like morphology did not increase the diversity of the more tightly adhering understory diatoms.

The last several examples notwithstanding, it is in many ways surprising that most predators do not become so proficient that they eliminate most of the species around them. In part, the explanation is that various refuges protect certain segments of the prey population from overexploitation: barnacles and mussels extend farther up the shore than do their various predators; corals living in cracks escape predation by asteroids; many reef fishes avoid their diurnal or nocturnal predators by being active at twilight or dawn (Hobson, 1968, 1974; Connell, 1972, 1975; Woodin, 1978). Many prey species, however, broadly overlap with their predators; adult gammarid amphipods may live among the stems of *Spartina* grasses in salt marshes, where they are more or less protected from small predatory fishes, but the juveniles are found in exposed places where they are vulnerable to predation (Vince et al., 1976; Van Dolah, 1978). Young *Tegula funebralis* snails in the intertidal zone of the Northeast Pacific characteristically live above the foraging zone of the sea star *Pisaster*, but adult snails overlap extensively with this predator (Paine, 1969).

Janzen (1970) and Connell (1972) independently proposed that efficient long-distance dispersal of seeds or other propagules, coupled with low adult densities, might reduce the searching effectiveness of potential enemies. Potentially preferred prey individuals then would be widely and sparsely distributed in the community and be difficult or time-consuming to locate. This argument was taken one step further by Atsatt and O'Dowd (1976), who suggested that toxic or repellent species protect neighboring individuals of otherwise vulnerable species by discouraging predation, or by making it unprofitable for a predator to search in their immediate vicinity. High local diversity could thus in itself lessen the exploitation of individual species by specialized predators.

These hypotheses and their supporting data refer largely to terrestrial plants and their fungal and insect pests, but there is no reason to believe that similar arguments would not apply to marine communities. Birkeland (1974), for example, has suggested that the unpredictability of successful larval settlement of sea pens (*Ptilosarcus*) prevents those predators that specialize on young individuals from overexploiting their food supply. Many cnidarian-associated molluscs, crabs, and other animals probably enjoy some measure of immunity from predation because of the close proximity of stinging nematocysts; and McLean and Mariscal (1973) showed that hermit crabs carrying *Calliactis* sea anemones on their shells are less vulnerable to predation by calappid crabs than hermits whose shells are not so endowed. Woodin (1978) has shown that shallowly buried polychaetes are less vulnerable to mortality caused by the sediment-disturbing activities of blue crabs (*Callinectes sapidus*) and horseshoe crabs (*Limulus polyphemus*) when they live close to the tubes of the large onuphid worm *Diopatra cuprea* than when they live away from these sediment-stabilizing structures.

Although predation and diversity seem to be interrelated by a complex series of ecological mechanisms, it must be stressed that high levels of diversity can be maintained and perhaps also achieved without predation's playing a significant role. Jackson and Buss (1975) suggest that chemically mediated competition between sessile animals (and plants?) may permit many species to coexist because they form dominance rings or networks in which no species is competitively dominant over all the others. For example, if species A is dominant over species B, and species B is dominant over species C, then a ring or network is formed if species C is competitively dominant over species A. Networks of this kind occur in cryptic epifaunal communities of tropical caves and crevices, where predation seems to be a minor contributor to mortality and of little significance in controlling local diversity (Jackson and Buss, 1975). Lang (1971, 1973) in the West Indies and Glynn (1974) in the Eastern Pacific have described a remarkable linear hierarchy in competitive dominance among scleractinian corals, based on the ability of dominant species to digest and arrest the growth of subordinate neighboring species. Mussids and other large-polyped corals, which are superior competitors in the digestive hierarchy, usually are inferior competitors with respect to growth rate and often cannot outcompete such fast-growing, branching forms as *Acropora* and *Pocillopora* (Lang, 1973). Even where linear hierarchies exist in one determinant of competitive ability, networks may still form because competitive success is determined by more than one factor; that is, competitive exclusion and consequent reduction in local diversity is unlikely when dominance networks exist (Jackson and Buss, 1975). It is not known how widespread such networks are, but there is no reason to believe that they should be rare or that networks should be limited to sessile marine species. Digestive interactions, for example, have been described in tropical sea anemones (Sebens, 1976), and chemically mediated inhibition of growth has been found when the encrusting brown alga *Ralfsia* encounters other algae as it grows in a mixed algal culture (Fletcher, 1975).

Interoceanic patterns in molluscan shell armor seem to be independent of regional variations in tides, wave regime, temperature fluctuations, and productivity; yet some patterns of distribution of corals and of other sessile photosynthesizers may be related to variations in productivity. On the whole, a positive correlation exists along shallow-water gradients between diversity and the development of antipredatory traits, but the degree of expression of these traits varies considerably among assemblages of high diversity. It therefore appears that high diversity is a necessary, but not a sufficient, condition for the evolution of heavily armed predators and of morphologically well-defended prey. Climatic stability, lack of physiological stress, and large area (or, more precisely, large effective population size) are usually associated with high diversity, but do not ensure it. Indeed, variations in area at the scale of biogeographical provinces cannot account completely for east-west patterns in either diversity or the expression of predation-related adaptations. Consequently, present-day physical and biological conditions are insufficient to explain observed patterns in skeletal architecture. It becomes critical to explore how high levels of diversity are attained in a community, and what factors lead to the increased emphasis on adaptations to biological interactions.

PART THREE

Geography and Evolution

CHAPTER 7

Extinction and Speciation

Wₕₑₙ we experimentally analyze communities rich in plant and animal species, we can isolate mechanisms through which the local coexistence of many species is made possible. This is what Paine (1966a, 1976b), Janzen (1970), Connell (1972), and others did when they suggested that preferential predation on a competitively dominant or common prey species makes available space that can then be occupied by less common or by competitively inferior species. Competitive networks in communities subject to low rates of predation and disturbance should also ensure high local diversity (Jackson and Buss, 1975). Several predator-mediated processes, moreover, encourage individuals of one species to be near to, or surrounded by, members of other species (Janzen, 1970; Atsatt and O'Dowd, 1976).

While these ecological processes are therefore responsible for maintaining or even increasing high local diversity, they also depend on a large, already existing, pool of available species. The question of how species accumulate thus is quite different from the question of how local diversity is maintained.

Three processes in combination determine the size of the species pool in a community or region. Speciation tends to increase diversity, while extinction reduces it; migration from one region to another may either enhance or depress species numbers. Speciation and extinction do not affect all species equally. For some, the physical and biological surroundings remain essentially constant through time (that is, the environment is perceived genetically as remaining the same); the status quo is maintained, and speciation and extinction are unlikely. For other species, an environmental change may result either in an adaptive change or in extinction.

Which measurable properties of species influence their genetic sensitivity to a change in surroundings? Do the conditions that promote speciation also encourage the evolution of predation-related traits? In this chapter I shall argue that species able to live in physiologically rigorous environments often are resistant to extinction and speciate little; species in physiologically more favorable environments, where biological interactions play a greater role, tend to show higher rates of both extinction and speciation.

Three Kinds of Species

In order to evaluate these questions and explore their implications, it will be convenient to recognize three types of species according to the environment in which they live and how they perceive it. This threefold division, first proposed by Van Valen (1971) and elaborated by Grime (1977) for vascular plants, distinguishes species that I shall here call opportunistic, stress-tolerant, and biotically competent. As in any such simplified scheme, the three extremes are connected by intermediates and do not necessarily constitute unambiguous realities; but this classification seems in many ways preferable to MacArthur and Wilson's (1967) twofold division into r- and K-selected species (the "r-K-ic" classification).

First, we may consider species that have a high rate r of population increase in the absence of overcrowding and that have a short life span. Often, these r-selected or opportunistic species have high dispersibility, reduced long-term competitive ability, and a propensity to occupy ephemeral or highly disturbed habitats. Organisms perceived to have a weed-like nature or to be of strictly seasonal occurrence belong to this category; they take advantage of temporarily favorable periods in an otherwise hostile environment (see also Gadgil and Solbrig, 1972; Abrahamson and Gadgil, 1973; Gaines et al., 1974; Diamond, 1974, 1975).

The second category of species includes long-lived forms highly tolerant to chronic physiological stress. Among vascular land plants, stress-tolerant species generally have slow growth rates and low reproductive potentials; they are found in areas deficient in water, soil nutrients, light, or heat. Even under conditions that are physiologically favorable to most organisms, stress-tolerant plants grow slowly; consequently they are poor competitors under most circumstances (Grime, 1977).

In the sea the adjective "stress-tolerant" can probably be applied to long-lived species in areas of severe desiccation (the high intertidal zone), low or variable salinity (estuaries, mangrove swamps, salt marshes), chronically low food supplies (caves and much of the deep sea), extreme perpetual cold (polar regions and the deep sea), and low oxygen levels (deep or sulfide-rich sediments and some deep-sea basins). Growth rates of most stress-tolerant marine species, especially those in the high intertidal and deep sea, are known to be slow; and rates of recolonization of denuded habitat are typically far lower (sometimes more than an order of magnitude less) than in the low intertidal and the shallow subtidal zones (Lawson, 1966; Paine and Vadas, 1969a; Grassle, 1977). Many high intertidal snails, and most polar and deep-sea animals, are characterized by reduction or loss of the planktonic dispersal stage. Instead, they have a few large, yolky eggs and direct development on the bottom (Thorson, 1950; Lewis, 1960; Ockelmann, 1965; Arnaud, 1974).

The third category comprises biotically competent species living in environments that are always or nearly always physiologically favorable. Typically, these species have long life spans, well-developed mechanisms for competition and for avoiding predation, and relatively high growth rates as juveniles; but their dispersibilities and reproductive potentials appear to vary widely. Marine environments where biotically competent species should predominate include open surfaces of reefs, the top few centimeters of sediment in shallow-water soft bottoms, and the lower zones in the intertidal belt.

The terms "K-selected" and "equilibrium" species (MacArthur, 1965; MacArthur and Wilson, 1967) have often been applied to the category here called "biotically competent" species. The former phrases embody the expectation that the carrying capacity (or equilibrium density) K of a given population is maintained at the highest possible level through effective competition with other species in the community. The choice of the phrase "biotically competent" carries with it a less strict interpretation: it does not stipulate which of several biological interactions are the major sources of selection or populational fluctuation.

In the literature much emphasis has been placed on the differences in life-history characteristics between r- and K-selected species. Opportunistic species are usually thought of as having short life spans, a single episode of reproduction early in life (semelparity), and a large investment of biomass into structures directly concerned with reproduction. Since opportunists are characteristic of ephemeral habitats, they are thought to be good colonizers with high dispersibility. By contrast, K-selected species are regarded as having long life spans, delayed and multiple reproductions (iteroparity), and a small investment of biomass directly concerned with reproduction. (For an incisive review of this topic see Stearns, 1976.) It should be noted, however, that the characteristics often attributed to K-selected species are very similar to the life-history features of stress-tolerant organisms, despite the likelihood that competition may be slow and perhaps feeble among individuals in the latter group (Van Valen, 1976). A good example of this similarity has been provided by Pitelka (1977), who studied three species of Californian *Lupinus*, a genus of leguminous land plants. The perennial herb *L. variicolor* allocates about 18 percent of its resources to reproductive structures and about 40 percent to roots. Although it is not found in climax vegetations, this species is successful in midsuccessional vegetations that occupy moist areas. The relatively short-lived shrub *L. arboreus*, which grows on harsh, dry, seaside hills, has a similar pattern of resource allocation, 20 percent going to reproductive structures and about 50 percent to stems during the second year of growth. This pattern contrasts with that of the weedy annual herb *L. nanus*; in this species 61 percent of the resources are channeled to reproduction, while only 3 to 4 percent is converted to roots.

It appears, therefore, that the life-history traits collectively thought to be favored by K-selection may be correlated with individual persistence, not with competitive ability. Since biotically competent and stress-tolerant species differ strikingly in their vulnerability to extinction and speciation (see below), it is essential to distinguish between them even if their life-history characteristics are identical in most respects.

An additional implication of the distinction between r- and K-selected species is that the rate r of population increase in the absence of overcrowding should be low when K is high, and vice versa; that is, K-selected species should generally have low population growth rates following a density-independent disturbance that reduces a population to levels far below the carrying capacity K (see Gadgil and Solbrig, 1972). This need not always be so, however. In many trees and marine organisms the number of seeds or eggs produced by an individual adult female is a direct function of body size: the larger the organism, the greater the number of potential progeny produced. This relationship has been established empirically for sea stars, crustaceans, and gastropods (Menge, 1974; Spight et al., 1974; Spight and Emlen, 1976; Reaka, 1978). Although there may be an upper limit of size beyond which fecundity declines, as has been shown for *Thais lamellosa* by Spight and Emlen (1976), it is evident that many large trees, cod, oysters, mussels, and other organisms with long life spans may produce vast numbers of seeds or eggs which, given an adequate food supply and low rates of predation, could all survive to adulthood and bring about a dramatic population explosion if the previous generation were decimated (Williams, 1975). Thus, the separation of species into those that are K-selected and that are r-selected, as deduced from reproductive criteria, often seems to break down.

This is not to say that the scheme adopted here is foolproof or always unambiguous. For instance, the North American white pine *(Pinus strobus)* is an early-successional, fast-growing, relatively r-selected opportunistic tree, which nevertheless is able to persist in later-successional forests (Horn, 1971). Dayton (1973b) has pointed out that *Postelsia palmaeformis*, a wave-resistant annual brown alga of the Northeast Pacific open-coast intertidal zone, is able to maintain itself as a permanent member of the community alongside such excellent perennial competitors as the mussel *Mytilus californianus* and the stalked barnacle *Pollicipes polymerus*. In highly productive parts of the tropical Eastern Pacific, Birkeland (1977) has shown that opportunistic barnacles and hydroids outgrow, smother, and therefore outcompete more K-selected (in this case biotically competent) reef corals. Moreover, small insects, parasites, and many other organisms have life spans of less than one year, yet they are abundantly found in physiologically unstressed environments and are in many respects biotically competent relative to truly opportunistic species. Probably the simple scheme outlined here will have to be modified to take into account such variations; still, for the present the threefold classification of species seems useful, since the various types differ in their susceptibility to predation, extinction, and speciation.

Susceptibility to Predation

Theoretical considerations and empirical evidence from both marine and nonmarine organisms suggest that opportunistic species are apt to have poor defenses against predators, while biotically competent species are often well defended. For flowering land plants, Levin (1975) has shown that species whose genetic systems allow for extensive recombination of genes often possess chemical defenses that are effective against insect and fungal pests; these species can genetically respond rapidly to changes in the nature or intensity of pest pressure. Short-lived weedy species, on the other hand, often possess genomes in which whole blocks of genes are transmitted intact to the next generation, with little or no recombination taking place. These species often exhibit weak structural or chemical defenses to potential grazers with a generalized diet, but may have evolved very specific deterrents designed to discourage highly specialized (species-specific) consumers (Cates and Orians, 1975; Feeny, 1975). Long-lived plants are certain to be found by a herbivore with a generalized diet; if such a plant is to survive and reproduce, it must defend itself against consumers through chemical or structural impediments. Feeny has argued that ephemeral species may not always be discovered by generalized herbivores, but that they will be the target of consumers specialized to feed on them; such plants should, so the argument goes, possess defenses that are highly effective against specialists. If weeds lack such antiherbivore devices, they may flood a temporary habitat with individuals during favorable times, develop and grow rapidly, then virtually disappear. In this way some species may be able to tolerate heavy predation by producing great quantities of offspring.

Among aquatic zooplankton (Allan, 1976), cladocerans and especially rotifers have higher reproductive potentials than do copepods and as a rule are more vulnerable to predation. Copepods reproduce sexually and tend to be dioecious; while cladocerans and rotifers discourage genetic recombination by being parthenogenetic. As with land plants, the more predation-resistant forms (copepods) are particularly characteristic of permanent habitats, including the sea; while the more weedy types (cladocerans and rotifers) are typical of temporary pools and ponds, and are represented by only a few species in nonestuarine marine habitats.

Porter (1973) has suggested that the early-successional phytoplankton in a New England lake are more easily and more completely digested by zooplankton than are the more slowly reproducing blue-green algae and dinoflagellates. These forms appear later in succession and may have a mucilaginous coat, a sturdy test, or a toxin that impedes or prevents digestion.

Fresh-water snails living in temporary or highly seasonal ponds or rivers typically have short life spans, high dispersibility as adults, and thin, predation-vulnerable shells (Clarke, 1969; Vermeij and Covich, 1978). Snails (especially prosobranchs) that predominate in more permanent standing or flowing waters are mostly long-lived and thicker-shelled and, unlike the species of pulmonate and prosobranch in ephemeral waters, are dioecious rather than hermaphroditic.

In the cryptic reef communities of Jamaica, solitary serpulid polychaetes, brachiopods, and bivalves are eventually overgrown and replaced by colonial sponges, bryozoans, and ascidians (Jackson, 1977). With the exception of the brachiopods, the solitary animals experience heavy predation; for example, Jackson finds that 50 percent of the mortality among bivalves (*Echinochama*, *Dimya*, and *Spondylus*) results from predation by drilling muricid gastropods. The colonial forms, on the other hand, experience very low rates of predation; their world is one of interference competition, and their weapons often include toxins. Jackson has proposed also that replacement of solitary by more permanent colonial organisms is a feature common to many subtidal open-surface and cryptic communities on hard substrata, particularly in clear tropical waters. The greater permanence of the colonial forms stems in part from their ability to invade temporarily open spaces adjacent to them by asexual growth and reproduction. Most sedentary solitary animals require sexual reproduction to accomplish this. Moreover, successful attack by a predator normally will be lethal for a solitary animal, but not for a colonial organism; the latter can often grow over or quickly repair the affected portion (Jackson, 1977).

Fast-growing ephemeral marine plants such as diatoms and the green algae *Ulva*, *Enteromorpha*, and *Chaetomorpha* are favorite food items for many intertidal herbivores; they appear early in marine succession and are typical of disturbed habitats (Castenholz, 1967; Doty, 1967; Dayton, 1971). Weedy algae are morphologically conservative and geographically very widespread in comparison to many of the larger algae of the lower and middle shore, among which chemical and structural defenses against herbivores are often developed to a high degree (Paine and Vadas, 1969b).

Many other examples of weedy animals with poorly developed antipredatory defenses could be cited. Among molluscs, small annual cerithiids and other gastropods living on algae and sea grasses rarely achieve morphological immunity to predation. Many tellinids, small mactrids, and other thin-shelled bivalves with short life spans may achieve very high population densities, but they are then quickly decimated by various means, probably including predation (Levinton, 1970; Levinton and Bambach, 1970). Annual tube-building worms are prominent in the diet of many predators, and only the more perennial forms exhibit morphological defenses against predation; thus, the tropical serpulid *Spirobranchus* has a thick-walled tube and well-developed operculum, while the large infaunal onuphid *Diopatra* has a tough long tube that may be able to transmit vibrations from potential attackers to the worm within (Vine, 1974; Brenchley, 1976).

There are, however, annual marine species that seem to be well defended against some, if not all, of their potential predators. Annual unshelled nudibranchs, which have been studied primarily in temperate waters, are protected against their enemies by a variety of mucus and acidic secretions (many dorids), by nematocysts in the dorsal cerata (especially common in aeolids), or by calcareous spicules in the skin (dorids). (For a review see Harris, 1973.) Typically, these opisthobranchs are associated with sessile invertebrate hosts such as sponges, hydroids, and bryozoans, whose abundance fluctuates seasonally. Just which animals prey on nudibranchs remains largely an open question; an analogy with chemically noxious annual land plants and the application of Feeny's (1975) ideas suggest that the chemical and dermal traits of nudibranchs may well be adaptations against specialized predators.

As a rule, stress-tolerant species seem to be little affected by herbivores or by predators, either because they live in environments where potential consumers cannot function, or because they have developed defensive mechanisms that are very difficult to circumvent. Bryophytes (mosses and liverworts), ferns, lycopods (club mosses), horsetails, and many gymnosperms are primitive plants nearly all characteristic of nutrient-deficient soils, severe cold, deep shade, or other physiologically marginal conditions where rates of photosynthesis and transpiration are low. Together with vascular plants in montane, alpine, and tundra environments, these primitive plants appear to have strikingly few herbivores (Corner, 1964; Ehrlich and Raven, 1964; Howard, 1969; Janzen, 1973a). Some of these stress-tolerant plants are known to possess a chemistry inimical to most potential attackers (as do ferns and many gymnosperms); others are festooned with spines, thorns, or greatly thickened and toughened leaves that are inedible and poor in nutrients for most herbivores (see, for instance, Howard, 1969; Mooney and Dunn, 1970).

Many of these structural and chemical devices are apparently dependent on very slow growth and on low rates of biomass replacement, and therefore are inappropriate for those biotically competent species that must grow rapidly in order not to be outdone by neighboring individuals. For example, heavily sclerotized leaves impede transpiration and gas exchange; this makes them well adapted to conditions where water, soil nutrients, or some other basic necessities for photosynthesis are in short supply, but poorly suited to most temperate and tropical forests where water is available in sufficient quantity for high transpiration rates (Mooney and Dunn, 1970; Givnish and Vermeij, 1976; Orians and Solbrig, 1977). For rapidly growing, biotically competent plants, other antiherbivore devices not in conflict with the demands of rapid growth and high rates of transpiration are called for. These traits may not always be as insuperable to potential insect and fungal enemies as are the toxins, waxes, or spines of many stress-tolerant species; however, speculation of this sort requires extensive investigation.

Janzen (1973b) has argued that cold temperatures in high-altitude and polar habitats greatly restrict the activities and the effectiveness of ectothermic herbivores; thus the interference and selective pressure from herbivores would be reduced in these, and perhaps also in other, stress-tolerant plants. Although experiments will have to be designed for testing this idea, it is interesting that antiherbivorous alkaloids are virtually unknown among the high-altitude vascular plants of New Guinea (Levin, 1976), and that high-altitude herbs in Scotland *(Trifolium repens* and *Lotus corniculatus)* lack the cyanogenic compounds often found in populations of these legumes near sea level (Brighton and Horne, 1977).

In Chapters 2, 3, and 4 I summarized the evidence, most of which is circumstantial, suggesting that predation on high-intertidal and deep-burrowing animals is less intense than on species at lower shore levels and at shallower depths in the sediment. Thus far, the only conclusive data come from barnacles (Connell, 1961a, b, 1970), but the trend is consistent with the pattern in land plants.

The blue-green algae (Cyanophyta) also are famous for occupying marginal habitats, including hot springs, supratidal rocks, and anoxic ocean basins. On the coast of Washington, blue-greens are rarely grazed by mid- and high-intertidal limpets and littorines, and the tropical Pacific snail *Turbo setosus* avoids certain species of Cyanophyta (Tsuda and Randall, 1970; Nicotri, 1977). Nor do fresh-water zooplankton as a rule feed on blue-greens (Porter, 1973). As with the land plants discussed above, a repellent chemistry is probably involved in these cases. Yet blue-greens are often grazed in unstressed marine environments. Fishes readily graze encrusting mats of blue-greens on the reef flat at Eniwetok Atoll in the Marshall Islands (Wiebe et al., 1975), and Garrett (1970) has shown that intense grazing by cerithiid and potamidid gastropods in the intertidal zone of the Bahamas restricts the formation of blue-green algal mats to the supratidal zone, where the snails do not occur.

Sclerosponges, brachiopods, certain bryozoans, and other sessile invertebrates found in caves, under overhangs, and in other dark habitats on tropical reefs grow very slowly and are subject to extremely low levels of pedation (Jackson et al., 1971; Lang et al., 1975; Jackson and Buss, 1975). Many of these animals possess chemical weapons whose primary function seems to be the prevention of overgrowth and consequent smothering by neighboring species; such defenses might equally interfere with predation. Shade-loving brachiopods in the Puget Sound area, according to C. W. Thayer, are surprisingly unexploited by predators despite their thin shell, perhaps because of unusual chemical deterrents.

The work of Onuf and colleagues (1977) on damage inflicted by herbivores to mangrove trees under different levels of nutrient supply suggests that stress is associated with low food content and consequently with reduced herbivore pressure. Trees *(Rhizophora mangle)* grown in soils enriched by bird guano grow more rapidly, have a higher nitrogen (and thus protein) content in their leaves, and must endure almost four times as much herbivory by insects as do individuals growing in soils not enriched by guano. It is not known whether reduced food content is a general property of stress-tolerant species, or what the underlying cause is for the relation between nutritional value and predation intensity. It does not seem unreasonable, however, that slow growth rates should accompany limited accumulation of protein.

Thus, although stress-tolerant species seem to be characterized by slow growth and relative immunity from attackers, the reasons for the immunity may vary. In many cases the chemical deterrents possessed by a stress-tolerant species may be difficult for potential enemies to overcome; they may be produced in small quantities, yet accumulate in large concentrations in the durable body. Together with these deterrents, physiologically marginal conditions reduce the effectiveness of predators and herbivores. Precisely why some stressed organisms have antipredatory deterrents while others do not is still a mystery. Different types of stress may impose varying limitations on the activity of predators and on the array of adaptations available to prey species for countering predation. Moreover, some predators and grazers, over geological time, may have been able to overcome certain types of stress better than they have responded to other types. In fact, what may have been predator-free habitats at one time may not have remained so throughout the history of life. This emphasizes the point that such words as "stressed" or "physiologically marginal" must always be used in a relative, and not in an absolute, sense. I would guess that the evolution of endothermy among mammal-like reptiles in the Permian period and perhaps among the dinosaurs in the Middle Triassic (Bakker, 1975) would have permitted predation in marginal environments where little or none had occurred previously; and that the evolution of plants with crassulacean night photosynthesis permitted a significantly greater occupation of habitats and a higher biomass of plants in very dry terrestrial environments. In the Early Paleozoic, all terrestrial environments would have been regarded as physiologically stressful to larger land plants, since no such primary producers were known until Late Silurian time. Clearly, the occupation and successful exploitation of stressed environments requires adaptations that evolve only rarely.

Extinction

In reviewing the groups discussed in the preceding paragraphs, I have been impressed that long-lived, stress-tolerant species are evolutionarily conservative and relatively immune from extinction. Groups that in geological periods occupied many environments and were highly diversified survive today as relicts or remnants in deep water, caves, oxygen-deficient sediments, low-salinity waters, and other physiologically marginal environments. Monoplacophoran molluscs, of which the deep-sea limpet-like genus *Neopilina* is the sole survivor, were represented in the shallow waters of the Early Paleozoic by a wealth of limpet-like and planispirally coiled forms. (For a review see Runnegar and Jell, 1976.) Glypheid crustaceans, the stem group of the lobster- and crab-like decapods, were common in shallow waters during the Jurassic, but are known in the Recent only by the genus *Neoglyphea*, found at a depth of 1,200 m in the Philippines (Forest et al., 1976). Hydrogen sulfide–rich, reduced, shallow-water sediments have an abundant biota of peculiar ciliates, turbellarians, gastrotrichs, nematodes, and gnathostomulid worms, which on morphological grounds have been judged to be very primitive (Fenchel and Riedl, 1970). Pleurotomariid gastropods, now found only in deep water, are primitive archaeogastropods that have changed little from Late Paleozoic and Early Mesozoic time, when these animals were represented by many species in shallow water. Articulate and inarticulate brachiopods were enormously common and diversified in Paleozoic and many Mesozoic marine assemblages; today they are predominantly found in tropical marine caves, inshore estuarine sediments, the deep sea, or high-latitude cryptic environments where oxygen levels are low (Rudwick, 1962). Sclerosponges, shade-loving demosponges with a well-developed skeleton of calcium carbonate, currently are found only in caves and on the deeper parts of West Indian and Indo-West-Pacific reefs, but in the Paleozoic and Mesozoic periods were important constituents of reefs in shallow water. Stalked crinoids were also abundant in Paleozoic and Mesozoic shallow-water communities, but are now found only in waters more than 100 m deep; they were replaced in shallower waters by comatulid crinoids, which are free to move around and attach themselves by means of specialized cirri (Meyer and Macurda, 1977). Blue-green algae and ancient groups of land plants also seem to conform to this pattern of distribution.

Hecht and Agan (1972) have noted that bivalve genera in the temperate Northwest Atlantic on the whole are more ancient than are genera of tropical West Indian bivalves, even though the tropical genera contain a larger number of species per genus and thus would be statistically less likely to disappear. For example, the average age of genera in cold-temperate New England is about 80 million years, while that of bivalve genera in subtropical Florida is only about 60 million years. A similar though less clear trend has been described for genera of planktonic and benthic Foraminifera

(Stehli et al., 1972; Durazzi and Stehli, 1972), but for unknown reasons there is no latitudinal trend in the geological age of foram species. The greater conservatism of cold-water as compared to warm-water genera in these and other groups (see also Nicol, 1967, for bivalves; and Hessler and Thistle, 1975, for isopods) is consistent with the ancient aspect of the biota in other stressful environments.

Not all species in stressed environments, however, are ancient. From analyses of the present-day deep-sea biota, it is evident that neogastropods, heterodont-hinged bivalves, and other animals of Late Mesozoic or Early Cenozoic vintage live side by side in the deep sea with protobranch bivalves, stalked crinoids, monoplacophorans, and other groups that have changed little since Paleozoic times (see Bruun, 1956; Menzies et al., 1961; Parker, 1961; Clarke, 1962; Nicol, 1967). Low extinction rates do not necessarily imply that species cannot evolve in physiologically stressed environments—or that animals cannot invade them from elsewhere.

Some so-called living fossils do not inhabit stressed environments. Thus *Nautilus*, the only surviving genus of the once highly diversified externally shelled cephalopods, is found at moderate and shallow depths in the tropical Western Pacific, where conditions are probably not particularly rigorous. Inarticulate brachiopods, the xiphosuran arthropod *Limulus* (horseshoe crab), and the mud-dwelling crustacean *Hutchinsoniella* are "living fossils" belonging to groups that apparently never contained many species; today these animals live in coastal waters that are not unusually stressful. Hobson (1974) noted that most primitive reef-associated teleost fishes are nocturnal or crepuscular in habit, while such advanced groups as chaetodontids, pomacentrids, scarids, labrids, and tetraodontids are primarily diurnal. Present-day forms of primitive mammals (insectivores, many primates) and lizards are also primarily nocturnal (Bakker, 1971). Comparatively few birds are active at night, but those that are belong to relatively primitive nonpasserine groups (see Morse, 1975). Among land vertebrates, early or primitive groups such as ratite birds, rhynchocephalian reptiles (tuatara), tenricid insectivores, and monotremes occupy islands or small continents that became isolated from larger land masses before the radiation of more advanced groups (Fooden, 1972; Jardine and McKenzie, 1972; Morse, 1975).

Why have these ancient groups been so resistant to extinction? Is there something about stressed environments, islands, and the night that permits species to persist for longer than average periods? One plausible answer is that environments which we think of as "physiologically favorable" are comparatively new, or at least more recently occupied by organisms, while stressful (particularly oxygen-poor) environments may be geologically very ancient. In their discussion on the primitive and ancient aspect of the biota in anoxic, sulfur-rich marine sediments, Fenchel and Riedl (1970) suggest that the most primitive monerans, protists, and metazoans originated and first diversified under conditions of extremely low oxygen tension, and that they have remained in those primeval, permanently reduced settings ever since. Rhoads and Morse (1971) make a similar case for primitive soft-bodied animals found in low-oxygen deep-sea basins. Although there is likely to be an element of truth in these speculations, it is doubtful whether all instances of conservatism in stressful environments can be ascribed to the persistence of obsolete organisms in geologically ancient environments, or that present-day ecological distributions of groups are similar to those in the past. For example, it seems unlikely that Paleozoic monoplacophorans and stalked crinoids were all light-shunning deep-sea creatures, or that well-lit, shallow-water marine environments did not exist in the Paleozoic when these groups were common.

Further evidence that groups living today in stressed habitats once occupied physiologically more favorable places comes from detailed biogeographical studies of fossil bivalves. The genus *Limopsis*, which Hallam (1977) recognizes as one of twenty-seven genera restricted to the tropics and subtropics in Jurassic time, at present is predominantly a polar group and has a particularly large number of species in Antarctic waters (Nicol, 1967). Other genera now restricted to cold waters but found abundantly in the Jurassic tropics and subtropics include *Musculus*, *Astarte*, and *Thracia* (see Hallam, 1976).

Some stressed environments may have become occupied only in comparatively recent time. None of the gastropod families characteristically found in the high intertidal zone today, for example, is known from before the Mesozoic. The bivalves that burrow deeply into oxygen-deficient sediments are mostly of Mesozoic or even Cenozoic vintage, and only a handful of short-lived genera occupied this habitat in the Paleozoic (Stanley, 1968, 1977). Yet, at least at the generic level, there has been little change in these groups since the time of their origin.

If a lineage persists morphologically unchanged over a long interval of time without becoming extinct, it externally perceives its physical and biological environment as remaining the same. In other words, there is nothing in the environment that demands significant change in external appearance or that threatens the lineage with extinction. This is not to say that change is altogether lacking; indeed, physiological and biochemical changes may occur over the course of time, but the external morphology is conservative and remains unaltered.

I believe that biological interactions are significant forces of environmental instability; they are instruments of both evolutionary change and extinction. The increased importance of biological interactions to the selective regime of a species could have two ways of increasing the sensitivity of a species to its environment. First, methods of predation and of competition may change with time, even if the physical environment remains substantially unaltered; such biological changes may call for new traits, and if these traits do not evolve, the population may face extinction. Second, adaptation to the pressures imposed by other organisms is often accomplished at the expense of tolerance to physical extremes. The environmental stress experienced as a matter of course by weeds or by stress-tolerant organisms may often be disastrous for biotically competent species and may result in their extinction (see also Futuyma, 1973; Horn, 1974; Jackson, 1974, 1977).

For example, generally long-lived, biotically competent animals succumbed to heat and desiccation on an outer-reef flat in Western Australia during a clear hot spell coincident with very low tides in the austral summer of 1959; more stress-tolerant inshore and high-intertidal species, including limpets and littorines, were hardly affected (Hodgkin, 1959). A similar pattern of selective mortality was noticed by Glynn and associates (1964) when a hurricane accompanied by high tides and heavy rain brushed the southwest coast of Puerto Rico. In June 1972 tropical storm Agnes caused a prolonged reduction in salinity in several rivers in estuarine Chesapeake Bay, with severe consequences for predatory gastropods (especially the oyster drill *Urosalpinx cinerea*), stenohaline echinoderms, and the thick-shelled arcid bivalve *Noetia ponderosa* (Andrews, 1973). Still other examples have been given by Glynn (1968) and Yamaguchi (1975b).

The morphological conservatism and geological longevity of stress-tolerant organisms, then, seems to be related to the normally minor role that biological interactions play in their selective regime. Both predation on, and competition among, stress-tolerant species are slow processes, and the extent to which these processes can change in nature and intensity is probably limited by physiological constraints on rates and methods of locomotion, muscular contraction, and skeletonization. Even if some of the animals in a community in a physiologically marginal habitat are not themselves constrained in body functions by the prevailing environmental conditions, their rates of feeding may often be curtailed by an inadequate food supply. Endothermic birds and mammals are found in various cold terrestrial and marine environments and may be important grazers (ptarmigans in the Arctic, llamas in the Andes) or predators (polar bears or wolves in the Arctic), but most of these endotherms migrate seasonally to other places or to other habitats, so that their exploitation of stressed organisms is only intermittent.

If weedy opportunists respond to changes in their biological environment in a numerical rather than in a genetic way, then they too should be relatively immune to extinction. Patches of suitable, ephemeral habitat are created with sufficient frequency in most communities, and their quality varies little. This may not be true, however, if the major cause of weed mortality is biological. In forests and reefs, for example, open space created by the falling of a tree or the breaking up of a coral-head may become occupied temporarily by weeds, which then experience heavy mortality as a result of predation or competition from adjacent, well-established, biotically competent organisms. In such cases weeds respond to the biological rather than to the physical environment, and evolution among them may be quite rapid. This topic demands careful investigation. In particular, it will be important to ascertain whether weeds that occupy ephemeral patches in communities dominated by biotically competent species the year round are in some way different from weeds that occupy seasonal or otherwise periodically bare habitats.

The susceptibility to extinction of biotically competent species should, all other things being equal, be greater than that of most stress-tolerant and many opportunistic species. Not only is their tolerance to occasional physical stress compromised, but they are vulnerable to changes in the nature and intensity of competition, predation, symbiosis, and parasitism (see also Futuyma, 1973).

Only a few circumstances may favor the long-term persistence of biotically competent species. First, prey species may "discover" defenses that are difficult or impossible for a potential enemy to circumvent. Gilbert (1971) has given a possible example of this phenomenon. The passionflower *Passiflora adenopoda* has developed hook-like hairs on its leaves, which puncture the integument of heliconiine caterpillars and kill them. Gilbert believes that this is an exceptionally effective device against caterpillars, and he speculates that *P. adenopoda* may have become liberated from the attacks of heliconiines. This does not, of course, mean that the plant is altogether free of herbivores, although further coevolution with heliconiines may no longer occur. Such defensive measures often exact a high price in metabolic energy, and they may seriously compromise important body functions; moreover, if the defense is ever breached, the species would probably experience a hair-raising decline that might result in rapid extinction.

In experiments with houseflies, Pimentel and Bellotti (1976) have shown that the possession of several defensive mechanisms is often far more effective in preventing counteradaptation than is a single deterrent. Again, energetic costs and functional compromises probably increase with the number of defensive gimmicks; yet it is plausible that a fortress is for a time immune from coevolutionary pressures.

A second stabilizing force, discussed by Atsatt and O'Dowd (1976), is the preservation of genes that confer resistance against predation as a result of the perhaps genetically based tendency for members of one species to be surrounded by members of other species. This local rarity reduces the searching effectiveness of a specialized predator, or of a predator with a search image for the rare species; furthermore, it is less likely that a predator will adapt genetically to overcome the structural or chemical deterrents in order to feed on the prey species. If, on the other hand, prey species occur in large monospecific aggregations, as is usually the case when man cultivates crops, prey genes that control defensive adaptations are more quickly neutralized by counteradaptation on the part of the prospective predator.

Although the above predictions about the susceptibility to extinction of the three types of species require extensive confirmation and testing, they are consistent with historical patterns of taxonomic change in various marine communities. Bretsky (1968) has suggested that inshore (and therefore presumably stressed) Paleozoic soft-bottom communities, dominated by inarticulate brachiopods, protobranch bivalves, and gastropods, changed less in taxonomic composition and were generally more conservative than were communities farther offshore dominated by articulate brachiopods, crinoids, and (toward the end of the Paleozoic) foraminifers. Similar conclusions were drawn by Boucot (1975) for nearshore and offshore Siluro-Devonian brachiopod assemblages. In fact, Silurian assemblages of deposit-feeding protobranchs were in many respects similar to Recent communities occupying the same soupy sediments (Levinton and Bambach, 1970, 1975); but shallow-water assemblages on firmer sediments today, dominated by crustaceans, siphonate bivalves, and neogastropods, are very different in composition from their Paleozoic equivalents in which crinoids, articulate brachiopods, and trilobites were common.

In his biogeographical study of Jurassic bivalves, Hallam (1977) noted that deep-water genera (including the still-living *Propeamussium*) and relatively euryhaline genera (including the living *Modiolus*, *Musculus*, *Isognomon*, *Chlamys*, and *Thracia*) were stratigraphically more persistent than were the often thick-shelled bivalves associated with reefs and with other shallow-water carbonate sediments. Some of these patterns are also apparent at the specific level. My calculations based on the stratigraphic ranges of European Jurassic bivalves given by Hallam (1976) indicate that deep-burrowing pholadomyoid and lucinacean bivalves lasted an average of 3.0 stages per species, while species that bored into limestone or other rock lasted an average of 4.0 stages. By contrast, reef-associated hippuritoids (including early rudists) lasted only 1.6 stages per species. In sediments believed to have been deposited in warm, shallow waters of normal marine salinity, deeply infaunal species survived an average of 3.3 stages, while shallowly infaunal and epifaunal bivalves each persisted for only 2.4 stages.

Several anomalies occur, however. Species interpreted by Hallam (1976) to have lived in brackish lagoonal waters (presumably a stressed environment) lasted only an average of 1.9 stratigraphic stages; thin-shelled and probably weedy epifaunal clams found in deoxygenated bituminous deep-water sediments persisted for a mean of 2.2 stages. These average stratigraphic intervals over which a species persists seem anomalously short for animals that are opportunistic or adapted to chronic stress.

For all the categories, however, there is a high degree of variation in the stratigraphic persistence of species, and it is not known how much the geological life span of a species would be increased if its stratigraphic range were known for Jurassic rocks throughout the world. Species-level work of this kind is crucial to further testing of the predictions about extinction offered here.

Probably the most impressive record of change in species composition and in architecture is found in what may be termed loosely the reef environment, where biotic interactions today are perhaps more refined than in any other marine setting. A reef is a biologically generated deposit or construction that accretes or maintains itself in the face of physical and biological erosion and that is elevated above the surrounding sea floor (Wray, 1971). The eventful history of the reef biota has been reviewed for the entire geological record by Newell (1971) and Wray (1971), and for the Paleozoic era in particular by Copper (1974). The first reefs, known from the Cambrian in the Early Paleozoic, were dominated by sponge-like members of the now extinct phylum Archaeocyatha. By Middle Ordovician time, after the demise of the Archaeocyatha, the primary reef-framework builders included rugose and tabulate corals, stromatoporoids, bryozoans (mostly trepostomes), and solenoporacean and other calcareous algae. During succeeding periods (Silurian and Devonian) reefs of this type flourished, but the end of the Frasnian epoch in the Late Devonian period brought a sudden end to this community. From modest beginnings in the Early Carboniferous (Mississippian) period, a new reef assemblage dominated by brachiopods, bryozoans, and sponges developed and flourished in the Permian; this too came to an abrupt end with a massive

episode of extinction at the end of the Paleozoic (Permo-Triassic boundary). In fact, several groups that at various times in the Paleozoic contributed importantly to the reef biota were completely extinguished: trepo-
stome bryozoans, rugose and most tabulate corals, most crinoids, and
productoid brachiopods. Mesozoic and Cenozoic reefs, dominated by
scleractinian corals and calcareous red and green algae, began their development in the Middle Triassic. From time to time large epifaunal bivalves
have been important contributors (and in the Cretaceous period consti-
tuted the major contributors) to the building of reefs or reef-like frameworks; good examples include the coral-like rudists (Hippuritacea), which
flourished during the Cretaceous but then disappeared with many other
animals at the end of that period (Kauffman and Sohl, 1974), and the Late
Tertiary and Recent giant clams (Tridacnidae), found on Indo-West-
Pacific reefs.

   In short, many of the taxa that have played critical roles in reef construction have been geologically ephemeral, and the biologically
"sophisticated" reef has changed radically over the course of earth history. Assemblages in environments where biological interactions have
been less important and not so lavishly elaborated have remained more
conservative in character and have been less ravaged by the major episodes of biotic extinction.

The Influence of Dispersal

   The degree to which individuals are dispersed relative to their parents
should influence the susceptibility of a population or lineage to extinction. As dispersibility and geographical range of a species increase, the
probability of total (or global) extinction of that species decreases even if
local catastrophes such as low tides, heavy fresh-water runoff, and
extreme temperatures cause local populations to be decimated (Levinton, 1970; Jackson, 1974). Since dispersibility varies widely within the
three groups of species, the relation between extinction and species type
is probably less simple and less uniform than was implied in the preceding section.

   Jackson (1974), for example, has found that tropical infaunal bivalves
that occupy inshore waters less than 1 m deep generally have broader
geographical ranges, larger contiguous populations, and a greater
average geological life span per species (Middle Pliocene) than do species living at depths from 1 to 100 m on the continental shelf, where
environmental conditions are more constant and physiologically less rigorous; the average geological age of the latter group is Pleistocene, and

many species are not known as fossils. The inshore bivalves, which include both weedy and long-lived stress-tolerant forms, are believed to have greater powers of dispersal than the deeper-water species. Other groups whose members tend to have large geographical ranges, great powers of dispersal, and therefore high resistance to total extinction, include weedy diatoms and green algae, many soft-bottom polychaetes, epifaunal fouling organisms, and regular echinoids.

High dispersibility usually has been associated with weediness, yet it seems equally characteristic of a surprisingly large number of tropical animals that have long adult life spans and whose environment would seem to be biologically challenging. Usually this wide dispersal is effected by a long planktonic larval phase (teleplanic larva); this has evolved independently in a number of decapod and stomatopod crustaceans (including the crab genus *Calappa*), gastropods *(Cymatium, Charonia, Tonna, Architectonica)*, bivalves *(Pinna)*, echinoderms, and polychaetes (see Thorson, 1961; Scheltema, 1971, 1977). Dispersal by means of teleplanic larvae seems often to be associated with long-term geological persistence (Shuto, 1974; Scheltema, 1977). For example, the gastropods *Cymatium parthenopeum, Architectonica nobilis*, and probably *Thais haemastoma* date from Early Miocene time, while *Cymatium nicobaricum* can be recognized in the Oligocene Caimito formation of Central America (see Scheltema, 1977).

It is particularly interesting that many marine predators have acquired remarkable powers of dispersal, wide geographical ranges, and therefore a low probability of total extinction in spite of usually low population densities. Among molluscivores, many species are circumtropical: the gastropods *Cymatium nicobaricum, C. muricinum*, and *C. pileare*; the puffers *Diodon hystrix* and *D. holacanthus*; the ray *Aetobatis narinari*; the stomatopods *Odontodactylus brevirostris, Pseudosquilla ciliata, P. oculata, Alima hyalina*, and *A. hieroglyphica*; and the crab *Calappa gallus* (Bigelow and Schroeder, 1953; Guinot, 1966; Randall, 1967; Manning, 1967, 1969; Scheltema, 1971). Many additional molluscivores are found throughout the Indian and Pacific oceans from the Red Sea and the east coast of Africa to the islands off western tropical America. These include the crabs *Calappa hepatica* and *Carpilius convexus* and the spiny lobster *Panulirus penicillatus* (Garth, 1973; George, 1974). Of the molluscivores distributed continuously from the Red Sea and East Africa to eastern Polynesia, examples of crabs *(Daldorfia horrida, Ozius guttatus, Eriphia sebana, Scylla serrata)*, gastropods *(Conus marmoreus, C. textile, C. pennaceus, Morula granulata)*, and octopods *(Octopus cyanea)* come to mind. Similarly wide distributions are found among coral predators: the sea star *Acanthaster planci*, the puffer *Arothron meleagris*, and the gastropods *Coralliophila violacea* and *Quoyula madreporarum* (Emerson, 1967; Chesher, 1972; Glynn et al., 1972). Large piscivorous sharks *(Carcharodon)*, barracuda *(Sphyraena)*, and some moray eels *(Gymnothorax)* are also renowned for their circumtropical or Indo-Pacific distributions.

The wide distribution of predators is often rivaled by that of their highly armed prey. Ready examples include the crustacean-defended coral *Pocillopora damaecornis*, the giant clam *Tridacna maxima*, and many species of the gastropod genera *Conus*, *Terebra*, *Mitra*, *Drupa*, *Morula*, and *Cypraea*.

Teleplanic larvae are rare or absent in cooler seas, but here dispersibility can be achieved aboard floating algae and ice even by invertebrates with direct development. An example is the Eastern North American oyster drill *Urosalpinx cinerea*, which may be dispersed as newly hatched snails on bits of grass or alga (Carriker, 1957). Many invertebrates have probably achieved a circumantarctic distribution by transport on kelps and ice (Arnaud, 1974). The Chatham Islands, east of New Zealand, have been colonized in part by species living with, and probably dispersing on, large algae (Knox, 1954).

Many biotically competent species, then, have achieved high powers of dispersal and acquired the potential for long stratigraphic persistence. If such species are important to a community, they could in principle lend an element of stability to such assemblages.

If high dispersibility confers a measure of immunity to extinction, then low dispersibility should increase the probability of global extinction when it is associated with a narrow geographical range. Data for this assertion are not yet available, but present evidence gives some reason to doubt that low dispersibility is invariably associated with greater vulnerability to extinction. The high intertidal zone is potentially an interesting habitat in which to carry out a test of this suggestion, since it is physiologically stressful for most snails, yet is discontinuous and subject to strong local atmospheric variations; many gastropods that inhabit the high shore have strikingly narrow geographical ranges compared to open-surface species at lower shore levels (Vermeij, 1972a, 1973b, c).

For example, six out of thirteen (46 percent) of the high intertidal snails along the limestone coast of Kenya are restricted to the Indian Ocean; low intertidal, near-shore gastropods (ten species) all appear to be widespread, and only one of fourteen outer-reef species is endemic to the Indian Ocean. In the Gulf of Aqaba (northern Red Sea), five of the eight high intertidal gastropods (63 percent) are endemic to the Red Sea; while only five of sixteen low intertidal species (31 percent) are endemic.

At least in Barbados, Hawaii, and the Red Sea, high intertidal snails usually have a reduced planktonic phase, or lack one entirely; suppression of this dispersal stage limits the dispersibility of the species as a whole (see Lewis, 1960; Kay, 1967). Underwood (1974), however, has not been able to detect differences in the length of planktonic stages between high- and low-shore snails in warm-temperate New South Wales.

It would be of great interest to know if high intertidal snails have shorter geological life spans than do lower-shore species, as would be predicted if dispersibility were the main determinant of stratigraphic persistence; or if they are long-lived taxa, as would be predicted if stress (in this case desiccation and extreme temperatures and salinities) is the primary source of persistence. Unfortunately, rocky-shore animals rarely fossilize, and the record of high intertidal snails in particular is spotty at best. Moreover, the relatively strong expression of predation-related adaptations among the high intertidal snails in two areas of the Indo-West-Pacific (see Chapter 5) suggests that predation may be a significant selective influence in the tropics, and that it is therefore potentially a source of narrower physiological tolerance and of extinction.

In both the tropical Atlantic and the Indo-West-Pacific, it has been noticed that gastropods associated with outer-reef or open-coast environments show less regionality in species composition, and are more widely and more continuously distributed, than are related species in lagoons or other inshore habitats where conditions are often quite different from one region to the next. In Brazil, for example, inshore gastropods are often endemic and are adapted to sand-scoured conditions that are relatively uncommon in other parts of the tropical Western Atlantic. Examples include the limpets *Acmaea subrugosa* and *Fissurella clenchi*, and the small mussel *Brachidontes solisianus*. Species living under stones or on the outer, unscoured reefs in Brazil are commonly found also on West Indian shores — for example, *Thais deltoidea*, *T. rustica*, *Trachypollia nodulosa*, and *Arca imbricata* (Vermeij and Porter, 1971). The fauna of the mainland coast of Queensland is so different from that of the adjacent Great Barrier Reef that Australian biogeographers have often regarded it as comprising a distinct province. Many of the species on the Great Barrier Reef, in contrast to mainland forms, have distributions that span the whole of the Indo-West-Pacific (Endean et al., 1956a, b; Bennett and Pope, 1960). Again, it is not known whether the inshore species have shorter or longer average geological life spans than do the animals associated with open-coast reefs; but in this case, the fossil record may be adequate to discover the answer.

It should be remarked here that some inshore gastropods with crawl-away larvae, and therefore with poor dispersibility, have been as geologically persistent as have some species with teleplanic, widely dispersing stages. The Recent Western Atlantic *Melongena melongena* and Eastern Pacific *M. patula* are very similar to each other and are virtually indistinguishable from the Early and Middle Miocene *M. consors* of the Gatun formation in Panama. The Western Atlantic *Cerithium lutosum*, with direct development, is probably indistinguishable from *C. harrisi* from the Pliocene of Trinidad (Houbrick, 1974b). It is not known what larval type the Western Atlantic *Vasum capitellus* possesses; but this species, which has a comparatively small southern Caribbean distribution, has persisted virtually unchanged since the Oligocene epoch (Vokes, 1966). The same can be said of the massive-shelled *Turbinella pyrum* of South India and Ceylon (Vokes, 1964; Kilburn, 1975). This species has a large, bulbous apex that strongly suggests the absence of a planktonic larval phase (Shuto, 1974). These examples illustrate that poorly dispersing lineages can persist for long periods of time, as do such wide dispersers as *Thais haemastoma*, *Architectonica nobilis*, and *Cymatium* species, all of which are known as fossils from the tropical American Miocene. A precise test of the significance of dispersibility in determining geological persistence must await the collection of more data on a large sample of both teleplanic and direct-developing species.

With this ignorance in mind, I conclude that dispersibility varies widely among species; it modifies the predicted sensitivity of the three types of species to environmental change. High dispersibility confers a certain stability on biotically competent species whose populations would otherwise be susceptible to physical calamities and even wholesale annihilation. The low dispersibility of certain biotically competent and stress-tolerant species may increase the vulnerability of these species to local perturbations, which in turn could result in total extinction. In other cases, however, low dispersibility has not interfered with the survival of species. Dispersibility is apparently only one of several factors that influence the way in which spatial and temporal variations are perceived, in a natural-selective sense, by organisms. What some of these factors are remains an important, intriguing, and largely unsolved problem.

Speciation

One of the necessary processes for the enrichment of biotas is speciation. Many of the conditions that increase the probability of extinction of a lineage also favor isolation, genetic change, and speciation. It seems clear that limited dispersibility may lead to the creation of many small isolated populations which, if they live under substantially different conditions, may undergo genetic differentiation and eventually become specifically distinct from the parent stock. For instance, geographical speciation has been more frequent among Eastern Pacific rocky-shore fishes, which are generally poor dispersers, than among species living on sandy or muddy bottoms (Rosenblatt, 1963, 1967). Among fresh-water gastropods, weedy pulmonates are conservative and speciate little, while permanent-water prosobranchs have low dispersibility and have radiated explosively into species flocks in various lakes and river systems (Clarke, 1969). Cichlid fishes in the Great African lakes show a striking propensity toward local differentiation, in contrast to their river-dwelling, better dispersed ancestors (Fryer and Iles, 1969, 1972). Brooks (1950) also saw a connection between poor dispersal, high diversity, and endemism among the species of Lake Baikal and other ancient fresh-water basins.

Climatic cycles and various geographical conditions have served to promote isolation and speciation. Vuilleumier (1971) and Vanzolini (1973) have summarized evidence for terrestrial plants and animals suggesting that populations expanded and contracted cyclically as climates in the Pleistocene alternated between wet (pluvial) and dry (interpluvial) phases. During the dry phases, forest populations contracted and became isolated from one another and genetic divergence became possible. Upon re-expansion during the wet phases, rain-forest populations came into contact again, but often remained genetically distinct by developing and perfecting antihybridization mechanisms. In South America these speciation-promoting cycles were apparently more effective than in tropical Africa, since more complex topography in South America ensured the dry-phase division of the rain forests into a larger number of isolated refugia in which local differentiation could take place (Vanzolini, 1973).

Speciation among terrestrial organisms in archipelagos has been promoted by cycles of isolation-invasion-isolation-reinvasion. Several colonizing individuals (the propagule) establish a population on an island; this population remains isolated from the parent stock for some time and may become genetically adapted to local conditions. A second wave of immigrants may arrive, perhaps sufficiently different to perceive the first isolate as a separate population. Such invasions and reinvasions depend not only on chance colonization, but also on alternating connection and separation between islands owing to the cyclical rise and fall of sea level. This and related topics have been extensively treated by Brooks (1950), Mayr (1963), and others.

In the sea climatic and sea-level cycles may also have played a key role in speciation of shallow-water organisms. Crane (1975), for example, has strong circumstantial evidence from morphological studies that tectonic events in the Philippines and Indonesia have promoted species formation in the fiddler-crab genus *Uca* through the creation, disappearance, and reappearance of geographical barriers.

Most of these cycles involve a contraction of the population followed by an expansion. Genetic variability during the expansion phase has been shown to be generally greater than for populations that have reached saturation (or equilibrium) levels. In an expanding population of the sciaenid fish *Bairdiella icistius*, a species introduced from the Gulf of California to the Salton Sea in southern California, a large number of individuals survived with abnormal jaws, anal fins, vertebral columns, and lateral lines; in fact, up to 23 percent of the fishes in the 1952 year-class were abnormal in some way in 1952. In subsequent years, when the population achieved saturation density, there was a twentyfold decline in the frequency of most abnormal traits (see Whitney, 1961). Better-known examples have been provided for butterflies by Ford (1965) and for small rodents by Krebs and associates (1973). In the vole *Microtus pennsylvanicus*, dispersing individuals from expanding populations are more heterozygous than are nondispersing individuals; moreover, certain alleles appear to be selected for in the expanding populations that are selected against when the population is in the peak and declining phases. (For a review see Krebs et al., 1973.) Thus, cycles of population expansion and decline may help create and preserve variation, as well as periodically divide populations into isolated units that are susceptible to potential speciation.

Isolation of populations, and therefore the likelihood of speciation, may have been greater at some times in the recent past than at others, even for marine species with teleplanic larvae. Fischer and Arthur (1977) have cogently argued that oceanic circulations may have been weaker and slower during the Eocene and Miocene epoch when diversity was very high, than in the Oligocene, Pliocene, and Recent, when diversity was generally lower. This hypothesis was inferred from the widespread occurrence of dark, laminated, organically rich, deep-sea sediments of Eocene and Miocene age. These sediments have been laid down in strongly reducing, oxygen-poor conditions and are comparatively rare in Recent open-ocean basins.

Potential isolation and speciation of populations with planktonically dispersed larvae may thus have occurred relatively often in the near geological past, especially in regions such as the Indo-West-Pacific and to a lesser extent the West Indies, where areas favorable for shallow-water organisms are separated as islands by stretches of deep-ocean bottom.

What other circumstances promote speciation? In the discussion on extinction I suggested that species are unlikely to change in external appearance if they perceive their environment as remaining the same. Even if several populations of a species become isolated from one another, they will become distinct only if they respond genetically to any changes in their surroundings (see also Ehrlich and Raven, 1969). Biological interactions seem to be an important source of environmental change, and many populations respond genetically to them. These interactions may be intraspecific (among individuals of the same species) or they may involve competition with or predation by members of other species (interspecific interactions). Potentially important intraspecific interactions that could influence selection include aggressive behavior among either males or females, sexual display, and signaling by means of specific morphological or behavioral patterns so that one individual recognizes another individual belonging to the same species. All these traits would be most important in species with relatively few offspring; here recognition of conspecifics is important in mating and internal fertilization, which often seem to accompany reduced litter size.

J. B. C. Jackson has pointed out to me that many groups with internal fertilization or with a complex behavioral repertoire do tend to have high speciation rates, while groups with external fertilization and little or no intraspecific behavior evolve slowly. Mammals, for instance, exhibit speciation rates as much as seven times higher than those of Late Tertiary bivalves (Stanley, 1973, 1977). Other animals with comparably high rates of speciation include ammonoids and other cephalopods, drosophilid flies, and cichlid and certain other teleost fishes (Simpson, 1953; Carson et al., 1970; Van Valen, 1974; Stanley, 1977). I would predict that stomatopods and crabs also speciate readily. On the other hand, regular echinoids, most archaeogastropods, and sponges should evolve more slowly.

Interspecific interactions which seem to be another important source of selection and of environmental change, often lead to coevolution of two or more species. Coevolution occurs when one species changes genetically in response to a genetic change in its competitor, predator, parasite, or symbiont (Ehrlich and Raven, 1964). Most well-documented cases of coevolution involve terrestrial organisms: flowering plants with their pollinating insects, bats, or birds; and plants with herbivorous insects and mammals (Ehrlich and Raven, 1964; Heithaus et al., 1974; Gilbert and Raven, 1975). For example, Benson and his co-workers

(1975) have been able to trace the reciprocal evolution between heliconiine butterflies and passionflowers of the family Passifloraceae in tropical
America. Heliconiine larvae feed exclusively on passionflower plants
and are the only butterflies found on these often toxic vines. The Benson group speculated that competitive exclusion from the larger, more
easily available species has promoted switches to smaller, more ephemeral members of the passionflower family. These plants in some cases
have evolved hairs that penetrate the integument of the larva (Gilbert,
1971), and in other instances have developed leaves with a pattern of
spots that mimics the presence of heliconiine eggs. This trick is probably
designed to inform a prospective egg-laying adult heliconiine that there
is no room at the inn, and that the insect would do well to go elsewhere
to find a "bed" of leaves on which to lay its eggs (Benson et al., 1975).

In much of lowland tropical America the weedy trees of the genus
Cecropia are protected from potential insect herbivores and from vines
and epiphytes by ants of the genus *Azteca*, which *Cecropia* "feeds" by
means of glycogen-rich bodies. Janzen (1973c), however, found that
Cecropia peltata in Puerto Rico is not protected by such ants and lacks
the glycogen-rich bodies; he speculates that strong pressure by herbivores and epiphytes on the mainland may have provided an ideal environment for the coevolution of *Cecropia* and *Azteca*, and that the
dissolution of this mutualistic relationship in Puerto Rico and on other
Caribbean islands is related to the plausible though as yet undemon-
strated reduction in herbivore pressure there.

Coevolution in the sea has almost certainly occurred between para-
sitic copepods and their invertebrate hosts, and there are other instances
of species-specific or genus-specific parasitism, mutualism, and symbiosis (Humes and Stock, 1973). A possible example is the protection of
the coral *Pocillopora* against *Acanthaster* predation through the activities of the crab *Trapezia* and the snapping shrimp *Alpheus* (Glynn,
1976). These commensal crustaceans seem to be to corals what ants are
to many land plants.

Coevolution may often result from, or be intensified by, selection
based on intraspecific interactions. Crabs, for instance, use the claws
not only for feeding, but also for display, ritualized combat, and all-out
fighting. Crabs that exhibit marked dimorphism in claw shape and size
are often aggressive, at least as males; whereas species whose claws show
little or no difference in size between males and females fight or display
less often and less vigorously (Warner, 1970; Hartnoll, 1974). If larger,
longer, or more powerful claws increase a male's attractiveness to a
female, or if they are instrumental in the elimination of competing

males, then greater claw size may be selected for in succeeding genera-
tions. Such selection inevitably influences the pattern of feeding and
perhaps the type and size of food available to the crab, and could initi-
ate or contribute to an episode of coevolution (Vermeij, 1977a). A
hypothesis of this kind is not, however, applicable to crabs whose dis-
playing claws are not used in feeding. The absurdly oversized master
claw of male fiddler crabs *(Uca)* is sometimes more than half the weight
of the entire body and is too clumsy to be used in feeding; but this claw
is important in sexual display and in species recognition (Crane, 1975).

In their work on stomatopods, Caldwell and Dingle (1975) concluded
that gonodactylids and other forms that inhabit rock or coral cavities are
more aggressive than are species that burrow in sand or mud. It is predominantly the gonodactylids that have perfected the hammering technique used both in fighting with other stomatopods and in obtaining and
pulverizing hard-shelled prey. The high levels of aggression may have
come about during the competition for cavities, which may be limited in
number and must be shared with other stomatopods as well as with such
diverse animals as crabs, molluscs, brittle stars, moray eels, and other
fishes. Since stomatopods in soft bottoms build their own burrows and
therefore are not likely to be limited by available hiding places, the
selective pressure to evolve hammering may have been much weaker
among these forms.

The situation in crabs is similar though slightly more complex (Vermeij, 1977a). Jeffries (1966) found that *Cancer borealis*, a relatively large-
clawed, hard-bottom species, is more aggressive than *C. irroratus*, a
smaller-clawed species more typical of soft bottoms. Xanthids and parthenopids possessing massive master claws are nearly all found on rocky
or bouldery bottoms. The only massive-clawed crabs found on muds or
sands are mangrove-associated intertidal forms, such as the xanthid
Eurytium and the portunid *Scylla*, or inshore species of *Menippe* and
Callinectes (*C. latimanus* in West Africa, *C. bocourti* in the Western
Atlantic). Space for these crabs that are limited to very shallow water
may be at a greater premium than for crabs with a wider depth
distribution.

Intraspecific and interspecific territoriality seems to be commonplace
among reef-associated pomacentrids and other fishes that require
shelters in areas of hard bottom; but it is rare among fishes living on
sands and muds (see Randall, 1967; Hobson, 1968, 1974). Intense competition for space, both within and between species, may have led to
greater effectiveness in overwhelming armored prey and may have
created coevolutionary pressures for the development of armor among

the besieged prey. This topic merits careful investigation. The possibility that the structure of the habitat itself could influence the intensity of coevolution should also be studied further, since it has a number of important ramifications for the development of communities.

The course that coevolution takes probably depends to a considerable extent on the adaptive options available to the organisms involved. For example, counteradaptation involving fleetness or fortress-like armor seems not to be an option available to cold-water ectotherms. Chemical warfare is possible at all latitudes, but Bakus and Green (1974) have suggested that the range of toxins that can be produced without undue metabolic sacrifice may be greatest in the tropics. In some environments morphological innovations might be tolerated that would be adaptively unacceptable elsewhere. Frazzetta (1970), for instance, has suggested that the morphological transition between typical boid snakes and their remarkable bolyerine descendants on the Mascarene Islands in the southwest Indian Ocean would have been improbable and adaptively unacceptable on a large island or continent where competitive and predatory pressures are believed to be greater; yet the bolyerine jaw apparatus is in many respects a better design than the ancestral type.

I have argued that, although dextral and sinistral coiling may be functionally equivalent in gastropods, the transition between these two coiling types is usually accompanied either by a mechanically weaker shell configuration (wide umbilicus) or by developmental abnormalities which in most marine environments would be purged through stabilizing selection. Even in the fresh-water prosobranch *Campeloma rufum*, Van Cleave (1936) found left-handed specimens only among individuals one year old or less; these deviants were evidently unfit relative to normal right-handed individuals. Nevertheless, left-handedness is a relatively common phenomenon among fresh-water and land gastropods, and it may be that selection against geometrically or developmentally deviant forms is less effective in fresh water and on land than in the sea. Why this should be so is unclear; predation by shell destruction may be less important for fresh-water snails than for most shallow-water marine gastropods (Vermeij, 1975; Vermeij and Covich, 1978). The important point is that a change in handedness is a potentially adaptive option available to many nonmarine gastropods, but generally unavailable to snails in the sea.

Coevolution could promote speciation by creating change and environmental instability. The selective pressures experienced by organisms undergoing coevolution change continually, because of the relentless selection for improved feeding efficiency of predators on the one hand and for improved predator avoidance by prey on the other (Rosenzweig, 1973). Unless one side or the other in the biological battle finds an insuperable weapon, coevolving organisms must genetically meet the changing demands of their environment or else face local or even total extinction. In the event of local extinction, a biological patchwork of isolated populations is created, and there is a potential for each of the isolates to become genetically distinct and adapted to the locally prevailing biotic conditions. Such fragmentation mediated by biological causes could thus lead to speciation, provided that isolates perceive their local environments as being different, and that migration from neighboring populations is insignificant.

MacArthur (1972) and Diamond (1973, 1975) have described many cases of disjunct distributions of species among the birds of tropical America and New Guinea. They surmise that these populations were once more continuously distributed, and that unspecified events caused the local extinction of intervening populations. Examples have also been found among marine species. In the Pacific *Strombus thersites* is found in the Ryukyu Islands south of Japan, in New Caledonia, and perhaps in the Society Islands, but not in any intervening localities. This heavy-shelled snail was apparently more widespread during the Pliocene epoch, when it occurred in Fiji (Abbott, 1960). An even better example, in which some local differentiation has occurred, may be seen in the *Strombus vomer* complex (Abbott, 1960). The nominate subspecies (*S. v. vomer*) is known today only from the Ryukyu Islands and from New Caledonia. Two other subspecies have been recognized, *S. v. iredalei* from northern Australia and *S. v. hawaiiensis* from the Hawaiian Islands. These isolated populations are the remnants of what in Pliocene times seems to have been a single, more widely distributed species.

It appears to me that coevolution can occur only under certain specified conditions of isolation coupled with small effective population size. Carson (1975) has pointed out that the genomes of individuals in large interbreeding populations are buffered to a remarkable degree against physical and biological variations in the environment. Continued immigration homogenizes the population genetically, and any mutations that might arise would be quickly swamped or purged as deviants, so that a genetic status quo is maintained. For example, the common sand-dwelling intertidal gastropod *Terebra felina* from the Indo-West-Pacific varies exceedingly little in external shell characters throughout its wide geographical range, yet different populations of this species are clearly exposed to strikingly different biological pressures. Some populations (such as one at Kunigami-Gun, Okinawa) show an average of as many as 2.7 repaired crab-induced injuries per shell, indicating intense predation pressure by calappid crabs; while other populations (such as one at Muti Island, Eniwetok) have an incidence of only 0.12 repaired injury per shell. Surveys of large samples from throughout the geographical ranges of many tropical molluscs suggest that such variations are the rule rather than the exception, even though there may be no detectable regional variation in shell traits. It is remarkable how insensitive the phenotype can be to what we, as humans, would interpret as selectively important differences in the environment.

In small, isolated populations, founder effects and reduced genetic contact with other members of the same species promote the breakdown of the genome's buffering capacity if the isolate occurs in an environment substantially different from that of the parent stock (Carson, 1975). Even rather small differences in the biological and physical environment may impose directional selection, which leads to the incorporation of new traits into the genetic makeup of the population. It may be in these small, isolated, though not necessarily peripheral populations that coevolution is possible, and not in the large stable populations characteristic of a species throughout most of its geographical range. New traits may eventually be incorporated by the species as the population in which they arose expands and intermingles with neighboring populations. Such centrifugal spreading of new traits has been postulated by Brown (1957), who has given numerous possible examples among land birds, insects, and reptiles. One possible case in the sea is the muricid gastropod genus *Nassa*. The Western Pacific species, *N. serta*, has numerous spiral cords on the shell; these are vestigial or absent in Hawaiian specimens and in the very closely related *N. francolina* from the Indian Ocean. The mangrove periwinkle *Littorina scabra* is represented in Singapore and vicinity by shells of two types, a smoothish form and a heavily corded form; in more peripheral parts of the Indo-West-Pacific, the species is less variable in sculpture and tends to be more like the smooth Singapore morph. The resemblance between the squat, smoothish shells of *L. scabra* from the northern Red Sea and from Hawaii is striking.

If coevolution occurs mostly in small populations, then we might expect to see reciprocal evolution more often in communities or regions characterized by high diversity than in areas less rich in species. If average body size and biomass are identical in two areas differing in diversity, then the average population size of any one species will be smaller in the more diverse area, and the probability that one of the populations is undergoing coevolution with another is higher. Selection should favor traits that confer resistance to predators if predation pressure on the local isolate is high. Again, this situation is more likely to obtain in the area with higher diversity (see Chapter 6). On the other hand, isolates in areas with low predation intensity should not develop or emphasize antipredatory features unless these happen to be appropriate to some unrelated environmental peculiarity or selective pressure.

Darlington (1959) and Briggs (1966, 1967a and b, 1974a) seem to have arrived at a similar conclusion, although they did not specify a plausible evolutionary mechanism. Both suggested that competitively superior animals evolve in high-diversity assemblages and that these animals subsequently emigrate into areas of lower diversity, competitively displacing less adept species as they do so. Briggs envisioned so-called evolutionary centers, the primary one located in the biotically very rich Indo-Malayan region and a secondary one located in the Caribbean Sea. If these are really evolutionary centers, then the rate of production of new taxa (species, genera and so on) should be higher there than elsewhere. Stehli and Wells (1971) have shown that the genera of scleractinian corals in the Indo-Malaysian part of the Indo-West-Pacific are greater in number and of younger average age (23 to 24 million years) than are the genera in such peripheral areas as Natal and Hawaii. Even in Natal, where the average age of a coral genus is 44 million years, the age of genera is less than in the richest parts of the Caribbean, where the average generic age of corals is 49 million years.

If average generic age is inversely correlated with rate of production of genera and species, as is commonly assumed (Simpson, 1953; Stanley, 1973, 1975a), then these data suggest that the speciation (or origination) rates among corals in the Indo-Malaysian region have been higher than those in other tropical regions. Origination rate will be the reciprocal of probability of extinction only if overall diversity remains constant (that is, originations minus extinctions equals zero). If generic age is positively correlated with number of species per genus, as might be expected on purely statistical grounds, then the trend for older genera to be more widely distributed is merely a biologically uninteresting artifact. These problems were not considered by Stehli and Wells (1971), but the statistical artifact may exist among some Indo-West-Pacific scleractinian genera.

The significantly older age of West Indian as compared to Indo-Malaysian genera, however, appears to be a real phenomenon, since West Indian genera are both smaller in number of constituent species and geographically more restricted in distribution. A second problem is that if tree-like, branching corals evolve more rapidly than do more massive or encrusting forms (Jackson, 1977), then the apparently slow origination rates of Caribbean as compared to Indo-West-Pacific corals may reflect nothing more than the comparatively low diversity of branching corals in the West Indies; that is, we cannot exclude the possibility that branching corals might evolve at much the same rate in the two regions. One complication here is that two genera of predominantly branching corals *(Porites* and *Acropora)* occur in both the Caribbean and the Indo-West-Pacific.

Further evidence for the existence of evolutionary centers is still scanty; and where fossil evidence is available, the problems in interpreting the data are similar to those with the scleractinian corals. The younger average age of tropical as compared to temperate bivalve genera in the Western Atlantic (Hecht and Agan, 1972) seems not to be a statistical artifact and may therefore indicate faster evolutionary rates in the tropics. Crane (1975) has concluded from comparative morphological studies that peripheral (and very widely distributed) Indo-West-Pacific species of the fiddler-crab genus *Uca* are more primitive and therefore presumably older than those restricted to the tectonically active archipelagos of the Philippines and Indonesia. In this case and others, however, evidence from fossils unfortunately is not available.

The world's richest marine biota today is found in the Indo-Malaysian region, encompassing the Philippines, Indonesia, New Guinea, the Solomons, and immediately adjacent areas. For example, the fauna of mangrove-associated gastropods in and around Singapore comprises at least twenty-five species, and is more than three times richer than that of the New Hebrides, Palau, or Madagascar (Vermeij, 1973c; Marshall and Medway, 1976). Up to twenty-seven species of shallow-water, reef-associated, comatulid crinoids are found at the northern end of the Great Barrier Reef, while in Palau and the Molluccas the number ranges from seventeen to twenty-two (Meyer and Macurda, 1977). The precise locality within the Indo-Malaysian area where diversity reaches its maximum probably varies for animals in different habitats and cannot be pinpointed in many cases because so little work has been done in this region; but I know of no taxonomic group in the Indo-West-Pacific that does not reach a peak in diversity in this region.

If coevolution is intense in the Indo-Malaysian area, and if predation pressure is particularly heavy there, we should see species with a distribution restricted to this region and with unusually well-developed predation-related traits. Moreover, I predict that such species will prove to be of comparatively recent origin. Paleontological studies are not yet sufficient to permit this last prediction to be evaluated properly, but the taxonomic literature on molluscs does provide some possible examples of well-armed species with a strictly Indo-Malaysian distribution. In the gastropod genus *Vasum*, for example, the extremely long-spined, sturdy *V. tubiferum* is restricted to the Philippines (Abbott, 1959). Many species of the genera *Lophiotoma* and *Turris*, which diverged in the Miocene epoch, are restricted to the Western Pacific arc (Powell, 1964). This is also true for most species of *Tectarius* and large representatives of the genus *Nerita* on the upper shore. All these shells are large, sturdy objects that place a certain stamp of boisterousness or superfluity on the Indo-Malaysian molluscan fauna.

I pointed out in Chapter 6 that, although there is a tendency for the most heavily armed molluscs and crustaceans to occur in areas of maximal diversity, there are instances in the peripheral regions of the Indo-West-Pacific where species that are highly predation resistant flourish in communities of relatively low diversity. In fact, I have been impressed that the incidence of repaired shell injuries may sometimes be very high in populations of the sand-dwelling *Terebra gouldi* from Hawaii (up to 9.5 per shell) and of *T. cerithina* from Tahiti (up to 4 per average shell) (see also Miller, 1975). If coevolution between predators and prey is common in such geographically marginal areas, then predation-related traits may often be at least as beneficial to isolated populations in the Central Pacific or Hawaii as to populations from the biologically richer Indo-Malaysian area, and the notion of an evolutionary center whose biotically superior species emigrate to more impoverished regions may be quite inaccurate. Conclusive evidence for the existence of evolutionary centers must come from the demonstration that competition or predation or both are more intense in the postulated center than in the area believed to be receiving rather than exporting species. It is not enough to show that diversity is higher in the postulated center of evolution, since it has not been conclusively proved that diversity and the evolution of "successful" species are causally interrelated.

The largely theoretical and speculative arguments outlined in the preceding paragraphs seem to be consistent with some patterns in the genetic variation of species as determined by chromosomal analyses and by electrophoresis. The latter method permits detection of differences in charge between two or more proteins (usually enzymes) produced by different alleles at a given gene locus. (For a review see Lewontin, 1974.) Genetic polymorphisms may arise for several reasons. Since sources of mortality may vary strikingly with age, some traits may be favored when the organism is young, while other traits may be favored at maturity. For example, desiccation is probably an important source of mortality

for young high-shore snails, whose surface-to-volume ratio is high; but it
may be a minor cause of death for larger adults. Predation and competition also are strongly size dependent. Levinton and Fundiller (1975)
have found that allele frequencies for various enzymes differ for adult
and juvenile *Mytilus edulis* from the same locality. Other examples of
this phenomenon have been reviewed by Battaglia (1975).

A second possible source of polymorphism is that heterozygous individuals may be more fit than homozygous ones. The physical and chemical environment may be so variable that more than one protein is
required to catalyze or regulate a reaction under the various conditions
encountered by the individual organism. For example, enzymes catalyzing reactions at cold temperatures may require a different conformation
from those catalyzing the same reaction under a warmer thermal
regime. Accordingly, it might be guessed that species with a broader
environmental tolerance, or species living under a greater variety of
conditions, would exhibit a greater degree of enzyme polymorphism
than would species whose tolerances are narrow (Bretsky and Lorenz,
1970). Some support for this prediction has come from work on leguminous flowering plants of the genus *Lupinus* (Babbel and Selander,
1974). Texan members of this genus that live in a variety of habitats
have greater electrophoretic variation than do species living in a more
restricted range of environments. Similarly, Levinton (1975) speculates
that the highest levels of electrophoretic polymorphism in the bivalve
genus *Macoma* are achieved in those species (*M. nasuta* and *M. eli-
mata*) in the Northeast Pacific that live under the greatest variety of
conditions as measured by water depth, sediment depth and type, and
salinity. In earlier work Levinton (1973) found that epifaunal and shallowly infaunal bivalves, which experience relatively variable conditions,
seem to have higher enzyme variation than do deep burrowers.

The related expectation that high-latitude organisms should be genetically more variable than tropical species has not been fulfilled, however.
In fact, available evidence points to an exactly opposite conclusion. (See
Valentine, 1976, for a review.) The calculated number of heterozygous
loci per individual in the Antarctic planktonic crustacean *Euphausia
superba*, for example, is only 5.8 percent; for *E. mucronata*, a species
found off the temperate coast of northern Chile, it is about 15.5 percent;
and for the tropical Eastern Pacific *E. distinguenda*, the figure is 21.1
percent (Valentine and Ayala, 1976). The mean number of heterozygous loci per individual in the horseshoe crab *Limulus polyphemus* is
low (5.7 percent), but is slightly higher for populations from the subtropical Gulf of Mexico than for those from the colder waters of Massachu-

setts and Virginia (Selander et al., 1970). As many as 20 to 22 percent of the loci may be heterozygous in individuals of the tropical giant clam *Tridacna maxima* (Ayala et al., 1973). Although comparisons across phyla are dangerous, it is noteworthy that the Antarctic articulate brachiopod *Liothyrella notorcadensis* has an average of only 3.9 percent of the loci polymorphic per individual (Ayala et al., 1975a).

This last example, together with Levinton's (1973) observations that deep-burrowing bivalves are genetically less variable than are species closer to the sediment-water interface, suggests in a preliminary way that electrophoretically detected enzyme variation may decline toward more stressful environments. If this is so, then we should expect a decline in polymorphism in an upshore direction among individuals or closely related species. Although such data should be easy to obtain, I am not aware that any exist.

The high genetic variability that has been observed in deep-sea asteroids, bryozoans, molluscs, and other animals would contradict the notion that polymorphism should decline with greater stress (see Gooch and Schopf, 1972; Ayala et al., 1975b). Up to 18 percent of the loci of deep-sea animals have been found to be polymorphic in an individual. Such high levels of variation in the deep sea and in the tropics have been attributed by Valentine (1976) to the high predictability of food that is postulated for these environments. He feels that adaptation to variable conditions in environments with unpredictable food supplies (such as shallow temperate waters) is achieved largely through plasticity of the phenotype, while in more predictable environments the adaptations are more genetic in character. Although plasticity certainly is characteristic of oysters, mussels, barnacles, and other sessile intertidal species on temperate coasts, as demonstrated by transplantation experiments (see, for example, Seed, 1968, 1969a), it seems equally clear that much phenotypic variation among cold-water snails is genetic in origin, and not closely tied to the predictability of the food supply. As examples, color variation in *Urosalpinx cinerea* from eastern North America has a genetic basis (Cole, 1975), and a polymorphism in chromosome number associated with differences in habitat preference has been described for some European populations of *Thais lapillus* (Staiger, 1957). Phenotypic plasticity should be characteristic of weeds, but not necessarily of long-lived organisms (Levin, 1975). Furthermore, no data are actually available on the predictability of food in the deep sea, and Dayton and Hessler (1972) have speculated that food may be much more patchy in distribution, and therefore much less predictable, for large deep-sea animals than for smaller species. Therefore I prefer an

alternative explanation that takes into account the number and importance of biological interactions experienced by an individual organism.

If coevolution promotes the retention and eventual spread of new
traits to keep pace with the changing biological environment, then a
third reason for genetic polymorphism may be that alleles are constantly
and slowly being replaced. Populations undergoing coevolution and
expanding after a period of isolation should display a high degree of genetic variability, while equilibrium or declining populations in environments that are perceived to remain constant over time should show
comparatively low levels of genetic variation. Indeed, it has been found
in plants and in some species of *Drosophila* that the incidence of chromosomal inversion and of genetic variation generally is higher in the
large populations of continents, where biotic pressures are thought to be
high, than in the smaller populations on biotically less demanding
islands. Moreover, variation tends to be greater in the center of a population than near its periphery (see Mayr, 1963; Lewontin, 1974; Levin,
1975; Johnson et al., 1975). D. E. Gill has suggested to me that populations in the center of a species' range may be expanding in the sense
that they export genes to more peripheral populations, while there may
be a net loss of genes in the stable or declining peripheral populations.
This would be especially plausible if turnover rates are lower at the
limits of a species' range because the species is living under physiologically marginal conditions.

Alternatively, the multitude of biological interactions and biological
settings in a physiologically favorable environment may demand accu-
rate chemical sensing; since the local biological surroundings of an individual are likely to vary widely in a species-rich biota, successful
coexistence with other species may depend on a large bank of minor
chemical variants on the same basic theme. Battaglia (1975) has
reviewed work on the harpacticoid copepod *Tisbe clodiensis* showing
that the fitness of this species in coexistence with *T. reticulatus* in culture vessels is greater for some enzymic variants than for others. I am
confident that similar examples will come to light in other organisms.
The general decline in genetic polymorphism toward the poles and from
shallow to deep layers in the sediment could then be reinterpreted as
reflecting a reduction in the importance of and variety in biotic interactions. Detailed work on members of the same or closely related species
under different biotic circumstances is badly needed to establish and
strengthen this tentative conclusion. Also needed are more data on
levels of variation in expanding populations compared with variation in
peak and declining populations.

It is now possible to integrate the arguments presented in this chapter with the conclusions of earlier sections of the book. Physiologically stressful environments are occupied for the most part by long-lived species that are little affected by predation and that are adapted to chronic stress; they are highly resistant to extinction and little prone to speciation. Weedy species, which take advantage of temporarily favorable conditions in stressed and benign environments alike, often are subject to intense predation; if they have wide geographical ranges and effective long-distance dispersal, weedy species should be relatively resistant to extinction and should speciate little. If, on the other hand, their environment is discontinuously distributed or subject to steep physical gradients, or if their mortality is predominantly biological, then these opportunists may have restricted distributions and their susceptibility to extinction and speciation may be quite high. Biotically competent predators and prey exhibit a wide variety of behavioral and morphological adaptations to predation; they are typically found in physiologically favorable environments and their susceptibility to extinction and to speciation is enhanced through the destabilizing influence of coevolution, which assures that selective pressures are continually changing. High dispersibility may confer a certain measure of resistance to extinction and may prevent genetic tracking of local biological changes that might otherwise result in coevolution.

CHAPTER 8

Patterns of Biotic History

T HE conclusions reached in Chapter 7 about the relative vulnerability to extinction and speciation of stress-tolerant, opportunistic, and biotically competent species may now be recast into statements concerning the permanence and regional variation of marine communities. Many of the present-day differences between marine biotas that occupy essentially similar climatic zones seem to be reflections of history; the course of development of communities may be similar, but large-scale disturbances apparently have affected some biotas far more severely and more frequently than they have others.

Long-Term Community Development

Habitats or regions barren of life must first be colonized by organisms that have dispersed from some already existing population (Van Valen, 1971). Depending on the character of the habitat, these pioneer species may be opportunistic or stress tolerant. As species accumulate in the new area from previously established biotas, they begin to interact with one another, and biological sources of mortality are likely to become increasingly important. Thus, selection may come to have an increasingly biological, and a decreasingly physical, component; and resident organisms will emphasize biological adaptations at the expense of physiological tolerances, within the constraints of the physical environment. The more physiologically favorable the environment, the greater the opportunity for biological specializations. Any biological patchiness in the developing community is favorable to the diversification of species, particularly of those that are well adapted to competition and predation. So long as episodes of wholesale extinction do not occur, the higher speciation rate of biotically competent species should result in a disproportionate increase through time in the number of such species. Moreover, biological interactions are likely to generate coevolutionary pressures that in turn lend an element of instability to the community. All these events are stimulated by expansions, contractions, and reexpansions of available habitat; fluctuations such as these must be large enough to affect population size, but small enough to prevent global extinction and to preserve refuges or sanctuaries where remnants of a population may persist for eventual reexpansion. Even though a local equilibrium in species number may become established, it need not imply evolutionary stagnation. If a large-scale disturbance does besiege a biologically sophisticated biota, then that biota is likely to suffer widespread extinction of its constituent species, and diversification and elaboration can take place once more after the crisis.

As viewed here, community development has many of the characteristics of plant succession, a process which, however, takes tens to hundreds of years rather than millions. When a new habitat is first colonized by plants, the first members of the community tend to be weedy herbs, fast-growing trees, or rock-dwelling mosses and lichens. In time, these are replaced by slower-growing, often more herbivore-resistant species (Cates and Orians, 1975), which over the long run are competitively superior to the early-successional plants. The invasion of these later plants in many cases is dependent on changes in the soil brought about by the earliest occupants. During this process of succession species number first increases, then as the so-called climax is reached, it decreases (at least in temperate forests) because of the predominance of a few competitively successful forms. Sudden changes in climate are often catastrophic for climax forests, but of little consequence for ephemeral early-successional plant assemblages. (For reviews and further discussion see Odum, 1969; Horn, 1971, 1974).

Similar successional patterns have been described for temperate fresh-water phytoplankton communities (Porter, 1973), fresh-water zooplankton communities (Allan, 1976), and marine communities of sessile organisms. For example, Grigg and Maragos (1974) have described a successional pattern for corals that colonize new lava flows in Hawaii. On rocky shores, diatoms tend to be the first colonists of barren rock, followed by ephemeral green algae, and later by more permanent algae, barnacles, and mussels (Castenholz, 1961; Doty, 1967; Paine and Vadas, 1969a; Dayton, 1971). Subtidally, early-successional organisms tend to be solitary (serpulid worms in cryptic areas, for instance), while later colonists are often more permanent sponges and colonial corals, hydroids, tunicates, and bryozoans (see Long, 1974; Jackson and Buss, 1975; Jackson, 1977). Succession among mobile animals has been little studied in marine environments; it is probably much less predictable than among sessile forms.

Good examples of long-term succession have come to light in studies of Paleozoic and Mesozoic marine assemblages. Ordovician soft-bottom communities in Quebec, for example, exhibited several cycles, during each of which an architecturally similar community of brachiopods and bryozoans developed from a simpler community dominated by fast-growing, apparently opportunistic, mud-floating brachiopods and infaunal deposit-feeding bivalves (Bretsky and Bretsky, 1975). The later-successional communities appear to have been destroyed by episodes of heavy sedimentation or by other, primarily physical, calamities. Mesozoic assemblages dominated by coral-like rudist bivalves passed through characteristic successional stages that culminated in tightly knit frameworks composed of a few large species (Kauffman and Sohl, 1974; Walker and Alberstadt, 1975).

The history of life has been beset with a number of major crises, the most catastrophic of which occurred at the end of the Ordovician, in the Late Devonian, at the Permo-Triassic boundary, and at the Cretaceous-Tertiary boundary. Fischer and Arthur (1977) believe that crises of greater or lesser magnitude occurred at intervals of about 32 million years; they speculate that the biotic crises corresponded to periods of strong climatic cooling, increased intensity of oceanic circulation, and reduced sea level. The communities most affected by these large-scale physical events are reefs and open-ocean plankton. After each episode of extinction, renewed diversification takes place. In the Tertiary period, Paleocene zooplankton communities were dominated by forams and other organisms with generally simple skeletons; most of the species were widespread, and certain apparently weedy forms and resting stages were extremely abundant. By Eocene time, these forms had given rise to, and were to some extent replaced by, more ornate types; these became largely extinct in the Oligocene, but by Miocene time another assemblage of ornate forms evolved (see Lipps, 1970; Cifelli, 1976; Fischer and Arthur,

1977). Valentine (1973) noted that Early Cambrian brachiopods, trilobites, and molluscs had relatively simple, unornamented skeletons; while bizarre coral-like brachiopods, complex fusilinid forams, and other ornate forms reached their Paleozoic zenith in the Permian. After the crisis at the end of the Permian, Triassic fossils were simple once again; toward the end of the Mesozoic, strange ammonoids and large coral-like rudist bivalves became common, only to disappear in the latest Cretaceous.

These cycles of long-term community development illustrate the importance of extinction as a selective filter, and of large-scale disturbance in arresting the elaboration of biological interactions and adaptations. Elton (1958) and Hutchinson (1959), in fact, suggest that terrestrial biotas in the Arctic and Boreal zones are actually being limited in diversification (and are thus kept in an early-successional stage) by frequent episodes of severe stress that cause local extinctions, while warm-temperate and tropical biotas achieve individuality because such disturbances are less frequent.

Superimposed on biotic cycles appear to be some long-term trends in the nature and intensity of biological interactions. Elsewhere I have argued that the intensity of crushing predation has increased in the course of geological time, especially since the Jurassic period (Vermeij, 1975, 1977c). Such skeleton-destroying predators as brachyuran crabs, lobsters, stomatopods, teleost fishes and rays arose in the Triassic or Jurassic and diversified in the Cretaceous. Drilling gastropods, first known in the Early Cretaceous, did not become common until the Late Cretaceous and Early Tertiary (Sohl, 1969). The first octopods are known from the Late Cretaceous, though Jeletzky (1965) thinks they may have arisen as early as the Triassic. Rock-destroying echinoids and teleosts radiated extensively in the later Mesozoic, especially the Cretaceous. The five-rayed scratches made on hard surfaces by the jaws (aristotle's lantern) of echinoids are known throughout the Mesozoic and Cenozoic, but did not become common in shallow-water settings until the Campanian in the Late Cretaceous (Bromley, 1975). Correspondingly, gastropods with narrowly elongate or toothed apertures and with heavy external sculpture did not become diversified until the Jurassic; indeed, only one genus of snail with a toothed outer lip *(Labridens)* is known from Paleozoic (Permian) rocks. Hermit crabs, which recycle snail shells after the latters' original inhabitants have died, arose in the Jurassic and may have increased the availability of shelled prey substantially. Sessile epifaunal brachiopods and stalked crinoids, which cannot reattach once dislodged from their site of settlement, gave way in warm shallow waters to byssally attached or cemented bivalves and to comatulid crinoids respectively in Late Mesozoic time (see also Meyer and Macurda, 1977; Stanley, 1977).

Possibly in connection with these epifaunal events, there was a large-scale invasion of soft bottoms by infaunal echinoids, gastropods, bivalves, and thalassinid crustaceans. This infaunalization of skeletonized taxa was well under way in the Jurassic and, like the epifaunal changes, was most strongly felt in shallow tropical waters. Another manifestation of the occupation of habitats below the surface was the evolution in the Jurassic of rock-boring bivalves. It is noteworthy that the biotic crisis at the end of the Mesozoic, and the less devastating biotic decline in the Oligocene, had no appreciable effect on this trend of increasing skeletal destruction by marine predators and grazers.

Trends such as these have evidently affected some habitats more than they have others. Biological interactions are limited in variety and intensity by physiological restrictions in stressful environments, and by spatial confinement in deep sediments and under boulders. In these environments coevolutionary pressures for the development of shell crushing and of resistance to it should be low; therefore, regional variations and rates of change in crushing-related architecture should be less in physiologically marginal or spatially confined habitats than in more open, physiologically more favorable environments. Correspondingly, the consequences of disturbances are less severe in stressed environments.

I suggested in the last chapter that competition for space may be much keener on rocky bottoms and in the intertidal zone than on most soft bottoms, and that such differences have had important consequences for coevolution between predators and their prey. Strife or ritual connected with the procurement or retention of shelters seems to be characteristic of many crustaceans and fishes inhabiting reefs and other rocky areas where they cannot make such hideaways themselves; but on most sandy or muddy bottoms well away from rocky areas, most large, active marine animals can make permanent burrows or hide by burying themselves temporarily in the sediment. I believe that this difference between hard-bottom and soft-bottom communities is reflected in the degree to which these communities display regional variations in predation-related architecture. If coevolutionary pressures from competition among larger skeleton-destroying predators are greater in rocky habitats than in sand or mud, and if these pressures affect the selection of prey, then biotic change might be more rapid on hard bottoms; similarly, a large-scale disturbance interfering with this biological elaboration might have a greater effect on the hard-bottom community than on one in soft sediments. Without disturbance, coevolution and the elaboration of biological interactions will proceed to a higher level of sophistication on hard bottoms. Thus, the biological contrast between two biotas that have been exposed to different frequencies or intensities of disturbance should be greater in reef and rocky-shore habitats than in most soft-bottom communities. Exceptions might occur if the soft sediments have become stabilized by the roots of mangroves or sea grasses, or when sand patches are found in reef settings. In the first case, burial is more difficult and permanent burrows may require defense. In the second case, predators requiring rocky shelters may be able to exploit organisms living in the sand patches.

It must be stressed that the greater regional and temporal variation of hard-bottom as compared to soft-bottom assemblages refers to the predation-related architecture of component species, and not necessarily to such other community properties as diversity or trophic structure. Woodin (1976) has demonstrated that Recent soft-bottom communities may be numerically dominated by any one of three functional groups — tube builders (polychaetes, amphipods, phoronids), suspension-feeding bivalves, or deposit feeders — and that adults of each of these three groups tend to exclude the larvae or adults of the two others. A striking regional variation in community composition and sediment characteristics is therefore often observed on soft bottoms even over comparatively short distances. The type of community present depends largely on which functional group is the first to settle as larvae in an area whose previous inhabitants succumbed to a local disturbance. Similarly, interactions between adults and larvae may create spatial variation in numerical dominance on rocky surfaces: some areas may be occupied by hydroids, others by tunicates, still others by bivalves, barnacles, bryozoans, or sponges (see Sutherland, 1974; Dayton et al., 1974).

Trans-American Comparisons

We may illustrate some of the foregoing ideas on changes in the architecture of species over time by examining the marine biotas on the two coasts of tropical America. Before the completion of the present Panama land bridge at least 3.5 million years ago, the faunas of the tropical Eastern Pacific and Western Atlantic were united into a single biogeographical entity (see Woodring, 1966, 1974; Saito, 1976). Since the isthmus was formed, however, the two biotas have diverged as a result of both extinction and speciation. Evidently, divergence has been greatest in environments where predation is thought to be important, and least in physiologically marginal environments.

One somewhat crude measure of the taxonomic similarity between a Caribbean and a physically comparable Eastern Pacific community is the proportion of cognate or geminate species. Cognate species are two or more geographically isolated forms that, morphologically speaking, have diverged only slightly (and in some cases not at all) from their common pre-Isthmian ancestor. In the Appendix I have compiled all the cognate species-pairs of molluscs known to me. The list is certain to be incomplete and to require revision; no consistent objective criteria of affinity were applied, and degree of similarity in nearly all cases is the considered opinion of systematists familiar with a given group. The compilation is probably least satisfactory for small gastropods, trochid and limpet-like archaeogastropods, turrids, opisthobranchs, and protobranch bivalves. I have excluded planktonic forms.

Some data on the frequency of cognate species of molluscs and other animals in various habitats are given in Table 8.1. The calculations for polychaetes, based on Fauchald's (1977) monograph, are to be regarded as preliminary at best; adequate collections have not been made from most areas and habitats in tropical America, and especially not from soft bottoms in the Eastern Pacific, and it is often difficult to assign a species unambiguously to habitat type. In calculating the frequency of cognates among polychaetes, I have considered only those species found both in the Western Atlantic and in the Eastern Pacific as cognates. The shortcomings described above for the Appendix probably bias the data on molluscs in Table 8.1 against the conclusions drawn in the next paragraphs.

It will be seen at once from Table 8.1 that shallow-water, soft-bottom assemblages of gastropods, decapod crustaceans, and stomatopods have diverged less on the two sides of the land bridge than have either high- or low-shore assemblages from rocky shores; that is, they have a higher frequency of cognate species. A similar conclusion was reached by Rosenblatt (1967) for Eastern Pacific fishes. Two groups do not conform to the trend. Hard-bottom polychaetes in the Eastern Pacific have proportionately more counterparts in the Atlantic than do soft-bottom species, but uncertainties in the data render this exception suspect. Bivalves as a group have diverged relatively little on the two sides of tropical America since the Late Pliocene, and I cannot detect differences in the frequency of cognate species between hard- and soft-bottom bivalve assemblages. The conservatism of bivalves has reared its head throughout this book, and I am unable to provide a convincing explanation for it. Stanley (1977) has suggested that bivalves are relatively opportunistic and that many of the hard-bottom forms in the tropics live in protected crevices or under ledges; but these apologies seem weak and insufficient.

The proportion of cognates seems especially high among deep-burrowing tellinid, semelid, mactrid, sanguinolariid, and solenid bivalves. With the exception of the Eastern Pacific cardiid genus *Lophocardium*, all tropical American genera with a posterior gape have cognate species in the Western Atlantic and Eastern Pacific (see Vermeij and Veil, 1978). The Lucinidae is the only family of deep-burrowing bivalves with a relatively low frequency of cognates. In the Eastern Pacific fifteen of the twenty mainland species are limited to Mexican waters, including Baja California and the Gulf of California (see Keen, 1971); Abbott (1974) lists twenty-four species in the tropical and subtropical Western Atlantic, most of which are common and widespread throughout the region. There are only seven cognate pairs listed in the Appendix. It may be that, for unknown reasons, large parts of the Eastern Pacific are unfavorable to lucinids. While dependence on grassbeds, which are almost absent in the Eastern Pacific, is a plausible explanation, other factors are certain to be involved.

On sandy beaches in Central America seven or eight taxa are so similar on the two coasts that they are regarded as disjunct populations of a single amphi-American species (Dexter, 1974). Examples among molluscs include the gastropods *Agaronia testacea*, *Mazatlania aciculata*, and *Terebra cinerea*, and the bivalve *Heterodonax bimaculatus*. Closely similar cognates are found in the bivalve genera *Donax* and *Tivela*, the anomuran mole crab *Emerita*, the brachyuran crabs *Ocypode* and *Arenaeus*, and the echinoid sand dollar *Encope*, as well as among various polychaetes, amphipods, and isopods. This similarity of species on beaches of the two coasts is in many ways surprising, since environmental conditions in the Atlantic and Pacific would seem to be markedly different. Most Eastern Pacific beaches (at least in Central America) are very wide and gently sloping; they are usually composed of fine, volcanic, land-derived sand, and the tidal range is large (more than 6 m in the Bay of Panama). In the Caribbean tidal amplitudes are rarely greater than 60 cm, and many beaches are composed of light-colored, often coarse, calcareous sand (see Glynn, 1972; Dexter, 1972, 1974, 1976). Atlantic beaches with volcanic or otherwise land-derived dark sand are found on the mainland coast of northern South America and on the volcanic Windward Islands and Leeward Islands in the eastern Caribbean. Since most Caribbean beach-dwelling animals have wide distributions, that encompass both volcanic and calcareous beaches, they can evidently tolerate a wide range of shore conditions associated with different sand types and tidal regimes. For example, most beach molluscs have dark shells on volcanic sand and light shells on white beaches; sculpture and overall shell shape are highly

Table 8.1   Incidence of cognate species in some tropical American assemblages.

Category	n	C	% C
Western Atlantic			
Grassbeds less than 1 m deep			
Bivalves, north coast of Jamaica	29	13	41
Gastropods, Cahuita, Costa Rica	7	5	71
Mangrove swamps			
Gastropods, Curaçao	5	3	60
Decapod crustaceans, Panama	17	10	59
Sandy beaches			
Fauna of Central America	33	14	42
Decapod crustaceans, Panama	8	4	50
Soft bottoms (general)			
Gastropods, Sanibel Island, Florida	19	8	42
Gastropods, Aruba	13	5	38
Gastropods, Amuay Bay, Venezuela	15	5	33
Bivalves, Amuay Bay, Venezuela	13	11	85
Bivalves, Sanibel Island, Florida	16	5	31
Uca	12	12	100
Squillid and hemisquillid stomatopods	39	13	33
Polychaetes, based on list from Panama	32	27	84
Hard bottoms (general)			
Gonodactylid stomatopods	18	4	22
Polychaetes, based on list from Panama	76	57	75
Decapod crustaceans, Panama	67	27	40
Bivalves, north coast of Jamaica	15	7	47
Undersurfaces of stones			
Gastropods, Pernambuco, Brazil	6	2	33
Gastropods, south coast of Curaçao	12	6	50
Gastropods, north coast of Jamaica	12	5	42
Open rocky surfaces			
Gastropods, Cahuita, Costa Rica	13	4	31
Gastropods, Fort Point, Jamaica	17	5	29
Gastropods, south coast of Curaçao	17	6	35
Gastropods, Amuay Bay, Venezuela	11	5	45
Gastropods, Pernambuco, Brazil	8	2	25
High intertidal zone			
Gastropods, Cahuita, Costa Rica	13	4	31
Gastropods, Fort Point, Jamaica	13	3	23
Eastern Pacific			
Mangrove swamps			
Gastropods, Mata de Limón, Costa Rica	9	2	22
Decapod crustaceans, Panama	20	10	50
Sandy beaches			
Fauna of Central America	52	14	26
Decapod crustaceans, Panama	17	4	24

Category	n	C	% C
Clean sand			
Gastropods, Venado Beach, Panama	35	12	34
Bivalves, Venado Beach, Panama	23	11	48
Soft bottoms (general)			
Uca	29	12	41
Polychaetes, based on list from Panama	53	27	51
Hard bottoms (general)			
Polychaetes, based on list from Panama	90	57	63
Decapod crustaceans, Panama	78	27	35
Undersurfaces of stones			
Gastropods, Venado Beach, Panama	12	7	58
Gastropods, Isla Taboga, Panama	13	5	38
Open rocky surfaces			
Gastropods, Playa de Panama, Costa Rica	20	3	15
Gastropods, Panama City, Panama	22	4	18
Gastropods, Venado Beach, Panama	21	6	29
Bivalves, Panama	15	6	40
High intertidal zone			
Gastropods, Playa de Panama, Costa Rica	8	2	25

n = total number of species.
C = number of cognate species.
% C = percentage of cognate species.
Source: Data in this table are based on species lists and distributional information given by Abele (1972, 1974) for decapods, Crane (1975) for *Uca*, Manning (1969) for stomatopods, Jackson (1973) for Jamaican grassbed bivalves, Dexter (1974) for sandy-beach fauna, and Fauchald (1977) for polychaetes. Other data on molluscs are based on my own collections and on the Appendix.

variable within some species (*Mazatlania aciculata*, for example). Thus regional variations in sand type and perhaps tidal regime promote polymorphism within species rather than differentiation among populations; and those differences that do exist between Western Atlantic and Eastern Pacific beaches appear to be irrevelant to the morphological features by which we recognize species.

As expected, the frequency of cognate species living under stones is considerably higher than that on open rocky surfaces, at least among gastropods. Examples of cognates under stones include gastropods in the genera *Diodora*, *Lucapina*, *Tegula* (*T. hotessieriana*), *Hipponix*, *Cheilea*, *Bursa* (*B. granularis*), *Bailya*, *Thala*, *Mitra*, and *Volvarina*; the chiton *Stenoplax*; and the bivalves *Arcopsis* and *Lima*. At least eleven species are morphologically indistinguishable on the two coasts and are therefore considered to be disjunct populations of the same amphi-American species.

On open rocky surfaces the frequency of cognate species is relatively low, and only two gastropods (*Cymatium pileare* and *Thais haemastoma*) are morphologically indistinguishable in this habitat on the Atlantic and Pacific coasts of tropical America. In fact, the casual observer is struck with the conspicuous differences in the composition of assemblages from open rocky sufaces on opposite sides of the isthmus. There are, of course, species that are nearly indistinguishable from cognate forms on the other coast. Good examples of this strong similarity may be seen in the genera *Chiton*, *Hemitoma*, *Fissurella*, *Tegula* (*T. viridula* and *T. verrucosa*), *Astraea*, *Cypraea*, *Thais* (*T. haemastoma*), *Purpura*, *Leucozonia*, and *Conus* (*C. mus* and *C. gladiator*, *C. regius* and *C. brunneus*). With the exception of *Haematoma*, *Astraea*, *Purpura*, and *Leucozonia*, most of the cognate pairs have relatively unornamented shells. For example, the two species-pairs of *Fissurella* listed in the Appendix have smoother shells than do the Western Atlantic *F. nodosa* and *F. barbadensis*, which have no close relatives in the Eastern Pacific. Other spinose Western Atlantic species without Eastern Pacific cognates include the fissurellid *Diodora listeri*; the littorinids *Echininus nodulosus* and *Nodilittorina tuberculata*; the cerithiid *Cerithium litteratum*; and the neogastropods *Thais deltoidea*, *Trachypollia nodulosa*, *Vasum capitellus*, and *V. globulus*. Eastern Pacific gastropods with well-developed knobs or spines and with no close Western Atlantic relatives include the archaeogastropods *Astraea buschi* and *Turbo saxosus*, the mesogastropod *Cerithium stercusmuscarum*, the neogastropods *Thais speciosa*, *T. triangularis*, *Cymia tectum*, *Neorapana* spp., *Vitularia salebrosa*, *Muricanthus radix*, *Anachis rugosa*, and *Leucozonia cerata*, and the pulmonate *Siphonaria gigas*. If the antipredatory criteria of Chapter 2 are applied to the open-surface rocky-shore gastropods of the Eastern Pacific, it is found that species without close relatives in the Western Atlantic are mechanically sturdier than are species with Atlantic cognates. In the Western Atlantic, there does not seem to be a significant difference in sturdiness between unique and cognate species.

One rather unexpected finding is that molluscs associated with grassbeds in the Western Atlantic very often have close relatives in the Eastern Pacific, where grassbeds are rare and *Thalassia* (one of the primary genera of sea grasses in the West Indies) is entirely absent. For example, the deep-burrowing lucinid bivalve *Codakia orbicularis* is commonly regarded as characteristic of grassbed environments in the West Indies (Jackson, 1972, 1973), yet its Eastern Pacific counterpart (*C. distinguenda*) apparently lives in unvegetated sediments. Western Atlantic gastropods living on grass-blades (*Tegula fasciata*, *Neritina virginea*, *Modulus modulus*, *Cerithium lutosum*, *C. atratum*, and *Columbella mercatoria*) generally are not restricted to this habitat and are seen also on boulders, algae, or even rocky shores; it is in these alternative habitats that the Eastern Pacific cognates of these species are found. (*C. mercatoria*, mechanically the sturdiest species, has no obvious close relative in the Eastern Pacific, however.)

A number of mangrove-associated molluscs and crustaceans also have close relatives on the other coast. Decapod examples may be found in the xanthid crab genera *Panopeus*, and *Eurytium*; the grapsid crabs *Aratus*, *Goniopsis*, *Pachygrapsus* and *Sesarma*; the ocypodid fiddler crab *Uca*; and the snapping shrimp *Alpheus* (Abele, 1972). Molluscan cognates occur in the genera *Littorina*, *Cerithidea*, *Thais*, *Mytella*, *Isognomon*, and *Protothaca*. The Eastern Pacific potamidid snail *Cerithidea montagnei* is found clinging to the trunks and branches of mangrove trees in Costa Rica, but in the Western Atlantic the closely similar *C. costata* is found only on mud flats and apparently never on trees (Vermeij, 1973c; Abbott, 1974).

This and the preceding examples again emphasize that although environmental regimes on the two coasts of tropical America may seem very different to us, at the morphological level they are perceived as virtually identical by some molluscan and other inhabitants. Of course, there may be biochemical and physiological differences between populations from opposite sides of the isthmus, but the morphological phenotype seems to be remarkably insensitive to regional variation in the environment. Yet it seems reasonable to suppose that many sources of selection, including predation, act directly and perhaps primarily on this morphological phenotype.

With the possible exception of *Thais trinitatensis* – *T. kiosquiformis* (a somewhat questionable species-pair), most of the cognate pairs among the mangrove molluscs seem to have relatively small, mechanically weak shells; possibly they are more opportunistic or more stress tolerant than are species that presently have no counterpart on the other coast. For example, *Littorina aberrans* (the Eastern Pacific counterpart of the Atlantic *L. angulifera*) is said to live high in the trees and to be zoned above the unique, large, or thick-shelled Eastern Pacific *L. fasciata*, *L. varia*, and *L. zebra* (Rosewater, 1972). The Eastern Pacific *Cerithidea montagnei* is zoned above the unique *C. valida* and *C. pulchra*, whose shells are both larger and thicker. Other examples of large Eastern Pacific mangrove-associated molluscs with no Atlantic cognates are found in the bivalve genera *Anadara* (*A. grandis* and *A. tuberculosa*) and *Laevicardium* (*L. elatum*). Similarly, the strongly sculptured Western Atlantic muricid *Chicoreus brevifrons* has no Eastern Pacific cognate.

This analysis suggests that species that are well adapted to resist crushing predation are less likely to have cognates on the opposite coast than are species with mechanically weaker shells. Other interesting trends emerge when cognates are compared to congeners that do not have cognates on the opposite coast. Using a list of forty-five molluscan genera that contain species having shell length greater than 10 mm, I find that thirty-five genera contain Eastern Pacific noncognates that are larger than any cognates in the same genus, while only about half that number (seventeen) contain Eastern Pacific noncognates that are smaller than any cognate. For instance, the genera *Cerithium, Nerita, Neritina, Planaxis, Tegula, Conus, Anadara, Ostrea,* and *Tagelus* (among others) contain unique Eastern Pacific species that are larger than any cognate congener; so do the crab genera *Ozius* and *Callinectes. Crepidula, Thais, Littorina,* and *Tagelus* (among others) contain unique Eastern Pacific species that are smaller than any congeneric cognates. In the Atlantic the situation is reversed: only thirteen genera contain unique Caribbean species that are larger than any cognates in the same genus, while more than twice that number (thirty) contain species that are smaller than any cognates. Thus, cognates are not a random sample of either the Eastern Pacific or the Western Atlantic biota.

In conclusion, communities with the most profound interoceanic variations in the expression of predation-related shell architecture are also the communities that seem to change most rapidly over time; that is, they have the lowest proportion of amphi-American cognates. We can conclude nothing about the evolutionary causes of these changes. For example, it is impossible to tell from a knowledge of the Recent faunas alone whether differential extinction or differential speciation is responsible for the biotic divergence on the two coasts of tropical America. To answer these questions, we must consult the fossil record.

Climatic History

The present-day biological differences among tropical marine regions summarized in Chapter 5 may now be viewed in a somewhat different light. The Indo-West-Pacific, and to a lesser extent the Eastern Pacific, have been more favorable to the evolution or retention of well-armed predators and strongly defended molluscan prey than has the tropical Atlantic. This statement invites several questions. First, can we show that the Pacific and Indian oceans have experienced less frequent and less calamitous large-scale disturbances than has the Atlantic? Second, when did the present-day interoceanic differences come about? Was the expression of predation-related traits once more uniform across the tropics than it is now, or have differences always existed?

It will not be possible to give unequivocal answers to these questions here. Our knowledge of the Tertiary fossil record is not yet detailed enough, particularly at the community level, to permit estimates of the degree of expression of antipredatory adaptations. (I am currently undertaking studies on this topic.) Nevertheless, there is evidence of biotically significant disturbance in the past, and some preliminary conclusions can be drawn from known patterns of extinction in the world oceans in geologically recent times.

Three types of disturbance seem to have influenced the history of marine benthic biotas in an important way: fluctuations in temperature, changes in sea level, and tectonic events. These disturbances are all somewhat interrelated, but their effects have not been uniform throughout the world. In general, they have caused biological disruption and fragmentation.

The cooling trend evident through much of the Cenozoic era and culminating in the glacial episodes of the Pleistocene may have had more severe effects in the Atlantic than in the other oceans. The nature and extent of this climatic trend, which shows many minor fluctuations, have been inferred largely from temporal changes in the geographical distribution of pollen grains and forams referable to present-day genera and species whose thermal tolerances are known. From such evidence and from geochemical inferences based on O_{18}/O_{16} ratios, Emiliani (1971) made some tentative estimates of the drop in average sea surface temperature during the Pleistocene glaciations. According to his interpretations, temperatures in the Caribbean might have been 7 to 8° C lower in the Pleistocene than they are today; the open equatorial Atlantic experienced a drop of 5 to 6° C, while the Pacific saw only a 3 to 4° C reduction in temperature.

Imbrie and colleagues (1971) inferred Pleistocene sea surface temperatures from the distributions and known thermal tolerances of whole assemblages of planktonic forams. According to their data, polar assemblages of planktonic Foraminifera never penetrated the Caribbean Sea, so that the sea surface temperatures there could not have dropped below 20° C. If this were true, then the drop in surface temperature in the Caribbean during glacial episodes in the Pleistocene could not have been more than 4 to 5° C (Imbrie et al., 1971).

A recent model of oceanic conditions during a Pleistocene glaciation some 18,000 years ago, proposed by the CLIMAP group (1976), predicts that tropical seas may not have been much cooler than they are at present, except in the tropical Eastern Atlantic where surface waters may have been as much as 5° C cooler than today. Based on the distribution of marine planktonic assemblages, the model further suggests that upwelling near the equator in the Pacific Ocean may have been more intense in the Pleistocene than it is now.

With all these predictions, however, it must be remembered that the inferred sea surface temperatures apply to open-ocean conditions; they may not always accurately reflect coastal or inshore thermal regimes. Moreover, the number of plankton core samples upon which the models and inferences are based is still small, and the geographical coverage not always adequate.

There is also controversy about whether the latitudinal extent of tropical water was reduced during the Pleistocene glaciations. The CLIMAP model suggests that, although isotherms in the polar and temperate zones were pushed toward the equator, the limits of the marine tropics remained more or less stationary except in the Eastern Atlantic. Again, however, it may be that inshore waters cooled more in winter than they do today, so that the extent of tropical waters may have become reduced for shallow-water benthic organisms. Today, for example, many West Indian molluscs and even some corals (such as *Siderastrea siderea* and *Solenastrea hyades*) are found offshore as far north as Cape Hatteras, North Carolina, well within the warm-temperate zone; the majority of inshore species, however, reach their northerly limits of distribution between Miami and Cape Canaveral, Florida (Cerame-Vivas and Gray, 1966; MacIntire and Pilkey, 1969).

If there were any reduction in the latitudinal extent of the tropics, it is likely that the Atlantic would have been more profoundly affected than the Pacific and Indian oceans. Not only is the Atlantic a comparatively small ocean, but it is surrounded by large continents which, during the Pleistocene glaciations, would have been substantially colder than they are today. The other two oceans are larger and hence were probably better buffered against terrestrial variations in climate and in average temperature.

One very important consequence of glaciation was the fluctuation in sea level that affected all the world's coasts. Four glacial periods are generally recognized in the Pleistocene, and each probably corresponded to a marked lowering in sea level. During the height of the last glaciation about 18,000 years ago, sea level is conservatively estimated to have been 85 m lower than it is at present (CLIMAP, 1976). The consequence of such variations in sea level could have been particularly severe for shallow-water, hard-bottom organisms. Interruption of continuous coral-reef construction may have reduced significantly the three-dimensionality of the reef, which provides many fishes, crustaceans, and molluscs with hiding places. Glynn and co-workers (1972), in fact, have argued that present reef frameworks in the Eastern Pacific and Caribbean are not more than 5,000 years old and that corals and other reef organisms grew in isolated mounds or on the bare rocky bottom during the Pleistocene. Whether reef construction was interrupted in the Indo-West-Pacific remains an open question.

Latitudinal changes in the positions of isotherms during the Pleistocene may well have contributed to the impoverishment of some temperate faunas, especially in the Western Atlantic. Glaciers are known to have advanced as far south as New York at 40° N, and the CLIMAP group (1976) estimates that sea surface temperatures there were some 18° C lower during the last glaciation than today. Natural rocky shores on the east coast of North America are absent between New York and southeast Florida, except for estuarine oyster beds. Any hard-bottom animal or plant incapable of tolerating prolonged glacial cold or turbid estuarine conditions would have been faced with the virtually complete disappearance of suitable habitat as they were pushed south of New England during episodes of glaciation. Possibly as the result of such events, the present Northwest Atlantic rocky-shore biota is little more than an impoverished version of the North European biota (Stephenson and Stephenson, 1954a, b). Thus such common European organisms as patellid limpets, intertidal trochids, and the high intertidal brown alga *Pelvetia canaliculata* are not found in eastern North America and have no ecological equivalents there. No such impoverishment is evident on boreal North American soft bottoms. Although some European species, such as the cockle *Cerastoderma edule*, are absent in North America, there are Northwest Atlantic species on soft bottoms with no obvious European counterparts (for instance, the moon snail *Polinices heros* and the sand dollar *Echinarachnius parma*).

Snails such as *Littorina littorea*, *L. "saxatilis,"* and *Thais lapillus* live both intertidally and subtidally in New England and eastern Canada, but are for the most part restricted to the intertidal zone in Great Britain. Presumably, predators and competitors which in Britain prevent occupation by these species below the low-water mark are less effective or absent in eastern North America (Newcombe, 1935; Stephenson and Stephenson, 1954a, b; Lewis, 1964). A parallel case of subtidal range extension may be seen in the brackish Baltic Sea, where the usually intertidal brown algae *Fucus vesiculosus* and *Ascophyllum nodosum* can be found to depths of 20 m (Segerstråle, 1957).

The destitute character of the cold-temperate biota of eastern North America is further underscored by the extraordinary success of *Littorina littorea* and *Carcinus maenas*, two rocky-shore animals apparently

Figure 8.1   *Littorina littorea* in North America and Europe.

a. *L. irrorata*, Tuckerton Meadows, New Jersey; lives in salt marshes, where it commonly ascends the stems of the grass *Spartina*

b. *L. littorea*, Tuckerton Meadows; lives on mud and sand at seaward edge of salt marshes

c. *L. littorea*, rock jetty at Belmar, New Jersey

d. *L. littorea*, rocky shore at Plymouth, south coast of England

e. *L. littorea*, salt marsh at Dawlish Warren, south coast of England

f. *L. littorea*, on sand at Squirrel Island, near Boothbay Harbor, Maine

g. *L. littorea*, rocky shore at Squirrel Island

L. littorea from salt marshes in New Jersey have convex whorls that are separated from one another by a well-defined suture. This suture is much less evident in *L. irrorata* (ranging from New Jersey to Texas) and in populations of *L. littorea* from Europe and North America north of Connecticut. (Photographs by F. Dixon.)

brought here from Europe. With the exception of a single specimen from a shell bed in southwestern Nova Scotia believed to be of Mid Wisconsin age (40,000 years old), *L. littorea* is unknown from the American Pleistocene (Wagner, 1977) and was not present south of Newfoundland and Labrador in historic time before 1840. The species evidently occurred in Iceland during the postglacial warm spell, and Kraeuter (1974) speculates that it could have arrived in North America through planktonic larval transport. When temperatures in the North Atlantic fell after the postglacial optimum, *L. littorea* may have survived in the vicinity of Northumberland Strait between New Brunswick and Prince Edward Island, where temperatures are higher than further south in the Bay of Fundy and Gulf of Maine. Berger (1977) has shown that populations of *L. littorea* between Cape Cod and Prince Edward Island display relatively little electrophoretically detectable variation (3 percent of loci are heterozygous per individual, compared to about 15 percent on the coast of Brittany in France). Moreover, the alleles in North American populations are radically different from those in France. Berger therefore believes that the American form has been geographically isolated from populations in Europe (at least those in France) for thousands of years.

Whatever the early history of the periwinkle in North America, there is little doubt that *L. littorea* became established in Nova Scotia by the mid-nineteenth century. It then quickly spread southward, reaching Atlantic City (New Jersey) by 1892, Ocean City (Maryland) by 1959, and Chincoteague (Virginia) by 1971 (Bequaert, 1943; Clarke, 1963; Kraeuter, 1974). The reasons for this rapid range extension remain obscure, but it may have been aided by the commercial activities of man. In any event, *L. littorea* is today one of the most abundant organisms on sheltered and semiexposed coasts of eastern Canada and New England, where it coexists with two other species of *Littorina*, *L.* "*saxatilis*" and *L. obtusata*, both of which are also known from Europe.

In the southern part of its North American range, *L. littorea* co-occurs in salt marshes with *L. irrorata*. The latter species is now found only as far north as the south coast of Long Island, New York, but in the late nineteenth and early twentieth centuries, it was still occasionally found on the coast of Connecticut in southern New England (Bequaert, 1943). The salt-marsh populations of *L. littorea* from New Jersey are characterized by a deep suture and markedly convex whorls in the spire, while more northerly populations from similar habitats have a barely perceptible suture and flatter-sided spire, features that are more reminiscent of *L. irrorata* (Figure 8.1). I do not know if this geographical variation in *L. littorea* is causally related to interactions with

L. irrorata, but the possibility cannot be discounted. No other intertidal species in eastern North America seem to have been affected or displaced by *L. littorea*.

The green crab *Carcinus maenas* may have been introduced from Europe to North America in ships, but the date is uncertain (see Christiansen, 1969). Before the twentieth century the species in North America was restricted to shores south of Cape Cod, but by 1920 it had reached Casco Bay in southern Maine. During the next thirty-six years, *C. maenas* spread gradually northward and eastward to northern Maine, New Brunswick, and Nova Scotia, probably aided by a general warming trend. After 1957, sea-surface temperatures declined, especially in winter, and populations of *C. maenas* were sharply reduced in the Bay of Fundy (Welch, 1968.) Nevertheless, *C. maenas* seems to be a well-established species and is a quantitatively important predator of commercial and other shellfish in the region (Welch, 1968; Ropes, 1968).

A third highly successful immigrant on the coast of the northeastern United States is the alga *Codium fragile tomentosoides*. This weedy species, first seen on Long Island in 1957 (Bouck and Morgan, 1957), is a European plant that grows abundantly in relatively sheltered situations on shells and other bits of rubble.

I view the impoverished character of the New England rocky-shore biota as resulting, at least in part, from the destructive effects of past glaciation. At the same time, however, it must be emphasized that present-day conditions are harsh; the region is cursed with severe winters, and temperatures fluctuate seasonally over a wide range (see Menge and Sutherland, 1976). It appears that the climate in the Pleistocene was even harsher than that prevailing in the Northwest Atlantic today.

The disastrous effects of glaciations may also have left their mark on the biota of Argentina in the Southwest Atlantic. CLIMAP (1976) estimates that there was a 6-degree northward shift of the polar front there 18,000 years ago as compared to the present time. This latitudinal shift may have blotted out most of the Patagonian rocky shores for year-round occupation by Argentinian hard-bottom forms and thus created precarious ecological conditions for them. Much of the coast in the vicinity of Buenos Aires and Uruguay is unsuitable for hard-bottom plants and animals, since it is characterized by wide stretches of sandy and estuarine habitats (Kuhnemann, 1972).

The Mediterranean Sea too has experienced large-scale extinctions since the Miocene, and especially in its eastern basin is occupied by a markedly impoverished biota. During the Pleistocene glaciations, temperatures in the western basin were probably some 9 to 10° C cooler than today, but temperatures in the eastern basin were not more than 1 to 2° cooler than at present (CLIMAP, 1976). The subtropical faunal elements contracted to the southern and eastern Mediterranean during each of at least three cold spells in the Pleistocene; at the same time, cold-tolerant North European forms such as *Mytilus edulis*, *Cyprina islandica*, and *Buccinum undatum*, invaded the western basin (Por, 1975). As sea level dropped during glacial periods, connection with the Atlantic Ocean via the Strait of Gibraltar became even more limited than it is today, and biotic exchange with other populations of subtropical and cold-tolerant species was limited or altogether impossible. Any extinctions resulting from excessive population declines often could not be compensated for by reinvasions.

The Pleistocene was only one of several biological setbacks suffered by the Mediterranean in Cenozoic time. The connection with the Red Sea was severed about 16 million years ago, during the early to middle Miocene; and a precipitous drop (60 m or more) in sea level in the late Miocene (Messinian), 6.5 to 5 million years ago, led to isolation of the Mediterranean from the Atlantic. This latter event, often referred to as the Messinian salinity crisis, resulted in the precipitation of something like 10^6 km^3 of salts in the Mediterranean (Adams et al., 1977).

The present impoverished character of the eastern Mediterranean biota may well explain why so many species native to the Red Sea and the Indo-West-Pacific have been able to invade that basin through the Suez Canal. This topic will be treated more fully in the next chapter, but it should be pointed out that animals from other geographical areas also have been successful invaders of the Mediterranean. The eastern North American blue crab *Callinectes sapidus*, for example, has become well established in the eastern Mediterranean (Holthuis and Gottlieb, 1958; Williams, 1974).

Other inland seas were strongly affected by fluctuating temperatures and by the raising and lowering of sea level in the Pleistocene. The Black Sea, connected today through the Bosporus Straits with the Mediterranean, periodically was cut off from the Mediterranean in the Pleistocene, and during such periods of sea-level reduction was occupied by brackish-water bivalves. At higher stands of the sea more stenohaline echinoderms (such as *Echinocyamus pusillus*) and molluscs (like the strombid *Aporrhais pespelicani*) have invaded the Black Sea (Por, 1975).

Similarly, the Baltic Sea from time to time has been isolated from the North Atlantic; during such periods the basin was invaded by freshwater organisms like the snail *Lymnaea peregra*. When connections with the North Sea were reestablished, these low-salinity species were replaced to varying degrees by more typically marine species, such as the snail *Littorina littorea*, the mussel *Mytilus edulis*, and the sea star *Asterias rubens*. Today the Baltic Sea is less saline than the North Sea, and various forms, including A. *rubens* and the brittle star *Ophiura albida*, are found only in the southwestern corner of the Baltic (Segerstråle, 1957; Por, 1975).

Other temperate coasts seem to have been much less affected by latitudinal shifts in climate and by fluctuations in sea level during the Pleistocene. On the Atlantic coast of Europe and in the North Pacific, rocky shores abound at all latitudes and there was no shortage of refuges for cold-water species during episodes of glaciation. Moreover, emergence of the Bering Strait and of the southern entrance to the Sea of Japan during low stands of sea level in the glacial periods prevented polar water from entering the North Pacific, and probably reduced the impact of temperature fluctuations on the rich biota of that region (CLIMAP, 1976). The North Pacific biota apparently was already richer than that of the North Atlantic before the onset of glaciation, possibly because the North Pacific is substantially older in a geological sense. The North Atlantic may not be older than Early Tertiary in age.

The contrasting histories of the various temperate regions may be appreciated further by examining what has happened to species introduced by man from other oceans. Many estuarine eastern North American molluscs, including *Crassostrea virginica*, *Geukensia demissa*, *Mercenaria mercenaria*, *Urosalpinx cinerea*, *Busycon canaliculatum*, *Crepidula fornicata*, and *Nassarius obsoletus*, have been introduced into Puget Sound and San Francisco Bay on the west coast of the United States. Several Japanese species (*Ocenebra japonica*, *Crassostrea gigas*, *Tapes philippinarum*, and the large brown alga *Sargassum muticum*) also have been introduced there. While many of these species have become abundant or even dominant in bays and estuaries, none has established populations on the biotically rich open coast in the fashion of *Littorina littorea* or *Carcinus maenas* in New England (Ricketts and Calvin, 1968). Furthermore, several foreign species have become established along the Atlantic coast of Europe; again, some of these have become abundant in bays and other sheltered areas, but none has penetrated to open rocky shores. Examples include the North American *Crepidula fornicata*, *Urosalpinx cinerea*, and *Mercenaria mercenaria*, and the New Zealand barnacle *Elminius modestus* (Lewis, 1964).

Events in the southern temperate zone are much less well understood, but it is doubtful that temperature fluctuations were as extreme as in the more land-dominated Northern Hemisphere. The crab *Car-*

cinus maenas apparently has established a population in South Australia, and there is some evidence that the New Zealand crab *Cancer novaezelandiae* also has become established in Australia (Stephenson, 1972; Nations, 1975).

Dell (1972) has summarized evidence that extinctions over the past 4 million years have seriously depleted the shallow-water plants and animals of Antarctica. He points out that most species inhabiting waters less than 30 m in depth have unusually wide depth distributions and may be found at bathal or even greater depths. Dell believes that four or more episodes of glaciation have been largely responsible for this decline; but Dayton and Oliver (1977) point out that the physical environment in shallow-water Antarctica is, except for its proximity to the sea surface, not very different from that in the present-day deep sea. It is unclear whether we should regard the Antarctic shallow-water benthos as impoverished, or whether it should be viewed as an extension of the deep-sea biota made possible by glaciation. It is interesting and puzzling that the few strictly shallow-water forms that still persist in the Antarctic, such as the echinoid *Sterechinus*, the isopod *Glyptonotus*, the gastropod *Neobuccinum*, and the bivalves *Laternula* and *Adamussium*, are all among the largest representatives of their respective groups in Antarctica.

Tectonic History

The Late Tertiary and Pleistocene climatic fluctuations and cooling were superimposed on a pattern of change in the positions of islands and continents. Evidence from plate tectonics and from the fossil record strongly suggests that the tropical biotas of today are far more isolated from one another, and are thus biologically more independent, than were the tropical biotas of the Cretaceous and Early Tertiary (Fell, 1967; Kauffman, 1973; Dana, 1975). Up to Late Eocene time, the extent of tropical faunal interconnection was so great that the tropical biotas can be profitably treated as representing a single unit, often referred to as the Tethyan fauna. Continental movements and episodes of orogeny (mountain building) slowly eroded the faunal connections and created land barriers. Throughout the Tertiary both North and South America migrated westward away from Eurasia and Africa; the Atlantic basin was broadened and dispersal of species from east to west was reduced (Fell, 1967; Dana, 1975). An apparent increase in biotic migration from the Indo-West-Pacific to the Western Atlantic during the Miocene, when tropical sea grasses first colonized the New World, was made possible by a warming of the waters around southern Africa; but this connection was short-lived (Brasier, 1975). Further north the contact between the Red Sea and the Mediterranean was broken by the development of a narrow land bridge during the Miocene (Por, 1971, 1975); and the Panama land bridge, now separating the Western Atlantic from the Eastern Pacific, was completed by Late Pliocene time, not more than 3.5 million years ago (Woodring, 1966; Saito, 1976).

The deep East Pacific barrier between the Indo-West-Pacific and Eastern Pacific (Ekman, 1953) has probably always been an effective deterrent to dispersal. Only in relatively very recent times (since the Late Pliocene) have reef-associated corals, molluscs, echinoderms, fishes, and a few coral-associated decapod crustaceans been able to invade the Eastern Pacific from the Central and Western Pacific (Emerson, 1967; Fell, 1967; Olsson, 1972; Garth, 1974). Dana (1975) has made the interesting suggestion that this eastward migration was made possible by the northward drift in the Central Pacific of the Line Islands. When by Late Pliocene time these islands had moved into the path of the eastward-flowing equatorial countercurrent, they could have acted as a source of widely dispersing teleplanic larvae of Indo-West-Pacific species. Before the Line Islands reached their present position some 6,500 km west of tropical America, larvae would have had to come from the Marshalls or the eastern Carolines, island groups too far away from the New World to act as sources for colonizing planktonic larvae from the Indo-West-Pacific. The alternative view, that species found today in both the Indo-West-Pacific and the Eastern Pacific have been long-term residents of the latter region, rests on the assumption that they survived episodes of Eastern Pacific extinction in relict offshore populations. Such an interpretation has been urged for the snails *Casmaria erinacea vibexmexicana* and *Heliacus caelatus planispira*; these forms are only subspecifically distinct from Indo-West-Pacific relatives, but are even closer to Tertiary tropical American shells (Abbott, 1968; Robertson, 1975). Similar arguments could be constructed for the coral genera *Pocillopora*, *Pavona*, *Leptoseris*, and *Porites*, since these are known as fossils in tropical America; but for most other Eastern Pacific species shared with the Indo-West-Pacific, there is no fossil record in tropical America. Many of these species, including the important reef coral *Pocillopora damaecornis* and the sea star *Acanthaster planci*, in the Eastern Pacific are indistinguishable from individuals in the Western Pacific. It thus seems best to regard these primarily reef-associated animals as comparatively recent immigrants to the New World tropics.

Weyl (1968) has theorized that the closing of the straits between the Caribbean and the Eastern Pacific had the effect of reducing the intensity of east-west oceanic circulation, and of increasing the north-south component of circulation. This change in oceanic circulation, which must have occurred gradually through the Tertiary as the connection between the tropical oceans in America became more tenuous, may have resulted in a proportionately greater transport of cold polar water to the tropics and of warm tropical water to the higher latitudes. In the polar regions higher evaporation rates eventually might have resulted in greater ice formation and in the growth of glaciers in the Pleistocene (Weyl, 1968). Thus, there may be a connection between tectonic events and climatic history.

The increasing isolation of the New World tropics from the Indo-West-Pacific during the Tertiary brought with it increased endemism, biotic fragmentation, and a high rate of extinction (Fell, 1967; Dana, 1975). The Tertiary extinctions seem to have affected the Eastern Pacific earlier than the Caribbean and may reflect an increasingly continental coastal environment there. For example, a number of coral genera, which in the Cretaceous and earliest Tertiary had circumtropical distributions, became extinct in the Eastern Pacific after the Eocene. These include *Astreopora*, *Goniopora*, *Hydnophora*, *Leptastrea*, *Montipora*, *Oulophyllia*, and *Platygyra*. By the end of the Oligocene, *Stylophora* had become extinct in the Eastern Pacific. All these genera are still represented on Indo-West-Pacific reefs, where they often contribute substantially to reef construction. With the exception of *Leptastrea* and *Montipora*, these genera persisted in the Caribbean until the Miocene; then they became extinct there also, together with *Pavona*, *Leptoseris*, *Plesiastrea*, and *Pocillopora*. Among the last four genera all but *Plesiastrea* are still found today in the Eastern Pacific, and all are known from the Indo-West-Pacific. Various coral genera now restricted to the Caribbean also lived at one time in the Eastern Pacific, but became extinct in that region at various times: *Montastrea* in the Early Cretaceous; *Manicina*, *Astrocoenia*, *Stephanocoenia*, and *Colpophyllia* in the Eocene; *Diploria*, *Dichocoenia*, *Eusmilia*, and *Siderastrea* in the Middle Pliocene (Dana, 1975).

Other groups with an Early Tertiary circumtropical distribution also became extinct in the New World. The poisonous sea urchin *Toxopneustes*, for example, is now found only in the Indo-West-Pacific and Eastern Pacific; in the Tertiary it occurred also in the Western Atlantic (Fell, 1967; Chesher, 1972). The mangrove-crab genus *Scylla*, now limited to the Indo-West-Pacific, may have occurred in tropical America during the Eocene, though the generic identity of the fossils in question requires confirmation (see Williams, 1974). Among gastropods, several relict genera and subgenera at present found only in the Indo-West-Pacific occurred in the Western Atlantic during the Tertiary (a relict group is one that in the

Early Tertiary had a circumtropical distribution, but that today is found in only one of the four major tropical regions):

Indo-West-Pacific: *Turbo* s.s., *Faunus*, *Terebralia*, *Rhinoclavis* s.s., *Campanile*, *Terebellum*, *Strombus* (*Labiostrombus*, *Canarium*, *Dolomena*), *Globularia*, *Oliva* (*Omogymna*), *Mitra* (*Dibaphus*)
Western Atlantic: *Siphocypraea* (*Muracypraea*), *Mitra* (*Dibaphimitra*)
Eastern Atlantic: *Tympanotonus*

The compilation above is taken from the works of Robertson (1957), Abbott (1960), Jung and Abbott (1967), Olsson and Petit (1968), Woodring (1973), and Cernohorsky (1976).

Giant clams (Tridacnidae) are today found only in the Western Pacific and Indian oceans, but several extinct genera that probably belong to this group are known from the Eocene of the New World and Europe. Shells resembling present-day *Hippopus*, a genus of giant clams that lie free on the substratum, have been found in the Miocene of Florida (Stasek, 1961; Rosewater, 1965). Thus, Tertiary extinctions of invertebrates affected the New World marine biota far more than they did the Indo-West-Pacific biota. In fact, only a handful of molluscan taxa with a circumtropical distribution in the Tertiary became extinct in the Indo-West-Pacific and survive today as relicts in the Caribbean.

This is not to say that extinctions or severe restrictions in geographical distribution did not take place in the Indo-West-Pacific. There is good evidence that many taxa now limited to Indonesia and the Western Pacific once had much larger distributions in the Indian and Pacific oceans. Today, for example, only two species of the giant-clam family Tridacnidae (*Tridacna maxima* and *T. squamosa*) are found living at Aldabra Atoll in the central Indian Ocean; but in the Pleistocene three additional species (*T. gigas*, *T. crocea*, and *Hippopus hippopus*) were also found there (Taylor, 1971). No species of the scorpion-shell *Lambis* is presently known from Hawaii, but *L. chiragra* occurred there in the Pleistocene era (Abbott, 1961). Ladd (1960, 1966, 1972, 1977) records numerous species from the Tertiary of atolls in the Central and Western Pacific, where these species no longer occur today. The genus *Clavocerithium*, for instance, now known only from New Guinea, has been found in the Pliocene of Guam and Java (see also Houbrick, 1975). The mangrove-associated *Terebralia sulcata* presently reaches Yap and Fiji in the Western Pacific, but has been found in the Upper Miocene of Eniwetok in the Marshall Islands; similarly, the mangrove-associated *Cerithidea obtusa*, limited today to the Indo-Malaysian area, is known from the Miocene of Saipan (Ladd, 1972). For many groups, then, the Indo-Malaysian area and the Western Pacific arc have served as refuges.

In examining the probably incomplete list of relict Indo-West-Pacific molluscs that in the Tertiary had a circumtropical distribution, I am not impressed that their commitment to armor is particularly great. Some of the relicts are clearly rather large animals: the thiarid *Faunus* and the potamidid *Terebralia* are both big for their families; *Campanile* is one of the largest cerithiids known; the Tridacnidae are the largest living bivalves. Members of the sand-dwelling genus *Harpa* may be relatively immune from attackers by virtue of being able to autotomize the posterior part of the foot (Rehder, 1973). On the other hand, two of the three relict subgenera of *Strombus (Labiostrombus* and *Canarium)* are composed of small species, and the same is true of the related relict genus *Terebellum*. The possibly relict nominate subgenus of *Turbo* has a medium-sized, smooth shell with a relatively elevated spire and on geometrical grounds seems less sturdy than do many Indo-West-Pacific species of *Turbo* that belong to other subgenera.

If the taxa that became extinct in the Western Atlantic and that persisted as relicts in the Indo-West-Pacific could be shown to be more heavily armored than the taxa that survived in the Western Atlantic, then the lesser expression of predation-related traits of modern Western Atlantic as compared to Indo-West-Pacific molluscs could be explained in part by differential extinction. Since this does not appear to be the case, extinction does not seem to be a primary cause for present-day interoceanic differences in shell architecture. The fossil record has preserved soft-bottom molluscs much more faithfully than it has species from rocky shores, however; and the possibility cannot yet be excluded that regional variations in the architecture of hard-bottom gastropods are the result of differential extinction.

The completion of the Panama land bridge in the Late Pliocene led to extinctions which, at least among molluscs, were far more widespread in the Western Atlantic than in the Eastern Pacific. From the work of Woodring (1966, personal communication) and others, I have compiled a list of sixty-two so-called Paciphile genera and subgenera of molluscs that have become extinct in the Western Atlantic but that still persist today in the Eastern Pacific (Table 8.2 and Figures 8.2 and 8.3). Only about eight genera and subgenera can be recognized as Caribphiles (persisting in the Western Atlantic but extinct in the Eastern Pacific).

Were it not for a few sanctuaries in the Western Atlantic, extinctions in that province would have been far more numerous than they actually were and the number of Paciphile taxa would have been very much greater. One such refuge, recognized by Petuch (1976), lies on the northern coast of South America in Venezuela and eastern Colombia. A number of grassbed- and sand-dwelling molluscs endemic to this area have changed

Table 8.2   Paciphile and Caribphile molluscan genera and subgenera.

Paciphiles
  Gastropoda
    Cyclostrematidae: *Woodringella*
    Neritidae: *Neritina (Clypeolum)*
    Vitrinellidae: *Vitrinella (Aepystoma)*, *Solariorbis (Hapalorbis)*
    Architectonicidae: *Heliacus (Astronacus)*
    Potamididae: *Rhinocoryne*
    Cerithiidae: *Rhinoclavis (Ochetoclava)*
    Calyptraeidae: *Trochita*, *Crepidula (Crepipatella)*
    Ovulidae: *Jenneria*
    Cypraeidae: *Cypraea (Basilitrona)*
    Naticidae: *Neverita (Glossaulax, Hypterita)*
    Cassidae: *Phalium (Echinophoria)*
    Tonnidae: *Malea*
    Thaididae: *Cymia*
    Muricidae: *Purpurellus*, *Vitularia*
    Columbellidae: *Parametaria*, *Strombina (Sincola)*
    Buccinidae: *Hanetia*, *Cymatophos*, *Northia*
    Olividae: *Oliva (Strephonella)*, *Olivella (Pachyoliva)*
    Mitridae: *Mitra (Cancilla, Dibaphus, Strigatella)*
    Harpidae: *Harpa*
    Cancellariidae: *Cancellaria (Euclia, Narona, Pyruclea)*, *Aphera*, *Perplicaria*,
      Trigonostoma (Extractrix)
    Terebridae: *Terebra (Panaterebra)*
    Turridae: *Pleurofusia (Cruziturricula)*, *Knefastia*, *Agladrilla* s.s.,
      Glyphostoma (Euglyphostoma)
    Acteonidae: *Rictaxis*
    Juliidae: *Julia*
  Scaphopoda
    Dentaliidae: *Dentalium (Tesseracme)*
  Bivalvia
    Nuculidae: *Nucula (Ennucula)*
    Arcidae: *Noetia* s.s., *Sheldonella*, *Anadara (Grandiarca, Tosarca)*
    Volsellidae: *Volsella*
    Pectinidae: *Pecten (Flabellipecten)*
    Anomiidae *(Placuanomia)*
    Crassatellidae: *Crassatella (Hybolophus)*
    Cardiidae: *Trachycardium (Phlogocardium)*, *Lophocardium*, *Mexicardia*
    Veneridae: *Agriopoma (Pitarella)*, *Pitar (Hyphantosoma)*, *Chione
      (Panchione)*, *Clementia*
    Mactridae: *Harvella*
    Corbulidae: *Bothrocorbula (Hexacorbula)*
Caribphiles
  Gastropoda
    Cypraeidae: *Siphocypraea (Muracypraea)*
    Cassidae: *Cassis*, *Sconsia*
    Turbinellidae: *Turbinella*

Bivalvia
  Pectinidae: *Pecten (Euvola)*
  Lucinidae: *Lucina (Phacoides, Callucina)*
  Cardiidae: *Dinocardium*

Source: Woodring (1966) originally listed forty-three Paciphile and four Caribphile taxa, but he has subsequently modified this compilation somewhat (Woodring, 1970; personal communication). The present table contains the forty-nine Paciphiles and seven Caribphiles now recognized by Woodring, as well as eight other Paciphiles and one Caribphile that have since come to light (see Olsson, 1956; Vokes, 1967, 1975a, b; Abbott, 1968; Cernohorsky, 1976; Bretsky, 1976).

very little since the Miocene: the muricids *Phyllonotus margaritensis* and *Calotrophon velero*, the cowry *Siphocypraea (Muracypraea) mus*, and the cone *Conus puncticulatus* (see also Radwin and D'Attilio, 1976). The area also boasts species of the sand-dwelling gastropod genera *Oliva*, *Ancilla*, *Voluta*, and *Strombina* not known from other parts of the Western Atlantic (Abbott, 1974). Meyer (1973a, b) has found that a number of large comatulid crinoids (such as *Tropiometra carinata carinata*, *Neocomatella pulchella*, and *Ctenantedon kinziei*) are limited to the coast of Colombia and Venezuela and the immediately adjacent islands of Curaçao, Aruba, and Bonaire.

Another region where many species have persisted that once had a more widespread Atlantic or tropical American distribution is Florida and the nearby coasts of the Bahamas, northern Cuba, and the Yucatan Peninsula of Mexico. Among tropical American gastropods native to this region and to the Eastern Pacific but with no living representatives in the West Indies or in Atlantic South America are *Pleuroploca gigantea*, *Ficus communis*, *Conus floridanus*, *C. villepini*, and *Calotrophon ostrearum*; the deep-burrowing *Anodontia philippiana* is an example among the bivalves. Woodring (1966) has already noted that many Paciphile taxa (*Ochetoclava*, *Cymia*, *Cymatophos*, and *Cancilla*, for example) may have persisted longer in Florida than elsewhere in the Western Atlantic. The possibly Pliocene Caloosahatchee Formation of Florida also contains a number of remarkable large, often well-sculptured gastropods in such extinct taxa as *Vasum (Hystrivasum)*, *Siphocypraea* s.s., and *Rhinoclavis (Cerithioclava)*.

A third, possibly less important, Western Atlantic refuge is the mainland coast of Brazil. Here the endemic large (15 cm long) gastropod *Turbinella laevigata* is the remnant of a lineage with a once much wider Western Atlantic distribution that included Florida (Vokes, 1964). Other tropical American genera whose Atlantic representatives are limited to the coast of

Figure 8.2 Dorsal *(left)* and apertural *(right)* views of some Paciphile molluscs. (Photographs by F. Dixon.)

a. *Mexicardia procera*, Venado Beach, Panama (sand)

b. *Jenneria pustulata*, Isla Taboga, Panama (rocks)

c. *Anadara (Grandiarca) grandis*, Mata de Limón, Costa Rica (mangrove swamp)

d. *Vitularia salebrosa*, Venado Beach (stones in sand)

e. *Hanetia dalli protera*, lower Gatun Formation, Middle Miocene, near Cativa, Panama (sand)

f. *Cymatophos veatchi*, lower Gatun Formation, Middle Miocene, near Cativa (sand)

g. *Strombina (Strombina) lessepsiana*, lower Gatun Formation, Middle Miocene, near Cativa (The subgenus *Strombina*, represented by about twenty species in the present-day Eastern Pacific, is not a true Paciphile, since it is represented in the lower Caribbean Sea by a single rare relict, *S. pumilio*. The related but smaller-shelled subgenus *Sincola* is a true Paciphile.)

h. *Northia pristis*, Venado Beach (sand)

the Guianas and Brazil are the bivalves *Micromactra* and *Mytella*, the scleractinian coral *Mussismilia*, and the gorgonian *Pacifigorgia* (see also Laborel, 1969). Mainland South American species with Pacific counterparts include the 30-cm-long *Strombus goliath*, the mangrove-associated 5-cm-long *Thais trinitatensis*, and the 4-cm-long predaceous sandy-beach olivid *Agaronia travassosi*. The tellinid bivalve *Strigilla (Simplistrigilla) surinamensis*, closely similar to the Eastern Pacific *S. strata*, is known only from off Surinam and Alabama (Boss, 1972a).

Figure 8.3 Dorsal *(above)* and apertural *(below)* views of some Caribphiles. (Photographs by F. Dixon.)
a. *Turbinella laevigata*, near Recife, Brazil (grassbeds)
b. *Siphocypraea (Muracypraea) mus*, Amuay Bay, Venezuela (grassbeds and rocks)
c. *Dinocardium robustum vanhyningi*, Sanibel Island, Florida (sand)

A common feature of all these supposed Western Atlantic refuges is that they lie along the shores of continents. In this respect they are more like the primarily continental shores of the Eastern Pacific and may differ from the waters surrounding most West Indian islands in being more turbid and perhaps in being richer in nutrients (Meyer, 1973a, b). This is especially true of the coast of western Venezuela and eastern Colombia. Waters along this desert coast are cold (apparently below 25° C) and turbid, and support dense assemblages of hard-bottom and soft-bottom molluscs, many of which are conspicuously large for their species. On the west coast of the Paraguana Peninsula of Venezuela, I found the muricid *Chicoreus brevifrons* up to 113 mm in shell length, compared with 84 mm on the coral-sand shores of nearby Aruba; the low intertidal buccinid *Pisania tincta* attains 35 mm in length in Paraguana, but only 23 mm on the south coast of Curaçao; the shallowly infaunal venerid bivalve *Chione cancellata* reaches a length of 41 mm in Paraguana, while on Aruba and Curaçao I have not seen specimens longer than 29 mm.

Some of the changes that have affected the Western Atlantic in the course of the Tertiary, then, may have been the consequence of decreasing continentality and declining nutrient supplies. In the Eastern Pacific the trend seems to have been in the reverse direction, judging from the widespread extinction of corals in that province (Dana, 1975). The primary Indo-West-Pacific refuge (Indo-Malaysian area) is also a more continental region than are most other parts of that large biotic province.

Most Paciphiles and all known Caribphiles are soft-bottom forms, so that it is difficult at present to form an opinion about their architectural sturdiness relative to taxa that have persisted on both coasts of tropical America. The few Paciphile rock-dwelling snails with coiled shells (*Cypraea, Basilitrona, Jenneria, Cymia,* and *Vitularia*) tend to have narrow apertures, a thickened outer lip in the adult stage (except in *Cymia*), and a relatively low or cryptic spire (except in *Vitularia*). The animals are not, however, notably large in size, for none exceeds 5 cm in length. Two additional Paciphile rock dwellers, *Crepidula (Crepipatella)* and *Trochita,* have limpet-like shells (Vokes, 1975b).

In contrast, many soft-bottom Paciphiles tend to be large in size, although some small-bodied forms are known. For example, the estuarine bivalve *Anadara (Grandiarca) grandis* is extremely large (up to 20 cm long) and has a massive, heavily ribbed shell. Other large soft-bottom Paciphiles include members of the *Conus patricius* and *C. fergusoni* groups (more than 10 cm long), and the species of *Rhinocoryne* (6 cm), *Rhinoclavis (Ochetoclava)* (5 cm), *Malea* (20 cm), *Mitra (Cancilla)* (6 cm), *Cancellaria (Narona)* (15 cm), *Terebra (Panaterebra)* (10 cm), *Placuanomia* (7 cm), *Pecten (Flabellipecten)* (7 cm), *Mexicardia* (10 cm), and *Bothrocorbula (Hexacorbula)* (2.2 cm). A few sand-dwelling Paciphiles are small compared with other members of their respective families or genera: *Oliva (Strephonella)* (1.2 cm), *Nucula (Ennucula)* (0.5 cm), *Trachycardium (Phlogocardium)* (2.3 cm), and *Harvella* (2.2 cm). Birkeland (1977) has already remarked that three of the seven Caribphiles are notably large animals: *Cassis* (10 to 22 cm), *Turbinella* (35 cm), and *Dinocardium* (15 cm).

Many of the Eastern Pacific and Western Atlantic extinctions thus have involved large species, and some species restricted today to the continental refuges are also large (*Pleuroploca*, *Turbinella*, and *Strombus goliath* in the Atlantic). This might suggest that large size, and perhaps other characteristics associated with resistance to predation, render a species more vulnerable to extinction, while small size favors survival. It must be emphasized, however, that many other taxa with large or sturdy shells have persisted on both coasts of tropical America and have wide distributions there: *Cypraea*, *Cymatium*, *Purpura*, *Melongena*, *Leucozonia*, *Vasum*, *Terebra*, *Lyropecten*, *Spondylus*, *Atrina*, *Codakia*, *Ventricolaria*, and *Perigylpta*, among others (see Appendix). It seems premature to conclude that differential extinction effected the architectural contrasts between the molluscan faunas of the Western Atlantic and Eastern Pacific.

If differential extinction cannot explain the present-day variations in shell architecture among tropical marine biotas, then we must ask whether the differences have arisen because of the evolution of especially well-armored shells in the Indo-West-Pacific. In Table 8.3 I have gathered together what is known of the times of origin and the antipredatory characteristics of genera and subgenera that are autochthonous to (that is, originated and have remained in) each of the four major tropical regions. In the list of autochthonous Indo-West-Pacific taxa are included many which in the Tertiary also occurred in Europe. I am keenly aware that this compilation is far from complete; I have tried to list only taxa that belong to families whose systematics have been recently revised and studied on a broad geographical scale. For some of the hard-bottom taxa the fossil record is dishearteningly inadequate, and especially in tropical America it is difficult to know whether some of the supposed autochthons might not have occurred in both the Eastern Pacific and the Western Atlantic before these regions became separated through the formation of the Panama land bridge.

The hard-bottom autochthonous Indo-West-Pacific taxa, which in some cases seem to have invaded parts of the Eastern Pacific very recently, in general are composed of highly predation-resistant species with strong sculpture, highly toothed or narrow apertures, and large adult body size. This seems to be particularly true of taxa that became differentiated in the Miocene and later (such as *Lambis*, *Drupa*, *Volema*, *Lunatica*, *Rochia*). Autochthonous groups in the Eastern Pacific are also remarkably resistant to crushing by puffers and spiny lobsters; they are in general thick shelled and low spired, but usually possess a thin lip and a regularly ovate rather than a narrowly elongate aperture. In fact, the convergence in shell form between *Thais* (*Vasula*) *melonis* and *Triumphis distorta*, which belong to

Table 8.3   Origin and predation-related traits of autochthonous gastropod genera and subgenera.

Group	Time of origin	Sc	Ap	SL
Indo-West-Pacific				
Tectus s.s.	Late Cretaceous	+	−	+
T. (Cardinalia)	Pliocene	−	−	−
Trochus s.s.	Miocene	−	−	−
T. (Rochia)	Pliocene	−	−	+
Thalotia	Miocene	+	−	−
Angaria	Neogene	+	−	−
Turbo (Lunatica)	Pliocene	+	−	+
Tectarius s.s.	?	+	−	−
Echininus s.s.	?	+	−	−
Telescopium	?	−	−	+
Rhinoclavis s.s.	Pliocene	−	−	−
R. (Proclava)	Eocene	−	−	−
Clavocerithium	Eocene	−	−	−
Pseudovertagus	Miocene	−	−	+
Longicerithium	?	−	+	−
Tibia	Eocene	−	−	+
Strombus (Doxander)	Miocene	−	−	−
S. (Laevistrombus)	Miocene	−	−	−
S. (Euprotomus)	Pliocene	+	+	−
S. (Conomurex)	Pliocene	−	+	−
S. (Gibberulus)	Pleistocene	−	+	−
Lambis	Pliocene	+	+	+
Apollon	Miocene	+	+	−
Nassa	Miocene	−	−	−
Drupa	Pliocene	+	+	−
Drupella	?	+	+	−
Azumamorula	?	+	+	−
Morula	?	+	+	−
Magilus	Eocene	−	−	−
Bedeva	?	−	−	−
Ergalatax	?	+	+	−
Lataxiena	?	+	+	−
Marchia	?	+	−	−
Haustellum	?	+	+	−
Naquetia	Oligocene	+	+	−
Takia	Early Miocene	+	−	−
Vexillum	?	+	+	−
Volema	Pliocene	+	+	−
Latirolagena	?	−	+	−
Tudicla	Cretaceous	+	−	−
Tudicula	Pliocene	+	−	−
Oliva (Parvoliva)	?	−	+	−
O. (Carmione)	?	−	+	−
O. (Galeola)	?	−	+	−

Group	Time of origin	Sc	Ap	SL
O. (Neocylindrus)	?	−	+	−
Cymbium	Eocene	−	−	+
Alcithoe	Eocene	+	−	+
Gemmula				
(Unedogemmula)	Eocene	−	−	−
Lophiotoma	Miocene	−	+	+
Turris	Miocene	−	+	+
Eastern Pacific				
Tegula s.s.	Miocene	−	−	−
Thais (Vasula)	Early Miocene	−	−	−
Bizetiella	?	−	−	−
Neorapana	Late Miocene	+	−	+
Triumphis	?	−	−	−
Opeatostoma	?	−	−	−
Western Atlantic				
Cittarium	Miocene	−	−	+
Turbo (Taeniaturbo)	Miocene	−	−	−
T. (Halopsephus)	Recent	−	−	−
Puperita	Miocene	−	−	−
Tectarius (Cenchritis)	?	−	−	−
Echininus (Tectininus)	?	+	−	−
Suturoglypta	Middle Miocene	−	+	−
Conella	Late Miocene	−	+	−
Vasum (Globovasum)	Early Miocene	+	−	−
V. (Siphovasum)	Early Miocene	+	−	−
Olivella (Dactylidia)	Miocene	−	+	−
O. (Macgintiella)	Miocene	−	+	−
Jaspidella	Oligocene	−	+	−
Eastern Atlantic				
Bolinus	Middle Oligocene	+	−	−
Jaton	?	+	−	−

Sc = presence (+) or absence (−) of strong sculpture.

Ap = presence (+) or absence (−) of narrowly elongate or toothed aperture.

SL = shell length greater (+) or smaller (−) than 100 mm.

Source: Data for this table have been compiled from the works of Knight et al. (1960) and Davies (1971); and from the monographs and articles of Rosewater (1972) for Littorinidae, Houbrick (1977) for Cerithiidae, Abbott (1960) for Strombidae, Emerson and Cernohorsky (1973) for Thaididae, Vokes (1971) and Radwin and D'Attilio (1976) for Muricidae, Olsson (1956) and Zeigler and Porreca (1969) for Olividae, Abbott (1959) and Vokes (1966) for *Vasum* and related genera, Powell (1964) for Turridae (Turrinae), Robertson (1957) for *Turbo*, Radwin (1977) for Columbellidae, and Woodring (1973a and b) for Thaididae.

the Thaididae and Buccinidae respectively, is striking. It may also be noteworthy that two Eastern Pacific autochthons (*Neorapana* and *Opeatostoma*) have labial spines. Few genera seem to have arisen and remained endemic to the Western Atlantic. Here again, however, shells that belong to the hard-bottom taxa (*Cittarium*, *Cenchritis*, *Tectininus*, and *Globovasum*) are relatively sturdy for their families, but have broadly ovate apertures with thin outer lips. *Cittarium pica* is known to be mechanically weaker than its Indo-West-Pacific autochthonous counterpart, *Trochus (Rochia) niloticus* (Vermeij, 1976). Another category of genera may be characterized as New World autochthons, having representatives only in the Western Atlantic and Eastern Pacific. Some of these (*Cyphoma*, *Macrocypraea*, *Cymia*, and *Leucozonia*) have quite sturdy shells, but many others, including *Tegula* (*Agathistoma*), *Calotrophon*, and *Eupleura*, would seem on morphological grounds to be relatively vulnerable to crushing. Experimental data are needed to verify all these claims.

Thus it can plausibly be argued that the greater shell sturdiness of rocky-shore molluscs in the Indo-West-Pacific and to a lesser extent the Eastern Pacific as compared to the Atlantic is caused not by differential extinction, but rather by the proliferation of well-armed taxa, especially in the Indian and Pacific oceans. Moreover, it seems that the most heavily armed forms are relatively recent productions arising in the Miocene or later. Vokes (1971) has already commented that sculpture and size have increased over time in various lineages of the gastropod family Muricidae. In the Paleocene, the family was composed almost entirely of small species less than 10 mm in shell length (for example, *Paziella* and *Poirieria*). Even in the Eocene, the largest members of the family (living in the Paris Basin in France) reached lengths of only about 50 mm. The very large (greater than 100 mm), highly spinose members of the family arose later: *Chicoreus* and *Phyllonotus* in the Late Oligocene, *Murex* and *Siratus* in the Early Miocene, and *Muricanthus* later in the Miocene (see Vokes, 1971; Radwin and D'Attilio, 1976). Abbott (1968) has pointed out that, although helmet shells (Cassidae) arose in the Eocene, the sturdy parietal shields and unresorbed former varices are characteristic only of forms that arose during or after the Miocene. Most Early Tertiary Conidae apparently had relatively tall spires and wide apertures; species with very low spires and extremely narrow apertures are known in Europe and North America from the Miocene on. The first species of *Vasum* (*V. humerosum* from the Upper Eocene of Louisiana) is relatively high spired compared to the Recent shallow-water members of that genus in tropical America and the Indo-West-Pacific (Vokes, 1966).

If the expression of predation-related traits among rocky-shore molluscs was more or less uniform across the tropics in the Early Tertiary, it was likely to have been low relative to the present. There was then a progressive increase until the Miocene, especially in the Indo-West-Pacific; since that time there seems to have been a slight faunal impoverishment, and many species have experienced range contractions and fragmentation of once continuous populations. On present evidence it seems that the tropical Atlantic biota never achieved the high level of predation-related adaptations attained in the Indo-West-Pacific. It must be stressed, however, that this is preliminary speculation and that more fossil data are required before any firm conclusion can be reached.

Indo-West-Pacific communities apparently were exposed to less frequent and less catastrophic disturbances during the Tertiary than tropical American communities in physically similar environments. As a result, Tertiary extinctions were more numerous in tropical America, and many taxa with a circumtropical distribution in the Early Tertiary are now limited to the Indo-West-Pacific. There is no convincing evidence that these relict taxa are morphologically sturdier than taxa that persisted in tropical America; therefore, differential extinction probably cannot account for the interoceanic differences in shell architecture between tropical American and Indo-West-Pacific communities. Rather, the postulated lower levels of disturbance in the Indo-West-Pacific may have permitted relatively uninterrupted evolution of molluscan shells highly resistant to crushing and to other destruction. In the Atlantic, which seems to have been profoundly affected by fluctuations in temperature and sea level during the Pleistocene, this evolution was evidently less spectacular — and possibly was subject to more frequent interruptions or even reversals. The Eastern Pacific biota seems to have been intermediate in all respects between the Atlantic and the Indo-West-Pacific.

More definitive statements must await the collection of additional data on fossil molluscs and other groups. In particular, we need to know what changes have taken place over time in predation-related shell architecture in the various oceans; and we must carefully analyze the architecture of taxa that become extinct in a given region compared to the architecture of groups that survive. Most of these aims critically depend on careful, world-wide systematic studies that cover both living and fossil species.

CHAPTER 9

Barriers and Biotic Exchange

W HEN biogeography is viewed from a historical perspective, as in Chapter 8, the story that unfolds is one of continual change in the relative isolation of biotas. While barriers are thrust up between once continuous seas in one region, other barriers break down or diminish in effectiveness and previously isolated biotas come into contact. In view of the impermanence of biogeographical barriers, it is of theoretical interest to predict the biological consequences of the removal or destruction of barriers. The problem takes on practical significance as well; since 1860 man has seriously reduced the effectiveness of one barrier (the Isthmus of Suez) and is contemplating the erasure of a second (the Central American isthmus).

In this chapter I shall examine several cases of recent biotic contact among marine biotas. I shall ask how species that have migrated from one biota to the other differ from those that have not taken part in the migration, and how this selection is related to conditions in the area of the connection and in the biotas themselves. Finally, I shall cautiously apply these findings to the question of what would happen if a sea-level canal were constructed across the Central American isthmus.

Hybridization and Migration

The two principal consequences of barrier removal are hybridization and migration. If a given barrier has been effective for only a relatively short time, its removal may reunite two isolates of a once continuous population; members of the two populations may still be interfertile and may produce viable hybrid offspring. If the interfertility is widespread, the area of contact between two recently joined biotas will contain many hybrid populations and may be referred to as a suture zone (Remington, 1967).

Many suture zones are known in the terrestrial biotas of North and South America (Remington, 1967; Vuilleumier, 1971). In the sea, the Sunda Shelf in Indonesia was dry land during the Pleistocene and effectively separated the Pacific Ocean from the Indian Ocean. This separation promoted speciation in a number of mangrove-associated crabs and other animals. When the Sunda Shelf became submerged with a postglacial rise in sea level, these pairs of incipient species came together, hybridized, and in many cases appear to have perfected antihybridization mechanisms so that they now coexist as fully separate entities (Macnae, 1968; Crane, 1975). Similar raising and lowering of barriers because of tectonic activity and climatically induced changes in sea level may have had similar consequences elsewhere in the Indo-Malaysian area, as well as in the Strait of Bab-el-Mandeb between the Red Sea and the Gulf of Aden, and in the area between the eastern and western basins of the Mediterranean.

The second principal consequence of barrier removal is migration between the joined biotas. The magnitude and the direction of migration depend to an important degree on the characteristics of the connection linking the biotas. Rarely is this connection penetrable to all species; in fact, what is a barrier to some organisms is no hindrance to migration for others. For example, the surface layer of low-salinity water in the Atlantic Ocean off the mouth of the Amazon is an effective barrier in separating a northern Caribbean fauna of light-loving corals and gorgonians from a smaller southern endemic fauna, but this same barrier does not restrict the distribution of shade-tolerant species that can apparently disperse through the more saline waters underlying the Amazon effluent (Laborel, 1969). A narrow strait of water is perceived by some tropical forest birds as an impenetrable barrier even though these animals may fly long distances over land to obtain food; but to the birds of more open environments, a strait is a trivial hurdle (MacArthur et al., 1972; Diamond, 1973, 1975). Terrestrial plants and animals living near the coasts of islands are less endemic and disperse more easily over stretches of open water than do the inhabitants of forests and uplands (Heatwole and Levins, 1972; Ricklefs and Cox, 1972; Lee, 1974; Diamond, 1975).

These examples illustrate the widely recognized fact that barriers, like extinction, act as selective filters. If stress tolerance is required to pass a given barrier, then there may be an upper limit on the biotic competence of species that migrate from one biota to another.

A second important limitation that prevents species from crossing a barrier and establishing a population in the biota on the other side is related to the biological conditions in both the donor and recipient biotas. This topic has received theoretical and empirical attention primarily from biogeographers working with mammals, birds, and fishes, all groups in which competition for food or space is widely thought to be the most important determinant of ecological coexistence among species (see, for example, MacArthur, 1972; Briggs, 1974a; Diamond, 1975). The only species that can gain a foothold in a recipient community are those exploiting resources that are untapped or only inefficiently exploited by species native to the recipient community. A community or biota in which all resources are utilized at the rate at which they are produced would therefore be highly resistant to invasion by species from the outside (MacArthur, 1972).

From a general principle of allocation MacArthur and others have argued that the most efficient utilization of a resource is by species specialized on that resource; species exploiting a wider array of resources (for example, a wider range of prey sizes) are less efficient at tapping any one resource and therefore unable to reduce that resource to as low a level as can a specialist. In communities made up of many potentially competing species, coexistence is made possible by the specialization of each species to a very limited range of the available resources; where fewer species coexist, each species in general exploits a wider array of resources. Therefore, species-rich communities should be more resistant to invasion than are more impoverished communities under the same physical regime (see Elton, 1958; MacArthur, 1972). Moreover, specialization on a limited range of resources is more feasible where environmental conditions are constant or highly predictable than in areas where availability of resources fluctuates unpredictably. Therefore, resistance to invasion should be greater for communities in predictable regimes than for those exposed to more stochastic variation (MacArthur, 1972). It was therefore not surprising to Elton (1958) that most of the strikingly successful invasions of introduced species into new biotas involve weeds in disturbed habitats.

Reasoning of this kind, based on communities of competing animals, provides an explanation for the common observation in biogeography that migration between two recently joined biotas is generally much greater in one direction than in the other, and that the rich biota tends to export species to the more impoverished biota (for reviews see Darlington, 1959; Briggs, 1974a, b). For instance, when at various times in the Late Tertiary and Pleistocene a land bridge between Asia and North America existed across the Bering Strait, the predominant migration of mammals and birds was from Asia to North America (Cracraft, 1973; Briggs, 1974b). When the interchange of mammalian faunas between North and South America in the Late Tertiary is considered overall, far more North American taxa invaded South America and diversified there than vice versa. Ungulates such as deer and camels, and carnivores such as bears, cats, and dogs, replaced notoungulates, litopterns, marsupials, and other groups that had evolved in isolation on the island continent of South America. The few South American taxa that invaded North America successfully (armadillos, ground sloths, porcupines) did so early during the faunal interchange (Late Miocene) and had no ecological equivalents in the North American mammal fauna (Webb, 1976).

In spite of the theoretical arguments of MacArthur (1972) and others, it remains to be established empirically that competitive relationships can account for the strong directionality of migration between biotas. For example, although competition seems to be an important determinant of fish community composition, especially among territorial reef fishes (Vine, 1974; Ehrlich, 1975; Kaufman, 1977; Sale, 1977), I know of no evidence to support the hypothesis that the competitive ability of the average fish species increases as the diversity of the community or biota in which that species evolved increases. In one of the few studies on interactions among closely allied reef fishes, Sale (1977) found that pomacentrids can coexist in territories that are often mutually exclusive, even though no species is able to monopolize space or food over any other.

Stomatopod crustaceans are another group of animals in which competition for space is believed to be important in community organization. They are known to establish interspecific dominance hierarchies, in which the competitive abilities of co-occurring species are linearly arranged (Caldwell and Dingle, 1975). The species at the top of the hierarchy is aggressively dominant over all the others and often interferes directly with the other species; the bottom species is subordinate to all other members of the hierarchy and is often relegated to spaces or foods that are not highly preferred. Hierarchies of this type are also known in terrestrial vertebrates, as well as in such marine groups as sea stars, corals, sea anemones, and even limpets (Lang, 1973; Stimson, 1973; Menge, 1974; Morse, 1974; Sebens, 1976).

It is possible that the competitive ability of the dominant species in a hierarchy increases as the number of species in the hierarchy becomes greater. At Eniwetok in the Marshall Islands, the gonodactylid stomatopod *Haptosquilla glyptocercus* is the aggressively dominant member of a three-species hierarchy; but on the west coast of Thailand, this same species, living in what is probably a larger assemblage of stomatopods, is dominated in agonistic encounters by *Gonodactylus chiragra* and *G. viridis* (Caldwell and Dingle, 1977). Two examples are known where a stomatopod species, recently introduced from a species-rich region to a species-poor island, was found to be dominant over any of the species in the island biota. In Hawaii, *Gonodactylus falcatus*, introduced from the Western Pacific in the early 1950s, was found to dominate *Pseudosquilla ciliata* in aggressive encounters and to have displaced this native species from areas of rubble to sandy bottoms (Kinzie, 1968). A similar displacement seems to have occurred in Bermuda, where *G. bredini* has become dominant over the native *P. ciliata* (Caldwell and Dingle, 1975).

At least five other explanations must be considered in a discussion of the preferred migration from species-rich to more impoverished biotas. First, the directionality may be nothing more than a statistical artifact. If migration is equally likely for each species in the two biotas, then migration should be more frequent from the rich to the poor biota simply because there are more species in the former area. In order to demonstrate that the migration actually occurs in a preferred direction, it must first be shown that the ratio of species migrating in one direction to the number migrating in the opposite direction is greater than the ratio of species numbers in the two areas.

Second, a donor region may contain species that perform ecological roles not performed by any species in the recipient biota. This explanation would seem to fit the establishment of *Littorina littorea* and of *Carcinus maenas* on the coast of New England, and the success of many estuarine species from Japan and eastern North America in the bays on the Pacific coast of the United States.

A third explanation for preferred migration is that species can become established in a recipient biota that is heavily exploited (and therefore undersaturated) as a result of large-scale harvesting by man. This possibility has been invoked by Christie (1974) in the case of the invasion of the Great Lakes by sea lampreys (*Petromyzon marinus*), smelt (*Osmerus mordax*), and other fish. Fishing pressures on native Great Lake coregonid whitefishes and salmonid trout became so heavy in the mid-twentieth century that a variety of fishes, which otherwise might not have found the necessary food for maintenance of permanent breeding populations, could become established.

Fourth, species in the donor region may be better protected against predators, and thus have more effective refuges from predation, than do species in the recipient biota. Since there is a trend for high resistance to predation to be associated with high diversity (Chapter 6), the average species in the richer biota may be more competent against predators than is the average species in the more impoverished community. As emphasized in Chapter 6, however, there are many exceptions to the general trend and diversity cannot serve as a reliable predictor of antipredatory prowess.

Finally, the connection between the biotas may pose fewer adaptive hurdles to species in a donor community than to those in the recipient biota. This explanation has been invoked by Webb (1976), who speculates that traversal of the Panama land bridge was easier for North American mammals adapted to open savannas than for forest-dwelling South American taxa.

In practice, of course, it is often difficult to distinguish among these explanations for the success of migrant species in new biotas. The case of *Trochus niloticus* will serve as an instructive example. This predation-resistant Indo-West-Pacific snail, indigenous to the Indo-Malaysian area, Melanesia, and Yap and Palau in Micronesia, was introduced to various Micronesian islands (Saipan, Guam, Truk, Ponape) and to the Society Islands from the 1920s on. (For a review see Gail and Devambez, 1958.) In most of these islands *T. niloticus* has become a common inhabitant of seaward reefs. It is not known whether it has ecologically replaced other species, or whether it has merely taken advantage of a niche previously ignored by other reef inhabitants. Exploitation by man of reef fishes and other organisms may have contributed to the successful establishment of *Trochus*; resistance to natural predation also may have played a role. No study of the ecological impact of *Trochus* has ever been undertaken, either in its native range or where it has been introduced.

Three Case Histories

Several episodes of marine biotic interchange have occurred in geologically recent time. These include migration through the Suez Canal (almost exclusively from the Red Sea to the Mediterranean), migration across the East Pacific barrier from the Western Pacific to the Panamic province, and migration via the Arctic Ocean from the North Pacific to the North Atlantic. In all these cases the biotic movement has been from the richer to the poorer biota; the degree to which one direction predominates over the other is far greater than would be predicted on the basis of diversity differences between the two biotas, except possibly in the Arctic migration.

Perhaps the most intensively studied biotic migration is the one through the Suez Canal, which connects the Mediterranean Sea on the north with the Red Sea on the south. (For a review see Por, 1971.) Before the Suez Canal was opened to navigation in 1869, there were hypersaline lakes and marshes in the Isthmus of Suez; during Pharaonic times, some permanent canals appear to have existed. Several euryhaline species had already penetrated from the Red Sea to the Mediterranean before the Suez Canal was created. These included the small potamidid gastropod *Pirenella cailliaudi* and the weedy intertidal mussel *Brachidontes variabilis* (Por, 1971, 1975).

With the creation of the permanent Suez Canal a major episode of biotic interchange began, which has resulted in a one-way migration of animals and plants from the Red Sea to the Mediterranean. Thus far, some thirty fishes (including three circumtropical species), twenty decapod crustaceans, and forty molluscs are known to have invaded the Mediterranean (Holthuis and Gottlieb, 1958; Lewinsohn and Holthuis, 1964; Ben-Tuvia, 1971b). These are indicated in Table 9.1. A very few animals, including the hypersaline-tolerant fish *Dicentrarchus punctatus*, have colonized the Red Sea from the Mediterranean (Ben-Tuvia, 1971a), but well-documented instances of this reversed migration are extremely rare.

Even today the waters of the Suez Canal, and especially those of the Bitter Lakes through which the canal runs, are hypersaline and therefore constitute a significant barrier to many potential migrants. Por (1975) has argued that species native to the Gulf of Suez, the hypersaline arm of the Red Sea at the southern extremity of the canal, were better adapted to withstand the physiological extremes in the canal and Bitter Lakes than were species native to the eastern Mediterranean, where salinity was kept at lower levels by the effluent of the Nile River. Thus, the virtual one-way migration to the Mediterranean could have resulted from the fact that the isthmian salinity barrier is less effective for Red Sea species than for Mediterranean taxa (Por, 1975).

Another explanation, championed by Topp (1969), holds that the impoverished biota of the eastern Mediterranean could absorb immigrant species more easily than could the biotically more saturated Red

Table 9.1   Red Sea molluscs that have migrated to the Mediterranean.

Gastropoda
  Haliotidae: *Haliotis pustulata*
  Fissurellidae: *Diodora rueppelli*
  Patellidae: *Cellana rota*
  Trochidae: *Minolia nedyma*
  Umboniidae: *Umbonium* cf. *vestitum*
  Alvaniidae: *Alvania orbignyi*
  Rissoinidae: *Rissoina bertelloti*
  Potamididae: *Pirenella cailliaudi*
  Cerithiidae: *Cerithium erythraoense, C. scabridum, Rhinoclavis (Proclava) kochi*
  Cypraeidae: *Cypraea caurica*
  Muricidae: *Aspella anceps, Murex tribulus*
  Thaididae: *Thais carinifera*
  Columbellidae: *Anachis savignyi*
  Nassariidae: *Naytiopsis granum*
  Fasciolariidae: *Fusinus marmoratus*
  Turbinellidae: *Vasum turbinellus*
  Opisthobranchia: *Notarchus indicus, Bursatella leachi savignyana*
  Siphonariidae: *Siphonaria kurracheensis* or *S. laciniosa*
Bivalvia
  Arcidae: *Arca* cf. *natalensis*
  Mytilidae: *Modiolus arcuatulus, M. auriculatus, M. glaberrimus, Brachidontes variabilis*
  Pteriidae: *Malleus regula, Pinctada radiata*
  Cardiidae: *Papyridea australe, P. papyracea*
  Veneridae: *Gafrarium pectinatum, Paphia textile, Clementia papyracea*
  Laternulidae: *Laternula subrostrata*
  Mactridae: *Mactra olorina*
  Gastrochaenidae: *Gastrochaena cymbium*

Source: This compilation is based on the findings of Barash and Danin (1972), Morrison (1972), Yaron (1972, 1976), and Mienis (1973).

Sea (see also Por, 1971; Briggs, 1974a). The impoverishment of the eastern Mediterranean came about because of Late Tertiary and Pleistocene changes in sea level, which periodically isolated this subtropical basin from the cooler Eastern Atlantic and western basin of the Mediterranean. Many subtropical and tropical species became extinct without being replaced by new forms. The Red Sea seems to have been less affected by such sea-level changes and today maintains contact with the rich tropical biota of the western Indian Ocean.

   Given the physiological limitations imposed on potential immigrants by the Suez barrier, we may still ask how the invaders differ from spe-

cies that have remained restricted to the Red Sea, and from species that
are native to the Mediterranean. At least two attempts have been made
to group the immigrants ecologically. Por (1971) believes that most of
the invaders fall into three ecological groups: (1) those associated with
boulders; (2) those living in beds of the sea grass *Halophylla stipulacea*;
and (3) animals with planktonic larvae. Groups that conspicuously have
not taken part in the migration are species whose larvae exhibit daily
vertical migrations in the water column and animals associated with
coral reefs (Por, 1971). Waters in the Suez Canal may be too shallow for
effective vertical migrations by larvae; and coral reefs, though present in
luxuriance in the Gulf of Aqaba to the east of the Sinai Peninsula, are
absent in the turbid and shallow Gulf of Suez on the west side of the
Sinai at the southern end of the canal (Loya and Slobodkin, 1971; Loya,
1972). The algae that have colonized the Mediterranean from the Red
Sea are ecologically less restricted; migrant plants include reef-associated species of *Caulerpa*, *Padina*, and *Sargassum*, as well as the sea
grass *H. stipulacea* (Lipkin, 1972). *Turbinaria*, another reef-associated
alga, has not yet been recorded from the southeastern Mediterranean.

Safriel and Lipkin (1975) have examined the ecological distribution of
twenty-one intertidal Red Sea immigrants and compared it with the distribution of 201 native species on the Mediterranean shore at Mikhmoret, Israel. According to their calculations, 76 percent of the
immigrants live in the low intertidal zone, while only 52 percent of the
native species live in that zone. Three of the presumed Red Sea immigrants (the red alga *Hypnea musciformis* and the green algae *Caulerpa
racemosa* and *C. mexicana*), all living in the low intertidal zone, are circumtropical in distribution and are known from the west coast of Africa
(John and Lawson, 1974). It is therefore at least possible that colonization of the eastern Mediterranean may have been from the Atlantic, not
from the Red Sea; or the species may even have been present in the
basin before the canal was opened. If so, then the Red Sea immigrants
are no more prone to live in the lower intertidal zone than are native
Mediterranean species. A definitive statement seems unwarranted at
present.

In any case, it appears that the high intertidal and littoral fringe in the
Mediterranean have not been colonized by Red Sea species (Barash and
Danin, 1972; Safriel and Lipkin, 1975). Among the possible reasons for
this absence of Red Sea elements on the high shore is the difficulty of
dispersal, which is particularly great for species with crawl-away larvae.
The dispersal problem is aggravated for rocky-shore animals by a 160 km
stretch of sandy beach on the Mediterranean coast between the north-

ern end of the Suez Canal and the first outcrop of rock to the east (Safriel and Lipkin, 1975). Another factor contributing to the poor success of high intertidal Red Sea immigrants is that, although water temperatures in the two seas are similar (Table 1.2), air temperatures are generally much lower year-round on the Mediterranean coast than on the shores of the northern Red Sea (Vermeij, 1972a; Safriel and Lipkin, 1975).

There are good reasons to believe that many Red Sea immigrants are either competitively superior to, or more resistant to predation than, native Mediterranean species; but it is equally clear that a large number of biotically competent Red Sea species have not taken part in the migration. Among migrant gastropods, potentially sturdy species include *Cerithium erythraoense*, *Cypraea caurica*, *Thais carinifera*, *Fusinus marmoratus*, and especially *Vasum turbinellus*. Many well-armed species, however, including members of ecological groups that have sent many invaders to the Mediterranean, have not yet been found in that sea. As an example, the large *Strombus tricornis*, a grassbed inhabitant tolerant of high salinity and known from the Suez Canal, is still unknown from the Mediterranean despite its thick sturdy shell. No species of Red Sea *Conus* has yet penetrated to the Mediterranean, despite the fact that the Mediterranean *C. ventricosus* has a thinner shell, a broader aperture, and a higher spire than most of the more sturdy Red Sea species. I have found that the incidence of such predation-related traits as toothed or elongate apertures, inflexible operculum, and strong sculpture is less than 10 percent among the rocky-shore gastropods on the Mediterranean coast of Israel. By contrast, rocky-shore assemblages of snails from the Gulf of Aqaba show a much higher frequency of the traits that confer resistance to crushing (Table 5.1). Yet relatively few Red Sea gastropods have colonized Mediterranean hard bottoms. The two Red Sea immigrants that seem to be common on the Mediterranean rocky coast of Israel (*Cerithium scabridum* and *Siphonaria* sp.) are not strikingly sturdy.

The lizard fish *Saurida undosquamis*, first seen in the Mediterranean in 1952, has successfully competed with, but not eliminated, the native Mediterranean hake *(Merluccius merluccius)*. Both fishes subsist to an important degree on small native and Red Sea fishes; indeed, the population reduction of *Leiognathus klunzingeri*, an earlier Red Sea migrant, may be causally related to the rapid rise in abundance of *S. undosquamis* after 1956 (Ben-Yami and Glaser, 1974). Today both the hake and lizard fish are exploited commercially along the coast of Israel.

Still another possible case of Red Sea biotic superiority has been suggested by Achituv (1973), who noted that the native Mediterranean sea star *Asterina gibbosa* has not been found on the Mediterranean coast of Israel since the Red Sea immigrant, A. *wega*, recently colonized that coast. If this indeed represents competitive exclusion, the mechanism underlying the superiority of A. *wega* remains unknown.

Several predatory swimming crabs of the family Portunidae have colonized the Mediterranean from the Red Sea, including the large *Portunus pelagicus* (10 cm broad) and the smaller *Thalamita poissoni*, *Charybdis helleri*, and *C. longicollis* (less than 5 cm broad) (Holthuis and Gottlieb, 1958; Lewinsohn and Holthuis, 1964). These crabs may be faster and more aggressive than native Mediterranean portunids (though this remains to be experimentally verified), but some large species, such as the 8-cm-wide *Thalamita crenata* and the 18-cm-wide *Charybdis erythrodactyla* have not traversed the Suez Canal. Several xanthids, including the 5-cm-broad, equal-clawed *Atergatis roseus*, have penetrated the Mediterranean; but massively clawed Red Sea natives such as *Carpilius convexus*, *Lydia tenax*, *Ozius rugulosus*, and *Eriphia sebana* have not. The case of *Eriphia* is particularly interesting. *E. verrucosa*, a Mediterranean species abundant on the coast of Israel, has smaller claws (ratios of claw height and thickness to carapace width for males respectively 0.415 and 0.290) than does *E. sebana* (ratios 0.527 and 0.332). Although *E. verrucosa* attains a larger size (up to 7.8 cm wide compared to 6.5 cm for *E. sebana*), this advantage is not sufficient to offset the larger cross-sectional area of the claws of *E. sebana* (Vermeij, 1977a). Possibly, dispersal of *E. sebana* and of other rocky-shore species is made difficult by the absence of suitable hard bottoms near the canal.

The Red Sea seems to have exported a number of weedy or stress-tolerant animals. These may be locally so abundant that they comprise an important part of the diet of both introduced and native Mediterranean predators. In their analysis of the stomach contents of the Mediterranean sparid fish *Sparus auratus* from the hypersaline lagoon of Bardawil on the north coast of the Sinai Peninsula, for instance, Barash and Danin (1971) found large numbers of such fragile Red Sea molluscs as *Pirenella cailliaudi*, *Rhinoclavis kochi*, and several species of *Modiolus*. Other examples of stress-tolerant Red Sea immigrants include the keyhole limpet *Diodora rueppelli* and perhaps the venerid clam *Gafrarium pectinatum*; these species are characteristic of hypersaline environments (Por, 1975).

We see that two factors may be primarily responsible for the one-way migration of species from the Red Sea to the Mediterranean through the Suez Canal: (1) the Mediterranean biota appears to be impoverished, while the Red Sea biota is not (or in any case is less so); and (2) Red Sea species may be better adapted to the salinity regime of the canal than are Mediterranean forms. Competitive or predation-related superiority seems to have played a much less important role and may not have been involved at all in the migration of hard-bottom molluscs. This could change if the salinity barrier in the Suez Canal and Bitter Lakes were to become less extreme with time, a possibility that may already have been realized (Por, 1975). My conclusions are broadly in accord with those of Topp (1969) and Por (1971, 1975).

Another case of interprovincial exchange that may be studied profitably is the relatively recent, and probably continuing, influx of Indo-West-Pacific species to the Eastern Pacific. This one-way migration, which may have begun in the Pliocene, has primarily involved organisms closely associated with coral reefs, though some sand dwellers (such as the crab *Calappa hepatica*) have taken part (Garth, 1965, 1974; Olsson, 1972; Dana, 1975). Largely from indirect evidence, it appears that most or all of the migrant species have long-lived teleplanic larvae that can successfully traverse the 6,500-km East Pacific barrier between the Line Islands in the Central Pacific and the islands off the west coast of tropical America (Emerson, 1967; Dana, 1975; Robertson, 1976).

As is evident from Table 9.2, most of the Indo-West-Pacific molluscs that have penetrated the Eastern Pacific are gastropods (Robertson, 1976). The majority of the snails are either architecturally very sturdy (*Cypraea, Drupa, Morula, Mitra, Conus, Terebra*) or they are associated with anthozoan cnidarians (*Pseudocypraea, Architectonica, Heliacus, Coralliophila, Quoyula, Magilus*). In Table 9.3 I compare the incidence of several predation-related shell traits among low intertidal Indo-West-Pacific gastropods from open rocky surfaces in the Line Islands, the Eastern Pacific offshore islands, and the mainland coast of tropical America. These data, compiled from faunal lists in the literature, are not strictly comparable to the data in Table 5.1 based on samples from specific localities; but there is excellent agreement between values for the incidence of apertural teeth, strong sculpture, and other antipredatory defenses in the Line Islands (Table 9.3) and values given for other Indo-West-Pacific areas (Table 5.1). Moreover, the inferred resistance to crushing is about the same in the Line Islands as among the Indo-West-Pacific migrants in the Eastern Pacific offshore islands; that is, the wide stretch of open ocean between the Indo-West-Pacific Line Islands and the Eastern Pacific offshore islands imposes little if any selective reduction in the incidence of antipredatory features among species. The Indo-West-Pacific faunal component that penetrates to the mainland

Table 9.2   Indo-West-Pacific molluscs in the Eastern Pacific. An asterisk before the name denotes species that have penetrated to the mainland of tropical western America.

Gastropoda
  Fissurellidae: *Diodora granifera*
  Neritidae: *Nerita plicata*
  Titiscaniidae: **Titiscania limacina*
  Architectonicidae: **Architectonica nobilis*, **Heliacus trochoides*
  Cerithiidae: *Cerithium nesioticum*
  Eulimidae: *Balcis thaanumi*, *B. vafra*
  Hipponicidae: **Hipponix antiquatus*, *H. pilosus*, *H. fimbriatus*
  Cypraeidae: *Cypraea arenosa*, *C. caputserpentis*, *C. helvola*, *C. depressa*, *C. maculifera*, *C. scurra*, *C. moneta*, *C. isabella*, *C. teres teres*, **C. t. pellucens*, *C. schilderorum*, *C. vitellus*, *C. rashleighiana*
  Ovulidae: *Pseudocypraea adamsoni*
  Cymatiidae: *Cymatium muricinum*, *C. nicobaricum*, **C. pileare*
  Bursidae: *Bursa cruentata*, **B. granularis*
  Thaididae: *Drupa morum*, *D. ricinus*, *Morula granulata*, *M. uva*
  Coralliophilidae: *Coralliophila violacea*, **Quoyula madreporarum*, *Magilus robilliardi*
  Melongenidae: *Pugilina lactea*
  Nassariidae: *Nassarius francolinus*
  Fasciolariidae: *Peristernia carolinae*, *P. thaanumi*
  Harpidae: *Harpa gracilis*
  Volutidae: *Voluta ancilla*, *V. deshayesii*
  Mitridae: **Mitra mitra*, *M. papalis*, *M. edentula*, *M. ferruginea*, *M. litterata*
  Conidae: *Conus chaldaeus*, **C. ebraeus*, **C. tessulatus*
  Terebridae: **Terebra affinis*, *T. crenulata*, *T. maculata*, **T. laevigata*
  Turridae: **Microdaphne trichodes*, **Spurilla alba*

Bivalvia
  Spondylidae: *Spondylus gloriosus*, *S. hawaiiensis*
  Ostreidae: *Pycnodonta hyotis*
  Lucinidae: *Codakia thaanumi*

Source: This compilation is taken from general works by Emerson (1967), Salvat and Ehrhardt (1970), Keen (1971), and Robertson (1976); and from monographs and notes by Emerson and Cernohorsky (1973), Rehder (1973), Cernohorsky (1976), and von Cosel (1977).

coast of tropical America, however, has a lower incidence of crushing-resistant traits than does either the Line Island fauna or the native Eastern Pacific biota. This reduction is especially evident with respect to apertural dentition and external sculpture. Thus, the region between the offshore islands and the mainland of western America imposes a more stringent adaptive barrier to Indo-West-Pacific molluscs than does the much wider East Pacific barrier.

Table 9.3 Incidence of predation-related traits among hard-bottom Indo-West-Pacific gastropods from the two sides of the East Pacific Barrier.

Region	Number of species	Percentage of species with—			
		Toothed apertures	Elongate apertures[a]	Inflexible operculum	Strong external sculpture
Line Islands	99	56	20	14	26
Eastern Pacific islands	31	65	26	6.5	19
Eastern Pacific mainland	5	20	20	0	0

[a] Ratio of length to breadth greater than 2.5.

With respect to size, most species that have crossed the East Pacific barrier to the Eastern Pacific islands and the tropical American mainland are small (less than 7 cm in longest dimension). The only exceptions are *Cypraea maculifera*, *C. schilderorum*, *Mitra mitra*, *Terebra crenulata*, and *T. maculata*; of these only *M. mitra*, the largest (15 cm long) and perhaps sturdiest species in the genus, has reached the mainland (von Cosel, 1977). Many larger (greater than 8 cm long) snails present in the Line Islands have never crossed the East Pacific barrier at all: *Turbo argyrostomus*, *Lambis chiragra*, *L. truncata sebae*, *Cypraea tigris*, *Cypraecassis rufa*, *Charonia tritonis*, *Bursa bubo*, *B. bufonia*, *Tonna perdix*, *Malea pomum*, *Thais armigera*, *Vasum armatum*, and *Pleuroploca filamentosa* (Kay, 1971). Generally speaking, then, Indo-West-Pacific migrants to the Eastern Pacific are resistant to crushing, but the largest and sturdiest species in the Line Islands have not taken part in the migration.

Much the same can be said for the decapod crustaceans that have invaded the Eastern Pacific from the west. At least three powerful molluscivorous decapods (the crabs *Calappa hepatica* and *Carpilius convexus* and the spiny lobster *Panulirus penicillatus*) have been found on one or more of the Eastern Pacific offshore islands. At least in the case of the crabs, however, these do not represent the largest or most powerful species in their respective genera. Indeed, such massively clawed or large crabs as *Eriphia sebana*, *Carpilius maculatus*, *Ozius guttatus*, *Scylla serrata*, *Daldorfia horrida*, *D. rathbunae*, and *Calappa calappa* have not yet been exported from the Indo-West-Pacific, despite their very wide distribution and apparent dispersibility throughout that large province.

Holthuis and Loesch (1967) noted that the Indo-West-Pacific spiny lobster *Panulirus penicillatus* coexists with the Eastern Pacific *P. grac-*

ilis in the Galapagos. In these islands the Indo-West-Pacific species appears to have displaced *P. gracilis* from its normally shallow-water habitat to slightly deeper water. Kent (1978), moreover, has shown that *P. penicillatus* has proportionately larger mandible teeth than does *P. gracilis*.

Other Indo-West-Pacific migrants to the Eastern Pacific are also well known for their predatory or grazing activities. They include the coral-eating sea star *Acanthaster planci*, the polyp-nipping smooth puffer *Arothron meleagris* (see Figure 4.6), and various parrotfishes, surgeon-fishes, and moray eels. The poisonous sea urchin *Toxopneustes variolatus* may be another recent addition to the Eastern Pacific fauna.

If the Indo-West-Pacific immigrants to the Eastern Pacific are indeed well adapted as adults to predation, it is surprising that so few of them have been successful in colonizing the mainland coast of tropical western America. Only fifteen of the sixty-two molluscs listed in Table 9.2 are found on the mainland; and of sixteen Indo-West-Pacific brachyuran crabs listed by Garth (1965) as occurring at Clipperton Island off the west coast of Mexico, only five are found on the continent (see also Garth, 1974). The two molluscivorous crabs *(Carpilius convexus* and *Calappa hepatica)*, for example, are known in the Eastern Pacific only from Clipperton Island. In spite of its possible competitive superiority over *P. gracilis*, the Indo-West-Pacific spiny lobster *Panulirus penicillatus* has never been found on the western American mainland. Indo-West-Pacific gastropods that penetrate to mainland tropical America are, if anything, less armored than local species, and in no instance can it be argued that a migrant *Terebra*, *Conus*, *Cymatium*, or *Cypraea* is better protected against predators than a local congener.

Even the species that have penetrated to the mainland have patchy or very restricted distributions there. For example, *Heliacus trochoides* thus far is known only from southwest Ecuador (Robertson, 1976), and *Mitra mitra* is known only from Isla Gorgona (Colombia) and from Costa Rica (von Cosel, 1977).

One possible reason why these Indo-West-Pacific species have not been more successful in colonizing mainland American shores is that the continental waters are usually more turbid and more productive than are the clear oceanic waters of the Line Islands and other Central Pacific atolls from which most of the migrants have come. Coral reefs are widely distributed along the American west coast, but perhaps these are oceanographically isolated from one another and from the main currents that transport larvae from the west. Surface currents flow from Panama southward to Ecuador and thence westward to the Galapagos, but very little return flow seems to occur (see Abbott, 1966).

It is interesting that no reverse migration has been recognized. Abbott (1966) suggested that the frequency of teleplanic larvae may be unusually low among shallow-water benthic organisms in western America. Although this remains to be demonstrated, it is possible that the East Pacific barrier is virtually insuperable to species of western America, while it is only partially effective against Indo-West-Pacific species. Moreover, the migration to the Eastern Pacific involved organisms living on and comprising coral reefs; they may therefore have filled an adaptive vacuum left after Eastern Pacific reefs were largely decimated in the Early and Middle Tertiary.

The migration from the Indo-West-Pacific to the Eastern Pacific leaves us with many problems. If competitive superiority and resistance to crushing predation played a role in the colonization of the Eastern Pacific by species from the Indo-West-Pacific, these biotic determinants were often overshadowed by factors related to the open-water barrier and to the exploitation of untapped or inefficiently tapped resources.

In the north temperate zone, periodic opening of the Bering Strait at times of high sea level and relatively mild climate in the Late Tertiary permitted interchange between the biotas of the North Pacific and the North Atlantic via the Arctic Ocean. (For a review see Briggs, 1970.) Prior to the Late Miocene, the North Pacific was biologically quite different from the Atlantic. Groups present in the North Pacific but absent in the North Atlantic apparently included sea grasses of the genera *Zostera* and *Phyllospadix*, the now extinct desmostylian mammals and the sirenian Steller's sea cow *Hydrodamalis*, otariids and related seals (sea lions and walruses), sea otters (*Enhydra*), and deep-burrowing bivalves of such genera as *Mya* and *Macoma*, as well as many groups of fish (Brasier, 1975; Dayton, 1975b; Lipps and Mitchell, 1976; Domning, 1976; Ray, 1976; Repenning, 1976a, b). When a connection between the two oceans was established, migration was predominantly from the Pacific to the Atlantic; groups taking part included walruses (*Odobenus*), the grass *Zostera*, and the bivalves mentioned above. Sea lions, sea otters, sea cows, the grass *Phyllospadix*, and doubtless many other Pacific autochthons were left behind.

It is difficult to estimate how directional the Arctic migration actually was. There is good evidence that phocid seals originated in the Atlantic during the Early Miocene and colonized the North Pacific by way of the Arctic (Briggs, 1970; Lipps and Mitchell, 1976; Ray, 1976). Not enough data are available to eliminate the possibility that the pattern of migration merely reflects the threefold difference in diversity between the North Pacific and the North Atlantic. In his thorough study of the history and systematics of North Pacific moon snails (Naticidae), Marincovich (1977) concludes that three lineages migrated from the Pacific to the Atlantic, whereas two migrated in the opposite direction. One of the Atlantic naticids of Pacific extraction is *Polinices (Euspira) heros*, also known under the name *Lunatia heros*. This species is the largest naticid in the cold-temperate Atlantic, much as its close relative, *P. lewisi*, is the largest naticid in the Northeast Pacific. In spite of this example of a biotically competent migrant, there is no compelling evidence that the Pacific invaders in the North Atlantic are biotically more competent than are species native to the Atlantic. Our knowledge of patterns of migration and of biological properties of species is so incomplete that scarcely any conclusions are warranted at this time.

The Panama Canal Problem

The practical importance of predicting the direction and magnitude of faunal interchange once a barrier has been removed is underscored by current proposals to build a sea-level ship canal across Central America, or to modify the present fresh-water canal by augmenting its source of water with sea water. At least four sometimes conflicting predictions have been made concerning the biological consequences of such a sea-level canal: (1) there would be no significant effects (Topp, 1969); (2) extensive hybridization and eventual evolution of isolating mechanisms would take place between pairs of species from the two oceans (Rubinoff, 1968); (3) faunal migrations would occur, with most species going from the Caribbean to the Eastern Pacific (Briggs, 1968, 1974a); and (4) faunal migrations would occur, with most species in rocky habitats and reefs going from the Eastern Pacific to the Caribbean (Vermeij, 1974b). I believe the first prediction to be wrong; the other three are probably correct, and it is our task to determine which types of organisms conform to each of these predictions.

If genetically similar (cognate) species from the Eastern Pacific and Western Atlantic were to come into contact through a sea-level canal across Central America, they would in many cases hybridize with one another and produce viable offspring. Rubinoff and Rubinoff (1971), for instance, have shown that hybridization and the production of fertile offspring occur readily when reciprocal crosses are made between Atlantic and Pacific species of *Bathygobius*, a genus of small tide-pool fishes. Although these species have diverged somewhat at the morphological level, they have not acquired mechanisms that prevent genetic exchange.

Hybridization between Atlantic and Pacific species should be more common in soft-bottom and cryptic communities than in assemblages from open rocky surfaces, since the frequency of cognate species is highest there (see Table 8.1). The only circumstance I can envision in which this prediction would not hold is if the entrances of the canal were located far from suitable cryptic, sandy, or muddy environments. This, however, is unlikely; the entrances of the present canal lie close to sand and mud flats and, for navigational reasons, the same is almost certain to be the case if a second canal were dug.

The direction and the magnitude of any biotic migration through a Central American sea-level canal are likely to differ among taxonomic groups as well as among organisms from different communities. Indeed, the contrasting opinions concerning migration through an Atlantic-Pacific connection seem to stem in part from a difference in orientation of the investigators who have taken sides in the debate. J. C. Briggs, I. and R. Rubinoff, and R. W. Topp have approached the subject primarily as ichthyologists; while my conclusions have been based largely on studies of molluscs and decapod crustaceans.

According to Briggs (1968, 1974a), the competitive ability of the average fish will be greater in the rich Caribbean fauna, comprising some 900 species, than in the more impoverished Eastern Pacific fauna of some 650 species. Given this presumed superiority of Caribbean fish species, Briggs predicted that these fishes could invade the Eastern Pacific and outcompete native western American species, while Caribbean fish communities would be relatively resistant to invasion. Possible exceptions of this pattern of migration would be Indo-West-Pacific fishes that have invaded the Eastern Pacific; these evolved in highly diverse communities and would therefore have a competitive edge not only over Eastern Pacific species, but over Caribbean ones as well. Included in this category are reef-associated fishes like scarids, zanclids, acanthurids, chaetodontids, and labrids, as well as some invertebrates. *Omobranchus punctatus*, an Indo-West-Pacific blenny not known from the Eastern Pacific, has already been found near the Caribbean entrance of the Panama Canal (McCosker and Dawson, 1975).

The pattern of interoceanic migration of fishes through the present fresh-water Panama Canal is consistent with Brigg's prediction, although competition may have had little to do with this migration. Five fishes are known to have traversed the canal from the Atlantic to the Pacific: the tarpon *Megalops atlanticus*, the gobies *Lophogobius cyprinoides* and *Barbulifer ceuthoecus*, and the blennies *Lupinoblennius dispar* and *Hypleurochilus aequipinnis*; only two have colonized the Caribbean from the Eastern Pacific: the yellow jack *Gnathanodon speciosus* and the goby *Gobisoma nudum* (McCosker and Dawson, 1975). All but one of these species (*B. ceuthoecus*) are highly tolerant of low salinity, and success in the new

ocean for most species has been modest. The Atlantic colonists to the
Pacific, for instance, are mostly restricted to the brackish Third Miraflores
Lock, a lake built in 1940 that is periodically exposed to Pacific water during
high tides. The walls of this lock are covered with the Eastern Pacific
Ostrea palmula, and many of the other animals there are native Eastern
Pacific forms typically associated with mangroves (such as the bivalve
Mytella falcata and the snail *Thais kiosquiformis*). This community of continually submerged oysters is rare or absent in the fully marine Pacific Bay
of Panama, which may account for the absence of Atlantic blennies and
gobies outside the Pacific entrance of the Panama Canal (Rubinoff and
Rubinoff, 1968). Competitive pressures may also contribute to this lack of
success, since the Eastern Pacific has a rich fauna of native blennies and
gobies. The Atlantic xanthid crab *Eurypanopeus dissimilis* too is known
from the Third Miraflores Lock (McCosker and Dawson, 1975) and like-
wise has not penetrated to the Bay of Panama.

Stomatopods constitute another group for which Briggs's predictions
might hold, although no empirical evidence has come to light on this subject. There are some sixty species of these aggressive crustaceans in the
tropical and subtropical Western Atlantic, and only about forty in the Eastern Pacific (Manning, 1969).

Competition for food or space is also likely to be important in the ecology
of such other large predators and grazers as rocky-shore crabs, echinoids,
lobsters, and certain gastropods. In contrast to fishes and stomatopods,
however, crabs and gastropods have slightly higher diversities on the
Pacific side of the isthmus than in the Western Atlantic. On the strength of
Briggs's arguments alone, then, transisthmian migration in these animals
might proceed primarily from the Pacific to the Atlantic. Moreover, Western American crabs of the genera *Eriphia*, *Menippe*, *Ozius*, and *Callinectes* grow to a larger adult size than do tropical Atlantic congeners; and
the Eastern Pacific *Ozius verreauxii*, *O. tenuidactylus*, and *Eriphia squamata* have proportionately larger crushing claws than do the only Atlantic
representatives of these genera, *O. reticulatus* and *E. gonagra* (Vermeij,
1977a). These differences may imply a competitive advantage of Eastern
Pacific as compared to Western Atlantic congeners, although experimental data are lacking. A similar but equally untested inference could be
drawn for fiddler crabs of the genus *Uca*; there are twenty-nine species on
the Pacific side and only twelve in the tropical Western Atlantic (Crane,
1975).

Although crabs as a group are slightly richer in species in the Eastern
Pacific than in the Western Atlantic, individual genera such as *Calappa*
and *Callinectes* show the reverse trend. *Callinectes bocourti*, one of nine

species of the genus in the Western Atlantic, has proportionately larger claws than do any of the three Eastern Pacific species. There is no Eastern Pacific counterpart of the large-clawed, reef-associated Atlantic *Carpilius corallinus* (Vermeij, 1977a).

The last several examples illustrate how difficult and perilous it is to infer properties of individual species from data on regional differences in diversity of whole groups; as a consequence, it is difficult to predict patterns of migration between two rich biotas such as the Panamic and the Caribbean. The complexity is compounded by the likelihood that animals of one major group are competing not only with members of their own group, but also with species in other taxa. For example, it seems certain that grazing grapsid crabs compete for food, if not for space, with grazing gastropods. Herbivorous echinoids are known to compete with pomacentrid and other fishes for food and other resources (see Kaufman, 1977). Fishes, stomatopods, crabs, and lobsters all crush hard-shelled prey and must probably compete for food with animals such as gastropods, octopods, and sea anemones, which use different methods to capture and exploit their prey.

Even if high diversity could be shown invariably to be associated with high competitive ability, preferential migration from a rich to a poor biota might have nothing to do with competition. Two further examples will serve to emphasize this point. All parties concerned with the biological effects of a Central American sea-level canal have worried about the colonization of the Atlantic by the Indo-Pacific sea star *Acanthaster planci* and the possible resulting devastation of West Indian coral reefs (see Porter, 1972a). There are thirty-seven species of asteroid in the tropical Eastern Pacific and only eighteen in the Western Atlantic (Chesher, 1972). From this large difference in diversity it might be expected that *Acanthaster* would have a competitive ability superior to that of any West Indian species, and that for this reason it should readily colonize the Western Atlantic. However, it is uncertain that *Acanthaster* would compete with any West Indian sea star, since it exploits a food resource (living coral) not efficiently exploited by any West Indian sea star or other animal. The endemic Panamic snail *Jenneria pustulata* is also an important consumer of Eastern Pacific coral (Glynn et al., 1972) and would probably do well in Caribbean reefs; yet there are only six ovulids in tropical western America, while there are about ten in the tropical Western Atlantic (Keen, 1971; Abbott, 1974).

Another potentially serious biological consequence of a Central American seaway would be the Atlantic colonization by the Indo-Pacific pelagic sea snake *Pelamis platurus*. This problem has been investigated by Rubinoff and Kropach (1970), who showed with trials in large salt-water tanks that potential Pacific predators of the sea snake very rarely attack and eat this poisonous reptile, while naive Caribbean predators can be provoked to attack and eat it. Even though the rate of consumption of snakes was some-times quite high (one *Lutjanus* snapper ate twenty-two snakes in thirty-one days), all the predatory fishes were eventually bitten either before or after ingesting a snake, and all died. Rubinoff and Kropach speculate that while the expansion of *Pelamis* into the Atlantic might be slow at first because of predation by fishes, intense selection would favor avoidance reactions on the part of the fishes, and little could stand in the way of eventual establishment of *Pelamis* in the tropical Atlantic.

Resistance to predation, as in the case of *Pelamis*, could be an important trait permitting a species to invade a new biota, especially if the immunity were more complete there than in the biota where the species evolved. I have predicted a predominantly Atlantic-ward migration of low intertidal rocky-shore snails on the basis of the greater resistance to crushing of Pacific species (Vermeij, 1974a). Shells of Eastern Pacific species become immune to crushing at a smaller size than do those of ecologically compa-rable, taxonomically related Western Atlantic species (see also Reynolds and Reynolds, 1977; Palmer, 1978). If juveniles from both coasts grow at about the same rate, then the period that Pacific individuals spend in the predation-vulnerable phase would be shorter than that spent by Caribbean individuals, and the probability of being eaten would be lower. This out-come should hold on either coast. It should thus prove easier for Eastern Pacific snails to colonize the West Indies than for Caribbean species to invade the Panamic province.

One assumption of this argument, of course, is that crushing constitutes a quantitatively significant form of gastropod predation. Although there is abundant circumstantial evidence that this is so (see Chapter 2), no quanti-tative estimates of the impact of crushing have been made. Much of the argument also hinges on the assumption that juveniles from both oceans grow at the same rate. If it could be shown that West Indian snails grow faster as juveniles and achieve immunity by reaching the critical size in the same time as, or more quickly than, Eastern Pacific species, then the Pacific advantage would be canceled or even reversed. Such a difference in growth would result if animals less committed to architectural defense as juveniles can grow more rapidly than animals that from the beginning must incorporate fortifications in their shells. It is therefore crucial to know how rapidly juvenile snails grow, to what extent architectural fea-tures that confer resistance to crushing are already incorporated in the early stages of shell growth, and how much the incorporation of armor in the juvenile shell depresses growth rate. Virtually no data of this kind are available for Caribbean and Eastern Pacific snails.

Except perhaps in *Conus* and a few other genera, most juveniles of snail species that as adults have heavily fortified shells grow rapidly to maturity. Juvenile cowries *(Cypraea)*, for instance, have fragile shells with thin outer lips; the thick shell, slit-like aperture, and massive toothed outer lip are developed only in the adult stage after spiral growth has ceased. Other shells with thickened or toothed outer lips *(Strombus, Thais, Drupa, Columbella, Nassarius)* are known to have a similar shell development; often the fragile juveniles inhabit deep crevices, bury themselves in sand, or live in some other environment where they are unlikely to be found by shell-destroying predators (Randall, 1964). Cymatiid and muricid shells are known to grow rapidly as juveniles and to cease growth for long periods as adults. During these stationary phases the outer lip thickens into a varix (Laxton, 1970; Spight and Lyons, 1974). Juvenile snails often compromise sturdiness in favor of rapid growth through a stage in their life cycle when they are exposed to heavy predation; and differences in growth rates of species that vary in sturdiness as adults may not be large.

Adult longevity may also influence species success. If Caribbean snails have longer individual life spans than their Eastern Pacific counterparts, and if they are comparatively immune from predation for most of their adult life, then an individual native to the West Indies can produce more offspring in its lifetime than a Pacific individual whose fecundity per unit time is the same. Although no comparative studies have been done on molluscan longevities or fecundities in the two oceans, I doubt that West Indian snails have longer life spans than their Eastern Pacific relatives.

In Chapter 5 I noted that snails in soft-bottom and cryptic assemblages do not show the interoceanic differences in predation-related traits seen in open-surface snails from hard bottoms. This might imply that migration of soft-bottom and cryptic species could occur in both directions through the canal. For statistical reasons, however, migrations toward the Caribbean might still be slightly more frequent, since the species-pool of snails is somewhat larger in the Eastern Pacific than in the Western Atlantic (see Table 8.1). Furthermore, the widespread extinctions of sand- and mud-dwelling molluscs in recent geological time in the Caribbean but not in the Eastern Pacific may have impoverished the Caribbean fauna somewhat and made it more vulnerable to invasion.

One soft-bottom community, however, may well be expected to migrate to the Eastern Pacific against the prevailing trend. Sea grasses such as *Cymodocea, Halodule, Halophylla,* and *Thalassia* are abundant throughout the Caribbean, but are virtually absent in the Eastern Pacific; the genus *Thalassia* is wholly unknown from tropical western America (Earle, 1972a). The reasons for this absence are unknown, but the Eastern Pacific probably does not constitute a uniformly unfavorable environment for sea grasses. From the abundance of *Thalassia* and *Halodule* on the productive and turbid west coast of the Paraguana peninsula of Venezuela, I conclude that the high nutrient levels and limited light penetration typical of much of the Eastern Pacific need not be obstacles for the establishment of sea grasses.

None of these predictions takes into account the influence of the sea-level canal as a selective barrier; yet, from the Suez and trans-Pacific migrations, it is evident that the barrier may well exercise primary control over the magnitude and direction of the migration, and over the types of organisms able to take part in the migration. The biotically most competent species in a biota often seem unable to traverse a barrier, probably because they tend to be intolerant of physiological stress. Turbid or low-salinity water might, for instance, create an insuperable obstacle for many typically marine species, especially for those that normally reside on coral reefs. Chronically anoxic conditions on the bottom might inhibit species typical of well-oxygenated waters from traversing a canal. Such conditions, it has been suggested to me, might be created by dumping large quantities of cornstalks into the canal and might prompt us to rename it the "*Zea*-level" canal. However, the more continental nature of the Eastern Pacific province compared to the tropical Atlantic should render Eastern Pacific species generally more tolerant of turbidity, reduction in salinity, high temperature, and other adversities in the canal. This would be particularly true for soft-bottom and intertidal species.

In the event that species from one side of the isthmus do establish breeding populations on the opposite side, it is important to ask how widespread they will become. One clue may lie in the geographical distributions of species that have cognates on the other coast. As I pointed out in Chapter 8, many such species have surprisingly restricted ranges in the Caribbean and Eastern Pacific, and many are limited to the presumably more productive shores of the continents. *Tegula cooksoni*, the Eastern Pacific cognate of the widespread Atlantic *T. fasciata*, is restricted to the Galapagos; *Acmaea semirubida*, Pacific relative of the widespread A. *pustulata* of the Atlantic, is limited to the Gulf of California and adjacent parts of Mexico; the venerid bivalve *Pitar dione*, the Atlantic cognate of the Pacific *P. lupanarius*, is a strictly continental species that ranges from Texas at least as far south and east as Venezuela (Keen, 1971; Abbott, 1974). If any such species were to establish populations on the opposite side, they would likely remain in a relatively restricted region. On the other hand, of course, many cognate species have extremely wide distributions. This is particularly true of reef-associated animals. There is thus no reason to believe that well-armed Pacific rock-loving gastropods would remain limited in distribution once they established breeding populations in the Caribbean.

In short, the creation of a sea-level canal across the Central American isthmus would, I believe, lead to extensive hybridization and possible introgression between pairs of closely related species (Rubinoff, 1968), and to migrations of animals and plants between the Atlantic and Pacific. The Caribbean biota would probably import more species than it would export for at least three reasons: (1) the West Indian fauna is in many respects slightly or considerably less rich than that of the tropical Eastern Pacific; (2) Eastern Pacific species of crabs, gastropods, and perhaps other groups seem to be better adapted to predation involving skeletal destruction than are Caribbean forms; and (3) the canal barrier is likely to be less effective against Eastern Pacific species than against most Caribbean forms. Some Pacific-ward migration is to be expected among sea grasses and perhaps fishes and stomatopods; and the teleplanic dispersal of many reef-associated Western Atlantic animals may also permit easy colonization of the shores of tropical western America.

In the foregoing chapters I have tried to show that a study of adaptation can lead to hypotheses and conclusions about the organization of benthic marine communities of organisms. The morphology and behavior of individual species comprising a community can often give us hints about the type and intensity of predation and competition, not only in communities of living organisms, but also in fossil biotas. Moreoever, they give us insight into the long-term dynamics of biological processes, and the manner and speed of community change. Although much of this book has been devoted to a consideration of molluscs and their enemies, I believe that a similar approach will be fruitful for other groups, both in the sea and on land.

I leave the reader with many unsolved problems and unanswered questions. We need comparative data on the sources and intensities of predation on single species at various representative localities in the geographical and microgeographical distribution of that species. We need to know more about productivity in relation to competition and predation. Much more work is required on the role of chemical defenses and warning coloration in the biology of marine animals, and on geographical patterns in these traits. Our understanding of why some species perceive their environment as invariant while others see it as patchy and heterogeneous is still meager. Studies are needed on the relation between the stratigraphical longevity of species and their adaptive response to the environment. Perhaps most important of all, we need to understand more about how resources are allocated to competing functions in an organism, and to what extent the principle of allocation limits adaptive choice.

Most of these unresolved issues demand a greater insight into the properties and natural history of individual organisms and species. I believe that significant advances in our understanding of biological communities will not come from studies on abstract group measures such as diversity or community stability, but rather from work on the measurable effects and properties of the organisms that comprise living and fossil communities.

APPENDIX

REFERENCES

INDEX

Appendix

Atlantic-Pacific cognates among molluscs on opposite coasts of tropical America

The listing below has been compiled from various general works (Olsson, 1961; Radwin, 1969; Keen, 1971; Abbott, 1974) and from a number of monographs and articles on Fissurellidae (Farfante, 1943), Truncatellidae (Clench and Turner, 1948), Architectonicidae (Robertson, 1975), Cerithiidae (Houbrick, 1974b), *Crepidula* (Hoagland, 1977), Naticidae (Marincovich, 1977), Strombidae (Abbott, 1960), Cassidae (Abbott, 1968), Cymatiidae (Beu, 1970), Thaididae (Clench, 1947), Muricidae (Radwin and D'Attilio, 1976; G. E. Radwin, personal communication), Columbellidae (Radwin, 1977), Vexillidae (Maes and Raeigle, 1975), Olividae (Olsson, 1956; Burch and Burch, 1964; Zeigler and Porreca, 1969), Mitridae (Cernohorsky, 1976), Conidae (Old, 1975), Mytilidae (Soot-Ryen, 1955), Pinnidae (Turner and Rosewater, 1958), Tellinidae (Boss, 1966, 1968, 1969, 1972a), Semelidae (Boss, 1972b), Pholadidae (Turner, 1954, 1955), and Lucinidae (Bretsky, 1976).

Western Atlantic	Eastern Pacific
Scissurellidae	
Scissurella crispata	*S. crispata*
Haliotidae	
Haliotis pourtalesii	*H. dalli*
Fissurellidae	
Emarginula tuberculosa	*E. tuberculosa*
Lucapina limatula	*L. callomarginata*
Hemitoma octoradiata	*H. hermosa*
Diodora cayenensis	*D. inaequalis*
Fissurella nimbosa	*F. virescens*
F. angusta	*F. microtrema*
Acmaeidae	
Acmaea pustulata	*A. semirubida*
Trochidae	
Tegula hotessieriana	*T. hotessieriana* (= *maculostriata*)
T. viridula	*T. verrucosa*
T. fasciata	*T. cooksoni*
Cyclostrematidae	
Anticlimax schumoi	*A. willeti*
Turbinidae	
Turbo castanea	*T. squamifera*
Astraea caelata	*A. turbanica*
Arene tricarinata	*A. balboai*
Neritidae	
Nerita fulgurans	*N. funiculata*
Neritina virginea	*N. luteofasciata*
Littorinidae	
Littorina angulifera	*L. aberrans*
Truncatellidae	
Truncatella bilabiata	*T. bilabiata*
Alvaniidae	
Alvania auberiana	*A. inconspicua*
Caecidae	
Caecum pulchellum	*C. diminuta*
Rissoinidae	
Rissoina fischeri	*R. effusa*
Vermetidae	
Stephopoma myrakeenae	*S. pennatum*
Turritellidae	
Vermicularia fargoi	*V. frisbeyae*
Architectonicidae	
Architectonica peracuta	*A. placentalis*
A. nobilis	*A. nobilis*
Heliacus architae	*H. architae*
H. cylindricus	*H. bicanaliculatus*
H. bisulcatus	*H. mazatlanicus*
H. perrieri	*H. perrieri*

Western Atlantic	Eastern Pacific
Epitoniidae	
Sthenorytis pernobilis	*S. dianae*
Amaea retifera	*A. ferminiana*
Eulimacea	
Mucronalia nidorum	*M. nidorum*
Niso aeglees	*N. aeglees*
Hipponicidae	
Hipponix antiquatus	*H. antiquatus*
Capulidae	
Capulus ungaricus	*C. ungaricus*
Calyptraeidae	
Crucibulum auricula	*C. personatum*
Cheilea equestris	*C. equestris*
Crepidula aculeata	*C. aculeata*
C. marginalis	*C. marginalis*
C. maculosa	*C. excavata*
Modulidae	
Modulus modulus	*M. disculus*
M. carchedonius	*M. catenulatus*
Potamididae	
Cerithidea costata	*C. montagnei*
Batillaria minima	*B. mutata*
Planaxidae	
Planaxis nucleus	*P. obsoletus*
Cerithiidae	
Cerithium lutosum	*C. menkei*
C. atratum	*C. uncinatum*
Litiopa melanostoma	*L. melanostoma*
Xenophoridae	
Xenophora conchyliophora	*X. robusta*
Strombidae	
Strombus pugilis, alatus	*S. gracilior*
S. goliath	*S. galeatus*
Lamellariidae	
Lamellaria perspicua	*L. perspicua*
Eratoidae	
Erato maugeriae	*E. panamensis*
Ovulidae	
Cyphoma gibbosum,	*C. emarginatum*
intermedium	
Naticidae	
Natica cayennensis	*N. broderipiana*
N. marochiensis	*N. chemnitzii*
N. floridiana	*N. idiostoma*
Polinices hepaticus	*P. otis*
P. lacteus	*P. uber*
Sinum maculatum	*S. debile*

Western Atlantic	Eastern Pacific
Cassidae	
Cypraecassis testiculus	*C. tenuis*
Phalium granulatum	*P. centiquadratum*
Casmaria atlantica	*C. vibexmexicana*
Morum oniscus	*M. tuberculosum*
M. dennisoni	*M. veleroae*
Cymatiidae	
Cymatium femorale	*C. tigrinum*
C. pileare	*C. pileare*
C. cingulatum	*C. wiegmanni*
C. parthenopeum parthenopeum	*C. p. keenae*
C. krebsii	*C. lineatum*
C. muricinum	*C. muricinum*
Distorsio constricta macgintyi	*D. c. constricta*
Bursidae	
Bursa corrugata	*B. corrugata*
B. granularis	*B. granularis*
Ficidae	
Ficus communis	*F. ventricosus*
Thaididae	
Thais haemastoma	*T. haemastoma* (= *biserialis*)
T. trinitatensis	*T. kiosquiformis*
Purpura patula patula	*P. p. pansa*
Coralliophilidae	
Coralliophila caribaea	*C. macleani*
Latiaxis dalli	*L. hindsii*
Muricidae	
Attiliosa philippiana	*A. incompta*
Aspella castor	*A. pollux*
Calotrophon ostrearum	*C. turrita*
Dermomurex elizabethae	*D. bakeri*
D. abyssicola	*D. cunninghamae*
Murex donmoorei	*M. recurvirostris*
Phyllonotus margaritensis	*P. regius*
P. pomum	*P. peratus*
Pteropurpura bequaerti	*P. centrifuga*
Eupleura caudata	*E. triquetra*
Evokesia grayi	*E. rufopunctata*
Trachypollia didyma	*T. lugubris*
Favartia alveata	*F. erosa*
Murexiella hidalgoi	*M. radwini*
M. levicula	*M. perita*
M. macgintyi	*M. humilis*
Muricopsis oxytata	*M. armata*
M. muricoides	*P. paxillus*

Western Atlantic	Eastern Pacific
Muricidae *(continued)*	
Pterotyphis pinnatus	*P. fimbriatus*
Talityphis perchardei	*T. latipennis*
Tripterotyphis triangularis	*T. lowei*
Columbellidae	
Anachis lyrata	*A. lyrata*
Mazatlania aciculata	*M. aciculata*
M. fulgurata	*M. fulgurata*
Alia unifasciata	*A. unifasciata*
Steironepion monilifera	*S. melanosticta*
Aesopus stearnsi	*A. sanctus*
Zafrona pulchella	*Z. incerta*
Mitrella ocellata	*M. ocellata*
Colubrariidae	
Colubraria testacea	*C. lucasensis*
C. lanceolata	*C. siphonata*
Buccinidae	
Bailya parva	*B. anomala*
Cantharus multangulus	*C. panamicus*
Melongenidae	
Melongena melongena	*M. patula*
Nassariidae	
Nassarius guadelupensis	*N. myristicatus*
Vexillidae	
Thala foveata	*T. gratiosa*
Fasciolariidae	
Pleuroploca gigantea	*P. princeps*
Latirus distinctus	*L. centrifugus*
Leucozonia nassa	*L. knorrii*
Turbinellidae	
Vasum muricatum muricatum	*V. m. coestus*
Cancellariidae	
Trigonostoma tenerum	*T. bullatum*
Olividae	
Oliva reticularis	*O. spicata*
Olivella floralia	*O. drangai*
O. nivea	*O. gracilis*
Agaronia testacea, travassosi	*A. testacea*
Mitridae	
Mitra barbadensis	*M. effusa*
M. nodulosa	*M. inca*
M. swainsonii antillensis	*M. s. swainsonii*
Volutidae	
Lyria guildingi	*L. barnesii*
Marginellidae	
Prunum prunum	*P. sapotilla*

Western Atlantic	Eastern Pacific
Marginellidae *(continued)*	
P. apicinum	*P. woodbridgei*
Dentimargo aureocinctus	*D. eremus*
Persicula porcellana	*P. accola*
P. multilineata	*P. bandera*
P. muralis	*P. imbricata*
P. swainsoniana	*P. phrygia*
Volvarina avena	*V. taeniolata*
Cystiscus bocasensis	*C. politulus*
Granula lavalleeana	*G. polita*
Granulina ovuliformis	*G. margaritula*
Conidae	
Conus testudinarius	*C. purpurascens*
C. puncticulatus (= minimus)	*C. perplexus*
C. mus	*C. gladiator*
C. regius	*C. brunneus*
C. beddomei	*C. orion*
C. daucus	*C. virgatus*
C. clerii	*C. regularis*
C. floridanus	*C. regularis monilifer*
C. austini	*C. arcuatus*
C. villepini	*C. poormani*
C. centurio	*C. recurvus*
Terebridae	
Terebra dislocata	*T. dislocata*
T. taurina	*T. ornata*
Hastula cinerea cinerea	*H. c. fluctuosa*
Turridae	
Drillia gibbosa	*D. walteri*
Crassispira dysoni	*C. rudis*
Strictispira ebenina	*S. stillmani*
Opisthobranchia	
Bulla striata	*B. punctulata*
Volvulella paupercula	*V. catharia*
V. acuminata, persimilis	*V. cylindrica*
V. texasiana	*V. panamica*
Pleurobranchus areolatus	*P. areolatus*
Berthellinia quadridens	*B. quadridens*
Oxynoe antillarum	*O. panamensis*
Lobiger souverbii	*L. souverbii*
Melampidae	
Melampus coffeus	*M. carolianus*
Polyplacophora	
Ischnochitonidae	
Stenoplax limaciformis	*S. limaciformis*
Radsiella rugulata	*R. rugulata*

Western Atlantic	Eastern Pacific
Chitonidae	
Chiton tuberculatus	*C. virgulatus, stokesi*
C. marmoratus	*C. laevigatus*
Scaphopoda	
Siphonodentaliidae	
Cadulus tetraschistus	*C. quadriscissatus*
Bivalvia	
Nuculanidae	
Nuculana acuta	*N. hindsii*
Nucinella adamsi	*N. subdolus*
Arcidae	
Arca zebra	*A. pacifica*
A. imbricata	*A. mutabilis*
Barbatia candida	*B. reeveana*
B. tenera	*B. illota*
B. reticulata	*B. gradata*
Anadara brasiliana	*A. bifrons*
A. chemnitzi	*A. nux*
A. notabilis	*A. biangulata*
Arcopsis adamsi	*A. solida*
Noetia olssoni	*N. centrota*
Glycymeridae	
Glycymeris lintea	*G. spectralis*
Mytilidae	
Mytella strigata	*M. strigata*
M. guyanensis	*M. guyanensis*
Modiolus americanus	*M. americanus*
Amygdalum sagittatum	*A. pallidulum*
Gregariella chenui	*G. chenui*
G. divaricata	*G. ecuadoriana*
Crenella divaricata	*C. inflata*
Lithophaga aristata	*L. aristata*
Botula fusca	*B. cylista*
Pinnidae	
Atrina seminuda	*A. maura*
A. rigida	*A. tuberculosa*
Isognomonidae	
Isognomon bicolor	*I. recognitus*
I. alatus	*I. chemnitzianus*
Pteriidae	
Pteria colymbus	*P. sterna*
Malleus candeanus	*M. rufipunctatus*
Pinctada radiata	*P. mazatlanica*
Pectinidae	
Argopecten gibbus, nucleus	*A. circularis*
Lyropecten nodosus	*L. subnodosus*

Western Atlantic	Eastern Pacific
Plicatulidae	
Plicatula gibbosa	*P. spondylopsis*
Spondylidae	
Spondylus americanus	*S. princeps*
Limidae	
Lima lima	*L. tetrica*
Ostreidae	
Ostrea frons	*O. serra*
Anomiidae	
Pododesmus rudis	*P. cepio*
Crassatellidae	
Eucrassatella antillarum	*E. digueti*
Crassinella adamsi	*C. adamsi*
Dreissensiidae	
Mytilopsis leucophaetus	*M. zeteki*
Condylocardiidae	
Condylocardia panamensis	*C. panamensis*
Lucinidae	
Lucina amianta	*L. cancellaris*
L. muricata	*L. liana*
L. trisulcata	*L. lingualis*
Divaricella quadrisulcata	*D. eburnea*
Codakia orbicularis	*C. distinguenda*
Ctena orbiculata	*C. mexicana*
Anodontia philippiana	*A. edentuloides*
Ungulinidae	
Diplodonta semiaspera	*D. semiaspera* (= *discrepans*)
Chamidae	
Arcinella arcinella arcinella	*A. a. californica*
Erycinacea	
Lasaea adansoni	*L. adansoni*
Kellia suborbicularis	*K. suborbicularis*
Ensitelops protexta	*E. pacifica*
Cardiidae	
Trachycardium muricatum	*T. senticosum*
T. isocardia, egmontianum	*T. discors*
Papyridea soleniformis	*P. aspersa*
Trigoniocardia media	*T. guanacastensis*
Laevicardium laevigatum	*L. clarionense*
Veneridae	
Pitar circinatus	*P. alternatus, vinaceus*
P. dione	*P. lupanarius* (= *multispinosa*)
Gouldia cerina	*G. californica*
Cyclinella tenuis	*C. singleyi*
Tivela mactroides	*T. byronensis*
Dosinia discus	*D. ponderosa*

Western Atlantic	Eastern Pacific
Veneridae *(continued)*	
Chione mazyckii	C. guatulcoensis
C. subrostrata	C. crenifera
C. pubera	C. purpurissata
C. obliterata	C. obliterata
C. grus	C. squamosa
C. paphia	C. mariae
Protothaca pectorina	P. asperrima
Ventricolaria rigida rigida	V. r. isocardia
V. strigillina	V. magdalenae
Periglypta listeri	P. multicostata
Petricolidae	
Petricola pholadiformis	P. parallela
Cooperella atlantica	C. subdiaphana
Rupellaria typica	R. robusta
Mactridae	
Mactra brasiliana	N. nasuta
Mactrellona alata	M. alata
Labiosa lineata	L. anatina
Raeta plicatella	R. undulata
Tellinidae	
Tellina listeri	T. cumingi
T. laevigata	T. ochracea
T. squamifera	T. pristiphora
T. persica	T. fluctigera
T. americana	T. pacifica
T. cristallina	T. rhynchoscuta
T. aequistriata	T. reclusa
T. martinicensis	T. ulloana
T. juttingae	T. lyra
T. angulosa	T. eburnea
T. alternata	T. laceridens
T. punicea	T. simulans
T. nitens	T. inaequistriata
T. mera	T. meropsis
T. tampaensis	T. suffusa
T. exerythra	T. subtrigona
T. sybaritica	T. amianta
T. gibber	T. hiberna
T. iris	T. virgo
T. sandix	T. esmeralda
Strigilla pseudocarnaria	S. chroma
S. pisiformis	S. interrupta
S. mirabilis	S. lenticula
S. gabbi	S. disjuncta
S. producta	S. ervilia

Western Atlantic	Eastern Pacific
Tellinidae *(continued)*	
S. surinamensis	*S. strata*
Tellidora cristata	*T. burnetti*
Macoma extenuata	*M. siliqua*
Psammotreta brevifrons	*P. aurora*
Donacidae	
Donax denticulatus	*D. punctatostriatus*
D. vellicatus	*D. contusus*
D. cayennensis	*D. panamensis*
D. striatus	*D. carinatus*
Iphigenia brasiliana	*I. altior*
Sanguinolariidae	
Heterodonax bimaculatus bimaculatus	*H. b. pacificus*
Sanguinolaria cruenta	*S. bertini*
S. sanguinolenta	*S. vespertina*
Tagelus plebeius	*T. affinis*
Semelidae	
Semele nuculoides	*S. subquadrata*
S. proficua	*S. lenticulare*
S. purpurascens	*S. sparsilineata*
S. bellastriata	*S. pacifica*
Cumingia coarctata	*C. lamellosa*
Abra lioica	*A. tepocana*
Solenidae	
Ensis minor	*E. californicus*
Solen obliquus	*S. rudis*
Corbulidae	
Corbula knoxiana	*C. elenensis*
Gastrochaenidae	
Gastrochaena ovata	*G. ovata*
Pholadidae	
Martesia fragilis	*M. fragilis*
M. striata	*M. striata*
Pholas campechiensis	*P. chiloensis*
Barnea truncata	*B. subtruncata*
Diplothyra smithii	*D. curta*
Verticordiidae	
Verticordia ornata	*V. ornata*
Thraciidae	
Cyathodonta magnifica	*C. undulata*
Cuspidariidae	
Cardiomya costata	*C. costata*
Bathyarca orbiculata	*B. orbiculata*

References

Abbott, D. P. 1966. Factors influencing the zoogeographical affinities of Galapagos inshore marine fauna. In *The Galapagos*, ed. R. I. Bowman. Berkeley: University of California Press, pp. 108–122.

Abbott, R. T. 1959. The family Vasidae in the Indo-Pacific. *Indo-Pacific Mollusca* 1 (1):403–472.

———. 1960. The genus *Strombus* in the Indo-Pacific. *Indo-Pacific Mollusca* 1 (2):831–999.

———. 1961. The genus *Lambis* in the Indo-Pacific. *Indo-Pacific Mollusca* 1 (3):51–88.

———. 1968. The helmet shells of the world (Cassidae). *Indo-Pacific Mollusca* 2 (9):2–202.

———. 1974. *American seashells*. 2nd ed. New York: Van Nostrand Reinhold.

Abbott, R. T., and R. Jensen. 1967. Molluscan faunal changes around Bermuda. *Science* 155:687–688.

Abe, N. 1941. Ecological observations on a limpet-like pulmonate, *Siphonaria atra* Quoy and Gaimard. *Palau Trop. Biol. Sta. Stud.* 2:239–278.

Abele, L. G. 1972. Comparative habitat diversity and faunal relationships between the Pacific and Caribbean Panamanian decapod Crustacea: a preliminary report, with some remarks on the crustacean fauna of Panama. *Bull. Biol. Soc. Washington* 2:125–138.

———. 1974. Species diversity of decapod crustaceans in marine habitats. *Ecology* 55:156–161.

Abrahamson, W. G., and M. Gadgil. 1973. Growth form and reproductive effort in goldenrods (*Solidago*, Compositae). *Amer. Nat.* 107:651–661.

Achituv, Y. 1972. The zonation of *Tetrachthamalus oblitteratus* Newman, and *Tetraclita squamosa rufotincta* Pilsbry in the Gulf of Eilat, Red Sea. *J. Exp. Mar. Biol. Ecol.* 8:73–82.

———. 1973. On the distribution and variability of the Indo-Pacific sea star *Asterina wega* (Echinodermata: Asteroidea) in the Mediterranean Sea. *Mar. Biol.* 18:333–336.

Adams, C. G., R. H. Benson, R. B. Kidd, W. B. F. Ryan, and R. C. Wright. 1977. The Messinian salinity crisis and evidence of Late Miocene eustatic changes in the world ocean. *Nature* 269:383–386.

Addicott, J. F. 1974. Predation and prey community structure: an experimental study of the effect of mosquito larvae on the protozoan communities of pitcher plants. *Ecology* 55:475–492.

Adegoke, O. S. 1973. Mineralogy and biogeochemistry of calcareous operculi and shells of some gastropods. *Malacologia* 14:39–46.

Alexander, R. M. 1967. *Functional design in fishes*. London: Hutchinson and Company.

Allan, J. D. 1976. Life history patterns in zooplankton. *Amer. Nat.* 110:165–180.

Allan, J. D., L. W. Barnthouse, R. A. Prestbye, and D. R. Strong. 1973. On foliage arthropod communities of Puerto Rican second-growth vegetation. *Ecology* 54:628–632.

Allen, J. A. 1958. On the basic form and adaptations to habitat in the Lucinacea (Lamellibranchia). *Phil. Trans. Roy. Soc. London* (B) 241:421–484.

Ameyaw-Akumfi, C. 1975. The feeding biology of two intertidal hermit crabs, *Clibanarius chapini* Schmitt and *C. senegalensis* Chevreux and Bouvier. *Ghana J. Sci.* 15:29–38.

Andrews, J. D. 1973. Effects of Tropical Storm Agnes on epifaunal invertebrates in Virginia estuaries. *Chesapeake Sci.* 14:223–234.

Ansell, A. D. 1960. Observations on predation of *Venus striatula* (da Costa) by *Natica alderi* (Forbes). *Proc. Malacol. Soc. London* 34:157–164.

———. 1969. Defensive adaptations to predation in the Mollusca. *Proc. Symp. Mollusca* 2:487–512.

Ansell, A. D., and A. Trevallion. 1969. Behavioural adaptations of intertidal molluscs from a tropical sandy beach. *J. Exp. Mar. Biol. Ecol.* 4:9–35.

Ansell, A. D., P. Sivadas, B. Narayanan, and A. Trevallion. 1972. The ecology of two sandy beaches in South West India. II. Notes on *Emerita holthuisi*. *Mar. Biol.* 17:311–317.

Arnaud, P. M. 1974. Contribution à la bionomie marine benthique des régions antarctiques et subantarctiques. *Téthys* 6:567–653.

Arnaud, P. M., and J. C. Hureau. 1966. Régime alimentaire de trois teleosteens Nototheniidae antarctiques (Terre Adelie). *Bull. Inst. Océanogr. Monaco* 66 (1368):3–24.

Arnaud, P. M., and B. A. Thomassin. 1976. First records and adaptive significance of boring into a free-living scleractinian coral (*Heteropsammia michelini*) by a date mussel (*Lithophaga lessepsiana*). *Veliger* 18:367–374.

Arnold, J. M., and K. O. Arnold. 1969. Some aspects of hole-boring predation by *Octopus vulgaris*. *Amer. Zool.* 9:991–996.

Atapattu, D. H. 1972. The distribution of molluscs on littoral rocks in Ceylon, with notes on their ecology. *Mar. Biol.* 16:150–164.

Atsatt, P. R., and D. J. O'Dowd. 1976. Plant defensive guilds. *Science* 193:24–29.

Ayala, F. J., D. Hedgecock, G. S. Zumwalt, and J. W. Valentine. 1973. Genetic variation in *Tridacna maxima*, an ecological analog of some unsuccessful evolutionary lineages. *Evolution* 27:177–191.

Ayala, F. J., J. W. Valentine, T. E. Delaca, and G. S. Zumwalt. 1975a. Genetic variability of the Antarctic brachiopod *Liothyrella notorcadensis* and its bearing on mass extinction hypotheses. *J. Paleont.* 49:1–9.

Ayala, F. J., J. W. Valentine, D. Hedgecock, and L. J. Barr. 1975b. Deep-sea asteroids: high genetic variability in a stable environment. *Evolution* 29:203–212.

Babbel, G. R., and R. K. Selander. 1974. Genetic variability in edaphically restricted and widespread plant species. *Evolution* 28:619–630.

Bak, R. P. M., and G. van Eys. 1975. Predation of the sea urchin *Diadema antillarum* Philippi on living coral. *Oecologia (Berlin)* 20:111–115.

Baker, A. J. 1974. Prey-specific feeding methods of New Zealand oystercatchers. *Notornis* 21:219–233.

Bakker, R. T. 1971. Dinosaur physiology and the origin of mammals. *Evolution* 25:636–658.

———. 1975. Experimental and fossil evidence for the evolution of tetrapod bioenergetics. In *Perspectives of biophysical ecology*, ed. D. Gates and R. Schmerl. New York: Springer Verlag, pp. 365–399.

Bakus, G. J. 1964. The effects of fish-grazing on invertebrate evolution in shallow tropical waters. *Allan Hancock Found. Publ. (Los Angeles)* 27:1–29.

———. 1966. Some relationships of fishes to benthic organisms on coral reefs. *Nature* 210:280–284.

———. 1967. The feeding habits of fishes and primary productivity of Eniwetok, Marshall Islands. *Micronesica* 3:135–149.

———. 1969. Energetics and feeding in shallow marine waters. *Intern. Rev. Gen. Exp. Zool.* 4:275–369.

———. 1974. Toxicity in holothurians: a geographical pattern. *Biotropica* 6:229–236.

———. 1975. Marine zonation and ecology of Cocos Island, off Central America. *Atoll Res. Bull.* 179:1–9.

Bakus, G. J., and G. Green. 1974. Toxicity in sponges and holothurians. *Science* 185:951–953.

Bandel, K. 1974. Studies on Littorinidae from the Atlantic. *Veliger* 17:92–114.

Barash, A., and Y. Danin. 1971. Mollusca from the stomach of *Sparus auratus* fished in the lagoon of Bardawil. *Argamon, Israel J. Malacol.* 2:97–104.

———. 1972. The Indo-Pacific species of Mollusca in the Mediterranean and notes on a collection from the Suez Canal. *Israel J. Zool.* 21:301–374.

Barlow, G. W. 1961. Causes and significance of morphological variation in fishes. *Syst. Zool.* 10:105–117.

Bartonek, J. C., and J. J. Hickey. 1969. Selective feeding by juvenile diving ducks in summer. *Auk* 86:443–457.

Bastida, R., A. Capezzani, and M. R. Torti. 1971. Fouling organisms in the port of Mar del Plata (Argentina). I. *Siphonaria lessoni*: ecological and biometric aspects. *Mar. Biol.* 10:297–307.

Battaglia, B. 1975. Fouling communities and genetic diversification in the marine environment. In *Ecology of fouling communities*, ed. J. D. Costlow. Washington, D. C.: U.S. Government Printing Office, pp. 69–84.

Bennett, I., and E. Pope. 1960. Intertidal zonation of the exposed rocky shores of Tasmania and its relationship with the rest of Australia. *Austr. J. Mar. Fresh-W. Res.* 11:182–219.

Benson, W. B., K. S. Brown, Jr., and L. E. Gilbert. 1975. Coevolution of plants and herbivores: passion flower butterflies. *Evolution* 29:659–680.

Ben-Tuvia, A. 1971a. On the occurrence of the Mediterranean serranid fish *Dicentrarchus punctatus* (Bloch) in the Gulf of Suez. *Copeia* 4:741–743.

———. 1971b. Revised list of the Mediterranean fishes of Israel. *Israel J. Zool.* 20:1–39.

Ben-Yami, M., and T. Glaser. 1974. The invasion of *Saurida undosquamis* (Richardson) into the Levant Basin—an example of biologic effect of interoceanic canals. *Fish. Bull.* 72:359–373.

Bequaert, J. 1943. The genus *Littorina* in the Western Atlantic. *Johnsonia* 1 (7):1–28.

Berger, E. 1977. Gene-enzyme variation in three sympatric species of *Littorina*. II. Roscoff population, with a note on the origin of North American *L. littorea*. *Biol. Bull.* 153:255–264.

Beu, A. G. 1970. Mollusca of the subgenus *Monoplex* (family Cymatiidae). *Trans. Roy. Soc. N.Z., Biol. Sciences* 11:225–237.

Beurois, J. 1975. Étude écologique et halieutique des fondes de pêche et des espèces d'intérêt commerciel (langoustes et poissons) des îles Saint-Paul et Amsterdam (Océan Indien). *Com. Nat. Franc. Res. Ant.* 37:1–91.

Bigelow, H. B., and W. C. Schroeder. 1953. Fishes of the Western North Atlantic. Part II. Sawfishes, guitarfishes, skates and rays. *Mem. Sears Foundation Mar. Res.* 1:1–585.

Bingham, F. O. 1972. The mucus holdfast of *Littorina irrorata* and its relationship to relative humidity and salinity. *Veliger* 15:48–50.

Birkeland, C. 1974. Interactions between a sea pen and seven of its predators. *Ecol. Monogrs.* 44:211–232.

———. 1977. The importance of rate of biomass accumulation in early successional stages of benthic communities to the survival of coral recruits. *Proc. Third Int. Coral Reef Symp.* 1:15–21.

Birkeland, C., and B. Gregory. 1975. Foraging behavior and rates of feeding of the gastropod, *Cyphoma gibbosum* (Linnaeus). *Nat. Hist. Mus. Los Angeles County Sci. Bull.* 20:57–67.

Birkeland, C., D. L. Meyer, J. Stames, and C. L. Burford. 1975. Subtidal communities of Malpelo Island. *Smithson. Contribs. Zool.* 176:55–68.

Black, R. 1976. The effects of grazing by the limpet, *Acmaea insessa*, on the kelp, *Egregia laevigata*, in the intertidal zone. *Ecology* 57:265–277.

Bloom, S. A. 1975. The motile escape response of a sessile prey: a sponge-scallop mutualism. *J. Exp. Mar. Biol. Ecol.* 17:311–321.

———. 1976. Morphological correlations between dorid nudibranch predators and sponge prey. *Veliger* 18:289–301.

Borland, C. 1950. Ecological study of *Benhamina obliquata* (Sowerby), a basommatophorous pulmonate in Otago Harbour. *Trans. Roy. Soc. N.Z.* 78:385–393.

Boschi, E. E. 1964. Los crustáceos decápodos Brachyura del litoral Bonaerense (R. Argentina). *Bol. Inst. Biol. Mar. (Mar del Plata)* 6:1–100.

Boss, K. J. 1966. The subfamily Tellininae in the Western Atlantic. The genus *Tellina* (I). *Johnsonia* 4:217–272.

———. 1968. The subfamily Tellininae in the Western Atlantic. The genera *Tellina* (II) and *Tellidora. Johnsonia* 4:273–344.

———. 1969. The subfamily Tellininae in the Western Atlantic. The genus *Strigilla. Johnsonia* 4:345–368.

———. 1971. Critical estimate of the number of Recent Mollusca. *Occ. Papers on Mollusks, Mus. Comp. Zool.* 3 (40):81–135.

———. 1972a. *Simplistrigilla* in the Western Atlantic Ocean (Mollusca, Bivalvia). *Zool. Meded.* 46:25–28.

———. 1972b. The genus *Semele* in the Western Atlantic (Semelidae; Bivalvia). *Johnsonia* 5:1–32.

Bouck, G. B., and E. Morgan. 1957. The occurrence of *Codium* in Long Island waters. *Bull. Torrey Bot. Club* 84:384–387.

Boucot, A. J. 1975. *Evolution and extinction rate controls.* Amsterdam: Elsevier.

Braber, L., and S. J. de Groot. 1973. The food of five flatfish species (Pleuronectiformes) in the southern North Sea. *Netherlands J. Sea Res.* 6:163–172.

Branch, G. M. 1971. The ecology of *Patella* Linnaeus from the Cape Peninsula, South Africa. I. Zonation, movements and feeding. *Zool. Afr.* 6:1–38.

———. 1974. The ecology of *Patella* Linnaeus from the Cape Peninsula, South Africa. III. Growth-rates. *Trans. Roy. Soc. S. Afr.* 41:161–193.

———. 1975a. Ecology of *Patella* Linnaeus from the Cape Peninsula, South Africa. IV. Desiccation. *Mar. Biol.* 32:179–188.

———. 1975b. Intraspecific competition in *Patella cochlear* Born. *J. Anim. Ecol.* 44:263–282.

———. 1976. Interspecific competition experienced by South African *Patella* species. *J. Anim. Ecol.* 45:507–529.

Brasier, M. D. 1975. An outline history of seagrass communities. *Palaeontology* 18:681–702.

Breen, P. A. 1972. Seasonal migration and population regulation in the limpet *Acmaea digitalis. Veliger* 15:133–141.

Brenchley, G. A. 1976. Predator detection and avoidance: ornamentation of tubecaps of *Diopatra* spp. (Polychaeta: Onuphidae). *Mar. Biol.* 38:179–188.

Bretsky, P. W. 1968. Evolution of Paleozoic marine invertebrate communities. *Science* 159:123–133.

Bretsky, P. W., and S. S. Bretsky. 1975. Succession and repetition of Late Ordovician fossil assemblages from the Nicolet River Valley, Quebec. *Paleobiology* 1:225–237.

Bretsky, P. W., and D. M. Lorenz. 1970. An essay on genetic-adaptive strategies and mass extinctions. *Geol. Soc. Amer. Bull.* 81:2449–2456.

Bretsky, S. S. 1976. Evolution and classification of the Lucinidae (Mollusca; Bivalvia). *Palaeontographica Americana* 8 (50):219–337.

Briggs, J. C. 1966. Zoogeography and evolution. *Evolution* 20:282–289.

———. 1967a. Dispersal of tropical marine shore animals: coriolis parameters or competition? *Nature* 216:350.

———. 1967b. Relationship of the tropical shelf regions. *Stud. Trop. Oceanogr. Univ. Miami* 5:569–578.

————. 1968. Panama sea-level canal. *Science* 162:511–513.

————. 1970. A faunal history of the North Atlantic. *Syst. Zool.* 19:19–34.

————. 1974a. *Marine zoogeography*. New York: McGraw-Hill.

————. 1974b. Operation of zoogeographic barriers. *Syst. Zool.* 23:248–256.

Bright, T. J. 1970. Food of deep-sea bottom fishes. *Texas A & M Univ. Oceanogr. Stud.* 1:245–252.

Brighton, F., and M. T. Horne. 1977. Influence of temperature on cyanogenic polymorphisms. *Nature* 265:437–438.

Bromley, R. G. 1975. Comparative analysis of fossil and Recent echinoid bioerosion. *Palaeontology* 18:725–739.

Brooks, J. L. 1950. Speciation in ancient lakes. *Quart. Rev. Biol.* 25:30–60, 137–176.

Brooks, J. L., and S. I. Dodson. 1965. Predation, body size, and composition of plankton. *Science* 150:28–34.

Brown, J. H., and A. K. Lee. 1969. Bergmann's rule and climatic adaptation in woodrats (*Neotoma*). *Evolution* 23:329–338.

Brown, W. L., Jr. 1957. Centrifugal speciation. *Quart. Rev. Biol.* 32:247–277.

Bruce, A. J. 1976. Coral reef Caridea and "commensalism." *Micronesica* 12:83–98.

Bruun, A. F. 1956. The abyssal fauna: its ecology, distribution and origin. *Nature* 177:105–108.

Bullock, T. H. 1955. Compensation for temperature in the metabolism and activity of poikilotherms. *Biol. Revs.* 30:311–342.

Burch, J. Q., and R. L. Burch. 1964. The genus *Agaronia* J. E. Gray, 1839. *Nautilus* 77:110–112.

Burrows, M. 1969. The mechanics and neural control of the prey capture strike in the mantid shrimps *Squilla* and *Hemisquilla*. *Z. Vergl. Psychol.* 62:361–381.

Burton, P. J. K. 1974. *Feeding and feeding apparatus in waders: a study of anatomy and adaptations in the Charadrii*. London: British Museum of Natural History.

Caine, E. A. 1975. Feeding and masticatory structures of selected Anomura (Crustacea). *J. Exp. Mar. Biol. Ecol.* 18:277–301.

Caldwell, R. L., and H. Dingle. 1975. Ecology and evolution of agonistic behavior in stomatopods. *Naturwissenschaften* 62:214–222.

————. 1977. Variation in agonistic behavior between populations of the stomatopod, *Haptosquilla glyptocercus*. *Evolution* 31:220–223.

Canton, M., J. Bedard, and H. Milne. 1974. The food and feeding of common eiders in the St. Lawrence estuary in summer. *Canad. J. Zool.* 52:319–334.

Capapé, C. 1976. Étude du régime alimentaire de l'aigle de mer, *Myliobatis aquila* (L., 1758) des côtes Tunésiens. *J. Cons. Int. Explor. R. Mer* 37:29–35.

Carriker, M. R. 1951. Observations on the penetration of tightly closing bivalves by *Busycon* and other predators. *Ecology* 32:73–83.

————. 1957. Preliminary study of behavior of newly hatched oyster drills, *Urosalpinx cinerea* (Say). *J. Elisha Mitchell Sci. Soc.* 73:328–351.

Carriker, M. R., and D. Van Zandt. 1972. Predatory behavior of a shell-boring muricid gastropod. In *Behavior of marine animals*, ed. H. E. Winn and B. L. Olla. New York: Plenum Press, vol. 1, pp. 157–244.

Carriker, M. R., and E. L. Yochelson. 1968. Recent gastropod boreholes and Ordovician cylindrical borings. *U.S. Geol. Surv. Prof. Paper* 593B:B1–B23.

Carson, H. L. 1975. The genetics of speciation at the diploid level. *Amer. Nat.* 109:83–92.

Carson, H. L., D. E. Hardy, H. T. Spieth, and W. S. Stone. 1970. The evolutionary biology of the Hawaiian Drosophilidae. In *Essays in evolution and genetics in honor of Theodosius Dobzhansky*, ed. M. K. Hecht and W. C. Steere. New York: Appleton-Century Crofts, pp. 437–543.

Carter, R. M. 1968. On the biology and palaeontology of some predators of bivalved Mollusca. *Paleogeogr., Paleoclimatol., Paleoecol.* 4:29–65.

Castenholz, R. W. 1961. The effect of grazing on marine littoral diatom populations. *Ecology* 42:783–794.

———. 1967. Stability and stresses in intertidal populations. In *Pollution and marine ecology*, ed. T. A. Olson and F. J. Burgess. New York: Wiley, pp. 15–28.

Castro, P. 1976. Brachyuran crabs symbiotic with scleractinian corals: a review of their biology. *Micronesica* 12:99–110.

Cates, R. G. 1975. The interface between slugs and wild ginger. *Ecology* 56:391–400.

Cates, R. G., and G. H. Orians. 1975. Successional status and the palatability of plants to generalized herbivores. *Ecology* 56:410–418.

Cerame-Vivas, M. J., and J. E. Gray. 1966. The distributional pattern of marine invertebrates of the continental shelf off North Carolina. *Ecology* 47:260–269.

Cernohorsky, W. O. 1976. The Mitridae of the world. I. The subfamily Mitrinae. *Indo-Pacific Mollusca* 3 (17):273–528.

Chapin, D. 1968. Some observations of predation on *Acmaea* species by the crab *Pachygrapsus crassipes*. *Veliger* 11 (suppl.):67–68.

Chapman, C. R. 1955. Feeding habits of the southern oyster drill, *Thais haemastoma*. *Proc. Nat. Shell Fish Assoc.* 46:169–176.

Chesher, R. H. 1969. Destruction of Pacific corals by the sea star *Acanthaster planci*. *Science* 165:280–283.

———. 1972. The status of knowledge of Panamanian echinoids, 1971, with comments on other echinoderms. *Bull. Biol. Soc. Washington* 2:139–158.

Christensen, A. M. 1970. Feeding biology of the sea-star *Astropecten irregularis* Pennant. *Ophelia* 8:1–134.

Christiansen, B. O. 1965. Notes of the littoral fauna of Bear Island. *Astarte* 26:1–15.

Christiansen, M. E. 1969. Crustacea Decapoda Brachyura. *Marine Invertebrates of Scandinavia*, No. 2. Oslo: University Press.

Christie, W. J. 1974. Changes in the fish species composition of the Great Lakes. *J. Fish. Res. Bd. Canada* 31:827–854.

Cifelli, R. 1976. Evolution of ocean climate and the record of planktonic Foraminifera. *Nature* 264:431–432.

Clarke, A. H., Jr. 1962. On the composition, zoogeography, origin and age of the deep-sea mollusk fauna. *Deep Sea Res.* 9:291–306.

———. 1963. Supplementary notes on pre-Columbian *Littorina littorea* in Nova Scotia. *Nautilus* 77:8–11.

———. 1969. Some aspects of adaptive radiation in Recent freshwater molluscs. *Malacologia* 9:263.

Clench, W. J. 1947. The genera *Purpura* and *Thais* in the Western Atlantic. *Johnsonia* 2 (23):61–91.

Clench, W. J., and R. D. Turner. 1948. The genus *Truncatella* in the Western Atlantic. *Johnsonia* 2 (25):149–164.

CLIMAP project members. 1976. The surface of the ice-age earth. *Science* 191:1131–1137.

Cody, M. L. 1968. On the methods of resource division in grassland bird communities. *Amer. Nat.* 102:107–147.

———. 1970. Chilean bird distribution. *Ecology* 51:455–464.

Coe, W. R. 1946. A resurgent population of the California Bay–Mussel, *Mytilus edulis diegensis*. *J. Morphol.* 78:85–105.

Coe, W. R., and D. L. Fox. 1942. Biology of the California Sea–Mussel (*Mytilus californianus*). I. Influence of temperature, food supply, sex and age on the rate of growth. *J. Exp. Zool.* 90:1–30.

Cole, T. J. 1975. Inheritance of juvenile shell colour of the oyster drill *Urosalpinx cinerea*. *Nature* 257:794–795.

Colman, J. S., and A. Stephenson. 1966. Aspects of the ecology of a tideless shore. In *Some contemporary studies in marine science*, ed. H. Barnes. London: Allen and Unwin, pp. 163–170.

Connell, J. H. 1961a. Effects of competition, predation by *Thais lapillus*, and other factors on natural populations of the barnacle *Balanus balanoides*. *Ecol. Monogrs.* 31:61–104.

———. 1961b. The influence of interspecific competition and other factors on the distribution of the barnacle *Chthamalus stellatus*. *Ecology* 42:710–723.

———. 1970. A predator-prey system in the rocky intertidal region. I. *Balanus glandula* and several predatory species of *Thais*. *Ecol. Monogrs.* 40:49–78.

———. 1972. On the role of natural enemies in preventing competitive exclusion in some marine animals and in rain forest trees. *Proc. Adv. Stud. Inst. Dynamics Numbers Populations* (Oosterbeek 1970):298–312.

———. 1975. Some mechanisms producing structure in natural communities: a model and evidence from field experiments. In *Ecology and evolution of communities*, ed. M. L. Cody and J. M. Diamond. Cambridge, Mass.: Belknap Press of Harvard University Press, pp. 460–490.

Connor, M. S. 1975. Niche apportionment among the chitons *Cyanoplax hartwegii* and *Mopalia muscosa* and the limpets *Collisella limatula* and *Collisella pelta* under the brown alga *Pelvetia fastigiata*. *Veliger* 18 (suppl.):9–17.

Cook, R. E. 1969. Variation in species density of North American birds. *Syst. Zool.* 18:63–84.

Coomans, H. E., and P. E. Clover. 1972. The genus *Rivomarginella* (Gastropoda, Marginellidae). *Beaufortia* 20:69–75.

Copper, P. 1974. Structure and development of early Paleozoic reefs. *Proc. Second Intern. Coral Reef Symp.* 1:365–386.

Corbet, P. S. 1961. The food of non-cichlid fishes in the Lake Victoria basin with remarks on their evolution and adaptation to lacustrine conditions. *Proc. Zool. Soc. London* 136:1–101.

Corner, E. J. H. 1964. *The life of plants*. New York: World.

Cosel, R. von. 1977. First record of *Mitra mitra* (Linnaeus, 1758) (Gastropoda: Prosobranchia) on the Pacific coast of Colombia, South America. *Veliger* 19:422–424.

Costa, H. Rodrigues da. 1962. Note préliminaire sur les peuplements intertidaux de substrat dur du littoral de Rio de Janeiro. *Rec. Trav. Sta. Mar. Endoume Bull.* 42:197–208.

Cracraft, J. 1973. Continental drift, paleoclimatology, and the evolution and biogeography of birds. *J. Zool. London* 169:455–545.

Crane, J. 1947. Eastern Pacific expeditions of the New York Zoological Society. XXXVIII. Intertidal brachygnathous crabs from the west coast of tropical America with special reference to ecology. *Zoologica* 32:69–95.

——. 1975. *Fiddler crabs of the world (Ocypodidae: genus* Uca). Princeton, N. J.: Princeton University Press.

Crenshaw, M. A., and J. M. Neff. 1969. Decalcification at the mantle-shell interface in molluscs. *Amer. Zool.* 9:881–885.

Crothers, J. H. 1968. The biology of the shore crab *Carcinus maenas* (L.). II. The life of the adult crab. *Field Studs.* 2:579–614.

Currey, J. D. 1975. A comparison of the strength of echinoderm spines and mollusc shells. *J. Mar. Biol. Assoc. U.K.* 55:419–424.

Dahl, A. L. 1964. Macroscopic algal foods of *Littorina planaxis* Philippi and *Littorina scutulata* Gould. *Veliger* 7:139–143.

Dahl, E. 1952. Some aspects of the ecology and zonation of the fauna on sandy beaches. *Oikos* 4:1–27.

Dana, T. F. 1975. Development of contemporary Eastern Pacific coral reefs. *Mar. Biol.* 33:355–374.

Dare, P. J., and A. J. Mercer. 1973. Foods of the oystercatcher in Morecambe Bay, Lancashire. *Bird Study* 20:173–184.

Darlington, P. J. 1959. Area, climate and evolution. *Evolution* 13:488–510.

Darnell, R. M. 1961. Trophic spectrum of an estuarine community, based on studies of Lake Pontchartrain, Louisiana. *Ecology* 42:553–568.

Davidson, P. E. 1967. The oystercatcher as a predator of commercial shell fisheries. *Ibis* 109:473–474.

——. 1971. Some foods taken by waders in Morecambe Bay, Lancashire. *Bird Study* 18:177–186.

Davies, A. M. 1971. *Tertiary faunas*, vol. 1 (rev. F. E. Eames). New York: American Elsevier.

Davies, P. S. 1966. Physiological ecology of *Patella*. I. The effect of body size and temperature on metabolic rate. *J. Mar. Biol. Assoc. U.K.* 46:646–658.

Dayton, P. K. 1971. Competition, disturbance, and community organization: the provision and subsequent utilization of space in a rocky intertidal community. *Ecol. Monogrs.* 41:351–389.

——. 1973a. Dispersion, dispersal, and persistence of the annual intertidal alga, *Postelsia palmaeformis* Ruprecht. *Ecology* 54:433–438.

——. 1973b. Two cases of resource partitioning in an intertidal community: making the right prediction for the wrong reason. *Amer. Nat.* 107:662–670.

——. 1975a. Experimental evaluation of ecological dominance in a rocky intertidal algal community. *Ecol. Monogrs.* 45:137–159.

——. 1975b. Experimental studies of algal canopy interactions in a sea otter–dominated kelp community at Amchitka Island, Alaska. *Fish. Bull.* 73:230–237.

Dayton, P. K., and R. R. Hessler. 1972. Role of biological disturbance in maintaining diversity in the deep sea. *Deep Sea Res.* 19:199–208.

Dayton, P. K., and J. S. Oliver. 1977. Antarctic soft-bottom benthos in oligotrophic and eutrophic environments. *Science* 197:55–58.

Dayton, P. K., G. A. Robilliard, and R. T. Paine. 1970. Benthic faunal zonation as a result of anchor ice at McMurdo Sound, Antarctica. In *Antarctic ecology*, ed. M. W. Holdgate. London: Academic Press, vol. 1, pp. 244–256.

Dayton, P. K., G. A. Robilliard, R. T. Paine, and L. B. Dayton. 1974. Biological accommodation in the benthic community at McMurdo Sound, Antarctica. *Ecol. Monogrs.* 44:105–128.

Dayton, P. K., R. J. Rosenthal, L. C. Mahen, and T. Antezana. 1977. Population structure and foraging biology of the predaceous Chilean asteroid *Meyenaster gelatinosus* and the escape biology of its prey. *Mar. Biol.* 39:361–370.

Dehnel, P. A. 1956. Growth rates in latitudinally and vertically separated populations of *Mytilus californianus*. *Biol. Bull.* 110:43–53.

Dell, R. K. 1972. Antarctic benthos. *Adv. Mar. Biol.* 10:1–216.

Demopulos, D. A. 1975. Diet, activity and feeding in *Tonicella lineata* (Wood, 1815). *Veliger* 18 (suppl.):42–46.

Deniel, C. 1974. Régime alimentaire des jeunes turbots *Scophthalmus maximus* L. de la classe 0 dans leur milieu naturel. *Cah. Biol. Mar.* 15:551–566.

Dexter, D. M. 1969. Structure of an intertidal sandy-beach community in North Carolina. *Chesapeake Sci.* 10:93–98.

———. 1972. Comparison of the community structures in a Pacific and an Atlantic Panamanian sandy beach. *Bull. Mar. Sci.* 22:449–462.

———. 1974. Sandy-beach fauna of the Pacific and Atlantic coasts of Costa Rica and Colombia. *Rev. Biol. Trop.* 22:61–66.

———. 1976. The sandy-beach fauna of Mexico. *Southwestern Naturalist* 20:479–485.

Dexter, R. W. 1962. Further studies on the marine mollusks of Cape Ann, Massachusetts. *Nautilus* 76:63–70.

Diamond, J. M. 1973. Distributional ecology of New Guinea birds. *Science* 179:759–769.

———. 1974. Colonization of exploded volcanic islands by birds: the supertramp strategy. *Science* 184:803–806.

———. 1975. Assembly of species communities. In *Ecology and evolution of communities*, ed. M. L. Cody and J. M. Diamond. Cambridge, Mass.: Belknap Press of Harvard University Press, pp. 342–444.

Dingle, H., and R. L. Caldwell. 1975. Distribution, abundance, and interspecific agonistic behavior of two mudflat stomatopods. *Oecologia (Berlin)* 20:167–178.

Dodd, J. R. 1964. Environmentally controlled variations in the shell structure of a pelecypod species. *J. Paleont.* 38:1065–1071.

Domning, D. P. 1976. An ecological model for Late Tertiary sirenian evolution in the North Pacific Ocean. *Syst. Zool.* 25:352–362.

Doty, M. S. 1967. Pioneer intertidal population and the related general vertical distribution of marine algae in Hawaii. *Blumea* 15:95–105.

Drinnan, R. E. 1957. The winter feeding of the oystercatcher (*Haematopus ostralegus*) on the edible cockle (*Cardium edule*). *J. Anim. Ecol.* 26:441–469.

Dunthorn, A. A. 1971. The predation of cultivated mussels by eiders. *Bird Study* 18:107–112.

Durazzi, J. T., and F. G. Stehli. 1972. Average generic age, the planetary temperature gradient, and pole location. *Syst. Zool.* 21:384–389.

Eales, N. B. 1949. The food of the dogfish, *Scyliorhinus caniculus* L. *J. Mar. Biol. Assoc. U.K.* 28:791–793.

Earle, S. A. 1972a. A review of the marine plants of Panama. *Bull. Biol. Soc. Washington* 2:69–87.

———. 1972b. The influence of herbivores on the marine plants of Great Lameshur Bay, with an annotated list of plants. *Nat. Hist. Mus. Los Angeles County Sci. Bull.* 14:17–44.

Ebling, F. J., J. A. Kitching, L. Muntz, and C. M. Taylor. 1964. The ecology of Lough Ine. XIII. Experimental observations of the destruction of *Mytilus edulis* and *Nucella lapillus* by crabs. *J. Anim. Ecol.* 33:73–83.

Ebling, F. J., A. D. Hawkins, J. A. Kitching, L. Muntz, and V. M. Pratt. 1966. The ecology of Lough Ine. XVI. The predation and diurnal migration in the *Paracentrotus* community. *J. Anim. Ecol.* 35:559–566.

Edington, J. M., P. J. Morgan, and R. A. Morgan. 1973. Feeding patterns of wading birds on the Gann flat and River Estuary at Dale. *Field Studs.* 3:783–800.

Edmondson, C. H. 1954. Hawaiian Portunidae. *Occ. Papers Bernice P. Bishop Mus.* 21 (12):217–274.

———. 1959. Hawaiian Grapsidae. *Occ. Papers Bernice P. Bishop Mus.* 22 (10):153–202.

———. 1962. Xanthidae of Hawaii. *Occ. Papers Bernice P. Bishop Mus.* 22 (13):215–309.

Edwards, D. C. 1969. Predators on *Olivella biplicata*, including a species-specific predator avoidance response. *Veliger* 11:326–333.

Edwards, R. R. C., and J. H. Steele. 1968. The ecology of 0-group plaice and common dabs at Loch Ewe. I. Population and food. *J. Exp. Mar. Biol. Ecol.* 2:215–238.

Edwards, R. R. C., J. H. Steele, and A. Trevallion. 1970. The ecology of 0-group plaice and common dabs in Loch Ewe. III. Prey-predator experiments with plaice. *J. Exp. Mar. Biol. Ecol.* 4:156–173.

Ehrlich, P. R. 1975. The population biology of coral reef fishes. *Ann. Rev. Ecol. Syst.* 6:213–247.

Ehrlich, P. R., and P. H. Raven. 1964. Butterflies and plants: a study in coevolution. *Evolution* 18:586–608.

———. 1969. Differentiation of populations. *Science* 165:1228–1231.

Ekman, S. 1953. *Zoogeography of the sea*. London: Sidgwick and Jackson.

Elton, C. S. 1958. *The ecology of invasions by animals and plants*. London: Methuen.

Emerson, W. K. 1967. Indo-Pacific faunal elements in the Eastern Pacific with special reference to mollusks. *Venus* 25:85–93.

Emerson, W. K., and W. O. Cernohorsky. 1973. The genus *Drupa* in the Indo-Pacific. *Indo-Pacific Mollusca* 3 (13):801–863.

Emiliani, C. 1971. The amplitude of Pleistocene climatic cycles at low latitudes and the isotopic composition of glacial ice. In *The Late Cenozoic glacial ages*, ed. K. Turikian. New Haven, Conn.: Yale University Press. pp. 183–197.

Endean, R., R. Kenny, and W. Stephenson. 1956a. The ecology and distribution of intertidal organisms on rocky shores of the Queensland mainland. *Austr. J. Mar. Fresh-W. Res.* 7:88–146.

Endean, R., W. Stephenson, and R. Kenny. 1956b. The ecology and distribution of intertidal organisms on certain islands off the Queensland coast. *Austr. J. Mar. Fresh-W. Res.* 7:317–340.

Estes, J. A., and J. F. Palmisano. 1974. Sea otters: their role in structuring near-shore communities. *Science* 185:1058–1060.

Farfante, I. P. 1943. The genera *Fissurella*, *Lucapina* and *Lucapinella* in the Western Atlantic. *Johnsonia* 1 (10):1–20.

Fauchald, K. 1977. Polychaetes from intertidal areas in Panama, with a review of previous shallow-water records. *Smithson. Contribs. Zool.* 221:1–81.

Feare, C. J. 1967. The effect of predation by shore-birds on a population of dogwhelks *Thais lapillus*. *Ibis* 109:474.

———. 1970. A note on the methods employed by crabs in reaching shells of dogwhelks *(Nucella lapillus)*. *Naturalist* 913:67–68.

———. 1971a. Predation of limpets and dogwhelks by oystercatchers. *Bird Study* 18:121–129.

———. 1971b. The adaptive significance of aggregation behaviour in the dog whelk *Nucella lapillus* (L.). *Oecologia (Berlin)* 7:117–126.

Feder, H. M. 1963. Gastropod defensive responses and their effectiveness in reducing predation by starfishes. *Ecology* 44:505–512.

Feeny, P. 1975. Biochemical coevolution between plants and their insect herbivores. In *Coevolution of animals and plants*, ed. L. E. Gilbert and P. H. Raven. Austin: University of Texas Press, pp. 3–19.

Fell, H. B. 1967. Cretaceous and Tertiary surface currents of the oceans. *Oceanogr. Mar. Biol. Ann. Rev.* 5:317–341.

Fenchel, T. M., and R. J. Riedl. 1970. The sulfide system: a new biotic community underneath the oxidized layer of marine sand bottoms. *Mar. Biol.* 7:255–268.

Figueiredo, M. J. de, and H. J. Thomas. 1967. *Nephrops norvegicus* (Linnaeus, 1758) Leach—a review. *Oceanogr. Mar. Biol. Ann. Rev.* 5:371–407.

Fischer, A. G. 1960. Latitudinal variations in organic diversity. *Evolution* 14:64–81.

Fischer, A. G., and M. A. Arthur. 1977. Secular variation in the pelagic realm. *Soc. Econ. Petrol. Mineral.* Special Publ. 25:19–50.

Fishelson, L. 1971. Ecology and distribution of the benthic fauna in the shallow waters of the Red Sea. *Mar. Biol.* 10:113–133.

———. 1974. Ecology of the northern Red Sea crinoids and their epi- and endozoic fauna. *Mar. Biol.* 26:183–192.

Fleming, T. H. 1973. Numbers of mammal species in North and Central American forest communities. *Ecology* 54:555–563.

Fletcher, R. L. 1975. Heteroantagonism observed in mixed algal cultures. *Nature* 253:534–535.

Foin, T. C. 1972. Ecological observations on the size of *Cypraea tigris* L., 1758, in the Pacific. *Proc. Malac. Soc. London* 40:211–218.

Fooden, J. 1972. Breakup of Pangaea and isolation of relict mammals in Australia, South America, and Madagascar. *Science* 175:894–898.

Ford, E. B. 1965. *Ecological genetics* (2nd ed.). London: Methuen.

Forest, J., and D. Guinot. 1962. Remarques biogéographiques sur les crabes des archipels de la Société et des Tuamotu. *Cah. Pac.* 4:41–75.

Forest, J., M. de Saint Laurent, and F. A. Chace, Jr. 1976. *Neoglyphea inopinata*: a crustacean "living fossil" from the Philippines. *Science* 192:884.

Foster, M. S. 1964. Microscopic algal food of *Littorina scutulata* Gould. *Veliger* 7:149–152.

Fotheringham, N. 1971. Field identification of crab predation on *Shaskyus festivus* and *Ocenebra poulsoni* (Prosobranchia: Muricidae). *Veliger* 14:204.

———. 1974. Trophic complexity in a littoral boulder field. *Limnol. Oceanogr.* 19:84–91.

Fraenkel, G. 1968. The heat resistance of intertidal snails at Bimini, Bahamas; Ocean Springs, Mississippi; and Woods Hole, Massachusetts. *Physiol. Zool.* 41:1–13.

Frank, P. W. 1965. The biodemography of an intertidal snail population. *Ecology* 46:831–844.

———. 1975. Latitudinal variation in the life history features of the black turban snail *Tegula funebralis* (Prosobranchia: Trochidae). *Mar. Biol.* 31:181–192.

Frazzetta, T. H. 1970. From hopeful monsters to bolyerine snakes? *Amer. Nat.* 104:55–71.

Fretter, V., and A. Graham. 1962. *British prosobranch molluscs.* London: Ray Society.

Fryer, G., and T. D. Iles. 1969. Alternative routes to evolutionary success as exhibited by African cichlid fishes of the genus *Tilapia* and the species flocks of the Great Lakes. *Evolution* 23:359–369.

———. 1972. *The cichlid fishes of the Great Lakes of Africa: their biology and evolution.* Edinburgh: Oliver and Boyd.

Furtado-Ogawa, E. 1970. Contribuição ao conhecimento da fauna malacológica intertidal de substratos duros do nordeste brasileiro. *Arq. Ciénc. Mar.* 10:193–196.

Futuyma, D. J. 1973. Community structure and stability in constant environments. *Amer. Nat.* 107:443–446.

Gadgil, M., and O. T. Solbrig. 1972. The concept of r- and K-selection: evidence from wild flowers. *Amer. Nat.* 106:14–31.

Gail, R., and L. Devambez. 1958. A selected annotated bibliography of *Trochus*. *South Pacific Commission Technical Paper* 111:1–17.

Gaillard, J. M. 1965. Aspects qualitatifs et quantitatifs de la croissance de la coquille de quelques espèces de mollusques prosobranches en fonction de la latitude et des conditions écologiques. *Mém. Mus. Nat. Hist. Natur. (n.s. A, Zoologie)* 38:1–155.

Gaines, M. S., K. J. Vogt, J. L. Hamrick, and J. Caldwell. 1974. Reproductive strategies and growth patterns in sunflowers *(Helianthus)*. *Amer. Nat.* 108:889–894.

Garrett, P. 1970. Phanerozoic stromatolites: noncompetitive ecologic restriction by grazing and burrowing animals. *Science* 169:171–173.

Garth, J. S. 1946. The littoral brachyuran fauna of the Galapagos archipelago. *Allan Hancock Pac. Exped.* 5:341–601.

———. 1960. Distribution and affinities of the brachyuran Crustacea. *Syst. Zool.* 9:105–123.

————. 1965. The brachyuran decapod Crustacea of Clipperton Island. *Proc. Calif. Acad. Sci.* (4) 33:1–46.

————. 1973. The brachyuran crabs of Easter Island. *Proc. Calif. Acad. Sci.* (4) 39:311–336.

————. 1974. On the occurrence in the Eastern tropical Pacific of Indo-West Pacific decapod crustaceans commensal with reef-building corals. *Proc. Second Intern. Coral Reef Symp.* 1:397–404.

George, R. W. 1974. Coral reefs and rock lobster ecology in the Indo-West Pacific region. *Proc. Second Intern. Coral Reef Symp.* 1:321–325.

George, R. W., and A. R. Main. 1968. The evolution of spiny lobsters (Palinuridae): a study of evolution in the marine environment. *Evolution* 22:803–820.

Gibb, J. 1956. Food, feeding habits and territoriality of the rock pipit *Anthus spinoletta*. *Ibis* 98:506–530.

Gibson, J. S. 1970. The function of the operculum of *Thais lapillus* (L.) in relation to desiccation and predation. *J. Anim. Ecol.* 39:159–168.

Gibson, R. N. 1969. Biology and behaviour of littoral fish. *Oceanogr. Mar. Biol. Ann. Rev.* 7:367–410.

————. 1972. The vertical distribution and feeding relationships of intertidal fish on the Atlantic coast of France. *J. Anim. Ecol.* 41:189–207.

Gilbert, L. E. 1971. Butterfly-plant coevolution: has *Passiflora adenopoda* won the selectional race with heliconiine butterflies? *Science* 172:585–586.

Gilbert, L. E., and P. H. Raven, eds. 1975. *Coevolution of animals and plants*. Austin: University of Texas Press.

Gislén, T. 1943. Physiographical and ecological investigations concerning the littoral of the northern Pacific. A comparison between the life conditions in the littoral of central Japan and California. *Lunds Univ. Arskr.* 39:1–63.

Givnish, T. J., and G. J. Vermeij. 1976. Sizes and shapes of liane leaves. *Amer. Nat.* 110:743–778.

Glynn, P. W. 1965. Community composition, structure, and interrelationships in the marine intertidal *Endocladia muricata-Balanus glandula* association in Monterey Bay, California. *Beaufortia* 12:1–198.

————. 1968. Mass mortalities of echinoids and other reef flat organisms coincident with mid day low water exposures in Puerto Rico. *Mar. Biol.* 1:226–243.

————. 1972. Observations on the ecology of the Caribbean and Pacific coasts of Panama. *Bull. Biol. Soc. Washington* 2:13–30.

————. 1973a. *Acanthaster*: effects on coral reef growth in Panama. *Science* 180:504–506.

————. 1973b. Aspects of the ecology of coral reefs in the Western Atlantic region. In *Biology and geology and coral reefs*, ed. O. A. Jones and R. Endean. New York: Academic Press, Vol. 1, pp. 273–324.

————. 1974. Rolling stones among the Scleractinia: mobile coralliths in the Gulf of Panama. *Proc. Second Intern. Coral Reef Symp.* 2:183–198.

————. 1976. Some physical and biological determinants of coral community structure in the Eastern Pacific. *Ecol. Monogrs.* 46:431–456.

Glynn, P. W., and R. H. Stewart. 1973. Distribution of coral reefs in the Pearl Islands (Gulf of Panama) in relation to thermal conditions. *Limnol. Oceanogr.* 18:367–379.

Glynn, P. W., L. R. Almodóvar, and J. G. González. 1964. Effects of Hurricane Edith on marine life in La Parguera, Puerto Rico. *Carib. J. Sci.* 4:335–345.

Glynn, P. W., R. H. Stewart, and J. E. McCosker. 1972. Pacific coral reefs of Panama: structure, distribution, and predators. *Geol. Rundschau* 61: 481–519.

Gonor, J. J. 1965. Predator-prey reactions between two marine prosobranch gastropods. *Veliger* 7:228–232.

———. 1966. Escape responses of North Borneo strombid gastropods elicited by predatory prosobranchs *Aulica vespertilio* and *Conus marmoreus*. *Veliger* 8:226–230.

Gooch, J. L., and T. J. M. Schopf. 1972. Genetic variability in the deep sea: relation to environmental variability. *Evolution* 26:545–552.

Gopalakrishnan, P. 1970. Some observations on the shore ecology of the Okha coast. *J. Mar. Biol. Assoc. India* 12:15–33.

Goreau, T. F., N. I. Goreau, T. Soot-Ryen, and C. M. Yonge. 1969. On a new commensal mytilid (Mollusca: Bivalvia) opening into the coelenteron of *Fungia scutaria* (Coelenterata). *J. Zool. London* 158:171–195.

Goreau, T. F., N. I. Goreau, and C. M. Yonge. 1971. Reef corals: autotrophs or heterotrophs? *Biol. Bull.* 141:247–260.

Goreau, T. F., J. Lang, E. Graham, and P. Goreau. 1972. Structure and ecology of the Saipan reefs in relation to predation by *Acanthaster planci* (Linnaeus). *Bull. Mar. Sci.* 22:113–152.

Goss-Custard, J. D. 1969. The winter feeding ecology of the redshank *Tringa totanus*. *Ibis* 111:338–356.

Gould, S. J. 1966. Allometry in Pleistocene land snails from Bermuda: the influence of size upon shape. *J. Paleont.* 40:1131–1141.

———. 1968. Ontogeny and the explanation of form: an allometric analysis. *J. Paleont.* 42, *Paleont. Soc. Mem.*, II:81–98.

Grassle, J. F. 1977. Slow recolonisation of deep-sea sediment. *Nature* 265:618–619.

Graus, R. R. 1974. Latitudinal trends in the shell characteristics of marine gastropods. *Lethaia* 7:303–314.

Green, R. H. 1968. Mortality and stability in a low diversity subtropical intertidal community. *Ecology* 49:848–854.

Greenwood, J. G. 1972. The mouthparts and feeding behaviour of two species of hermit crabs. *J. Nat. Hist.* 6:325–337.

Greenwood, P. H. 1974. The cichlid fishes of Lake Victoria, East Africa: the biology and distribution of a species flock. *Bull. Brit. Mus. (Nat. Hist.) Zool.* (suppl.) 6:1–134.

Griffin, D. J. G. 1971. The ecological distribution of grapsid and ocypodid shore crabs (Crustacea: Brachyura) in Tasmania. *J. Anim. Ecol.* 40:597–622.

Grigg, R. W., and J. E. Maragos. 1974. Recolonization of hermatypic corals on submerged lava flows in Hawaii. *Ecology* 55:387–395.

Grime, J. P. 1977. Evidence for the existence of three primary strategies in plants and its relevance to ecological and evolutionary theory. *Amer. Nat.* 111:1169–1194.

Guiler, E. R. 1959a. Intertidal belt-forming species on the rocky coasts of northern Chile. *Papers Proc. Roy. Soc. Tasmania* 93:33–58.

———. 1959b. The intertidal ecology of the Montemar area, Chile. *Papers Proc. Roy. Soc. Tasmania* 93:165–183.

Guinot, D. 1966. La faune carcinologique (Crustacea Brachyura) de l'Océan Indien occidental et de la Mer Rouge. Catalogue, remarques biogéographiques et bibliographie. *Mém. Inst. Fond. Afr. Noir* 77:1–352.

Habe, T. 1958. A study on the productivity of the Tanabe Bay. Part I. VII. Zonal arrangement of intertidal benthic animals in the Tanabe Bay. *Rec. Oceanogr. Works Japan*, spec. no. 2:43–49.

Hadfield, M. G. 1976. Molluscs associated with living tropical corals. *Micronesica* 12:133–148.

Hagen, H. O. von. 1970. Anpassungen an der spezielle Gezeitenzonen-niveau bei Ocypodiden (Decapoda: Brachyura). *Forma Functio* 2:361–413.

Hall, C. A. 1964. Shallow water marine climates and molluscan provinces. *Ecology* 45:226–234.

Hallam, A. 1976. Stratigraphic distribution and ecology of European Jurassic bivalves. *Lethaia* 9:245–259.

———. 1977. Jurassic bivalve biogeography. *Paleobiology* 3:58–73.

Hamai, I. 1937. Some notes on relative growth with special reference to the growth of limpets. *Sci. Rep. Tohoku Imp. Univ. Biol.* (4) 12:71–95.

Hamajima, F., T. Fujino, and M. Koga. 1976. Studies on the host-parasite relationship of *Paragonimus westermani* (Kerbert, 1878). IV. Predatory habits of some fresh-water crabs and crayfish on the snail, *Semisulcospira libertina* (Gould, 1859). *Ann. Zool. Jap.* 49:274–278.

Hamilton, P. V. 1976. Predation on *Littorina irrorata* (Mollusca: Gastropoda) by *Callinectes sapidus* (Crustacea: Portunidae). *Bull. Mar. Sci.* 26:403–409.

Hamilton, W. J. 1973. *Life's color code*. New York: McGraw-Hill.

Hancock, D. A. 1965. Adductor muscle size in Danish and British mussels and its relation to starfish predation. *Ophelia* 2:253–267.

———. 1972. The role of predators and parasites in a fishery for the mollusk *Cardium edule* L. *Proc. Adv. Stud. Inst. Dynamics Numbers Populations* (Oosterbeek, 1970):419–420.

Harger, J. R. 1972a. Variation in relative "niche" size in the sea mussel *Mytilus edulis* in association with *Mytilus californianus*. *Veliger* 14:275–282.

———. 1972b. Competitive co-existence: maintenance of interacting associations of the sea mussels *Mytilus edulis* and *Mytilus californianus*. *Veliger* 14:387–409.

Harris, L. G. 1973. Nudibranch associations. In *Current topics in comparative pathobiology*. New York: Academic Press, vol. 2, pp. 213–315.

———. 1975. Studies on the life history of two coral-eating nudibranchs of the genus *Phestilla*. *Biol. Bull.* 149:539–550.

Harris, M. P. 1965. The food of some *Larus* gulls. *Ibis* 107:43–51.

Harry, H. W. 1951. Growth changes in the shell of *Pythia scarabaeus* (Linné). *Proc. Calif. Zool. Club* 2:7–14.

Hartnoll, R. G. 1974. Variation in growth pattern between some secondary sexual characters in crabs (Decapoda Brachyura). *Crustaceana* 27:131–136.

———. 1976. The ecology of some rocky shores in tropical East Africa. *Estuarine Coastal Mar. Sci.* 4:1–21.

Hartwick, E. B. 1976. Foraging strategy of the black oyster catcher (*Haematopus bachmani* Audubon). *Canad. J. Zool.* 54:142–155.

Haven, S. B. 1971. Niche differences in the intertidal limpets *Acmaea scabra* and *Acmaea digitalis* (Gastropoda) in central California. *Veliger* 13:231–248.

————. 1973. Competition for food between the intertidal gastropods *Acmaea scabra* and *Acmaea digitalis*. *Ecology* 54:143–151.

Heatwole, H., and R. Levins. 1972. Biogeography of the Puerto Rican Bank: flotsam transport of terrestrial animals. *Ecology* 53:112–117.

————. 1973. Biogeography of the Puerto Rican Bank: species turn-over on a small cay, Cayo Ahogado. *Ecology* 54:1042–1055.

Hecht, A. D., and B. Agan. 1972. Diversity and age relationships in Recent and Miocene bivalves. *Syst. Zool.* 21:308–312.

Hedgpeth, J. W. 1969a. An intertidal reconnaissance of rocky shores of the Galapagos. *Wasmann J. Biol.* 27:1–24.

————. 1969b. Preliminary observations of life between tidemarks at Palmer Station, 64° 45 min S, 64° 05 min W. *Antarctic J. U.S.* 4:106–107.

Heithaus, E. R., P. A. Opler, and H. G. Baker. 1974. Bat activity and pollination of *Bauhinia pauletia*: plant-pollinator coevolution. *Ecology* 55:412–419.

Heller, J. 1975. Taxonomy of some British *Littorina* species with notes on their reproduction (Mollusca: Prosobranchia). *Zool. J. Linn. Soc.* 56:131–151.

————. 1976. The effects of exposure and predation on the shell of two British winkles. *J. Zool. London* 179:201–213.

Hemingway, G. T. 1975. Functional morphology of feeding in the predatory whelk, *Acanthina spirata* (Gastropoda: Prosobranchia). *Bull. Amer. Malacol. Union*:64–65.

Hendler, G. 1977. The differential effects of seasonal stress and predation on the stability of reef-flat echinoid populations. *Proc. Third Intern. Coral Reef Symp.* 1:217–223.

Hepper, B. T. 1957. Notes on *Mytilus galloprovincialis* Lamarck in Great Britain. *J. Mar. Biol. Assoc. U.K.* 36:33–40.

Heppleston, B. P. 1971. Feeding techniques of the oystercatcher. *Bird Study* 18:15–20.

Herrnkind, W. F., J. A. Van Derwalker, and L. Barr. 1975. Population dynamics, ecology and behavior of spiny lobsters, *Panulirus argus*, of St. John, U.S.V.I. (IV). Habitation, patterns of movement, and general behavior. *Nat. Hist. Mus. Los Angeles County Sci. Bull.* 20:31–45.

Hessler, R. R., and D. Thistle. 1975. On the place of origin of deep-sea isopods. *Mar. Biol.* 32:155–165.

Hiatt, R. W. 1948. The biology of the lined shore crab, *Pachygrapsus crassipes* Forskal. *Pacific Sci.* 2:133–213.

Hiatt, R. W., and D. W. Strassburg. 1960. Ecological relationships of the fish fauna on coral reefs of the Marshall Islands. *Ecol. Monogrs.* 30:65–127.

Hickman, R. W. 1972. Rock lobsters feeding on oysters (note). *N.Z. J. Mar. Fresh-W. Res.* 6:641–644.

Himmelman, J. H., and D. H. Steele. 1971. Foods and predators of the green sea urchin *Strongylocentrotus droebachiensis* in Newfoundland waters. *Mar. Biol.* 9:315–322.

Hoagland, K. E. 1977. Systematic review of fossil and Recent *Crepidula* and discussion of evolution of the Calyptraeidae. *Malacologia* 16:353–420.

Hobson, E. S. 1968. Predatory behavior of some shore fishes in the Gulf of California. *U.S. Dept. Interior Bur. Sport Fish. Wildlife Res. Rep.* 73:1–92.

————. 1974. Feeding relationships of teleostean fishes on coral reefs in Kona, Hawaii. *Fish. Bull.* 72:915–1031.

Hochachka, P. W., and G. N. Somero. 1973. Strategies of biochemical adaptation. Philadelphia: Saunders.

Hodgkin, E. P. 1959. Catastrophic destruction of the littoral fauna and flora near Fremantle, January, 1959. *Western Austr. Naturalist* 7:6–11.

Hodgkin, E. P., and C. Michel. 1961. Zonation of plants and animals on rocky shores of Mauritius. *Proc. Roy. Soc. Arts Sci. Mauritius* 2:121–145.

Holeton, G. F. 1974. Metabolic cold adaptation of polar fish: fact or artifact? *Physiol. Zool.* 47:137–152.

Holme, N. A. 1954. The ecology of British species of *Ensis*. *J. Mar. Biol. Assoc. U.K.* 33:145–172.

Holmes, J. C., and W. M. Bethel. 1972. Modification of intermediate host behavior by parasites. In *Behavioral aspect of parasite transmission*, ed. E. U. Canning and C. A. Wright. *Zool. J. Linn. Soc.*, suppl. 1:123–149.

Holthuis, L. B. 1959. The Crustacea Decapoda of Suriname (Dutch Guiana). *Zool. Verh. Rijksmus. Nat. Hist. Leiden* 44:1–296.

Holthuis, L. B., and E. Gottlieb. 1958. An annotated list of the decapod Crustacea of the Mediterranean coast of Israel, with an appendix listing the Decapoda of the eastern Mediterranean. *Bull. Res. Council Isr., Sec. B (Zool.)* 7b:1–126.

Holthuis, L. B., and H. Loesch. 1967. The lobsters of the Galapagos Islands (Decapoda, Palinuridae). *Crustaceana* 12:214–222.

Holthuis, L. B., and E. Sibertsen. 1967. The Crustacea Decapoda, Mysidacea and Cirripedia of the Tristan da Cunha archipelago, with a revision of the "*frontalis*" subgroup of the genus *Jasus*. *Results Norw. Sci. Exped. Tristan da Cunha (1937-1938)* 52:1–55.

Horn, H. S. 1971. *The adaptive geometry of trees*. Princeton, N. J.: Princeton University Press.

———. 1974. The ecology of secondary succession. *Ann. Rev. Ecol. System* 5:25–37.

Houbrick, R. S. 1974a. Growth studies on the genus *Cerithium* (Gastropoda: Prosobranchia) with notes on ecology and microhabitat. *Nautilus* 88:14–27.

———. 1974b. The genus *Cerithium* in the Western Atlantic (Cerithiidae: Prosobranchia). *Johnsonia* 5:33–84.

———. 1975. *Clavocerithium (Indocerithium) taeniatum*, a little-known and unusual cerithiid from New Guinea. *Nautilus* 89:99–105.

———. 1978. The family Cerithiidae in the Indo-Pacific. I. The genera *Rhinoclavis, Pseudovertagus, Clavocerithium*, and *Longicerithium* n.gen. *Indo-Pacific Mollusca*, in press.

Houbrick, R. S., and V. Fretter. 1969. Some aspects of the functional anatomy and biology of *Cymatium* and *Bursa*. *Proc. Malacol. Soc. London* 38:415–429.

Houvenagel, G. T., and N. Houvenagel. 1974. Aspects écologiques de la zonation intertidale sur les côtes rocheuses des îles Galapagos. *Mar. Biol.* 26:135–152.

Howard, R. A. 1969. The ecology of an elfin forest in Puerto Rico. VIII. Studies of stem growth and form and of leaf structure. *J. Arnold Arboretum* 50:225–267.

Hughes, D. A. 1966. Behavioural and ecological investigation of the crab *Ocypode ceratophthalmus* (Crustacea: Ocypodidae). *J. Zool. London* 150:129–143.

Hughes, R. N. 1970. Population dynamics of the bivalve *Scrobicularia plana* (da Costa) on an intertidal mud-flat in North Wales. *J. Anim. Ecology* 39:333–356.

Hulscher, J. B. 1973. Burying-depth and trematode infection in *Macoma balthica*. *Netherlands J. Sea Res.* 6:141–156.

Humes, A. G., and J. H. Stock. 1973. A revision of the family Lichomolgidae Kossmann, 1877, cyclopoid copepods mainly associated with marine invertebrates. *Smithson. Contribs. Zool.* 127:1–368.

Hunt, O. D. 1925. The food of the bottom fauna of the Plymouth fishing grounds. *J. Mar. Biol. Assoc. U.K.* 13:560–599.

Hutchins, L. W. 1947. The bases for temperature zonation in geographic distribution. *Ecol. Monogrs.* 17:325–335.

Hutchinson, G. E. 1959. Homage to Santa Rosalia or why are there so many kinds of animals? *Amer. Nat.* 93:145–159.

———. 1967. *A treatise on limnology*, vol. 2. New York: Wiley.

Imbrie, J., and N. G. Kipp. 1971. A new micropaleontological method for quantitative paleoclimatology: application to a late Pleistocene Caribbean core. In *The Late Cenozoic glacial ages*, ed. K. Turikian. New Haven, Conn.: Yale University Press, pp. 71–181.

Jackson, J. B. C. 1972. The ecology of molluscs of *Thalassia* communities, Jamaica, West Indies. II. Molluscan population variability along an environmental stress gradient. *Mar. Biol.* 14:304–337.

———. 1973. The ecology of molluscs of *Thalassia* communities, Jamaica, West Indies. I. Distribution, environmental physiology, and ecology of common shallow-water species. *Bull. Mar. Sci.* 23:313–350.

———. 1974. Biogeographic consequences of eurytopy and stenotopy among marine bivalves and their evolutionary significance. *Amer. Nat.* 108:541–560.

———. 1977. Competition on marine hard substrata: the adaptive significance of solitary and colonial strategies. *Amer. Nat.* 111:743–767.

Jackson, J. B. C., and L. Buss. 1975. Allelopathy and spatial competition among coral reef invertebrates. *Proc. Nat. Acad. Sci. U.S.A.* 72:5160–5163.

Jackson, J. B. C., T. F. Goreau, and W. D. Hartman. 1971. Recent brachiopod-coralline sponge communities and their paleoecological significance. *Science* 173:623–625.

James, B. L. 1968. The characters and distribution of the subspecies and varieties of *Littorina saxatilis* (Olivi 1792) in Britain. *Cah. Biol. Mar.* 9:143–165.

Janzen, D. H. 1970. Herbivores and the number of tree species in tropical forests. *Amer. Nat.* 104:501–528.

———. 1973a. Sweep samples of tropical foliage insects: effects of seasons, vegetation types, elevation, time of day, and insularity. *Ecology* 54:687–708.

———. 1973b. Rate of regeneration after a tropical high elevation fire. *Biotropica* 5:117–122.

———. 1973c. Dissolution of mutualism between *Cecropia* and its *Azteca* ants. *Biotropica* 5:15–28.

———. 1976. The depression of reptile biomass by large herbivores. *Amer. Nat.* 110:371–400.

Jardine, N., and D. McKenzie. 1972. Continental drift and the dispersal and evolution of organisms. *Nature* 235:20–24.

Jeffries, H. P. 1966. Partitioning of the estuarine environment by two species of *Cancer*. *Ecology* 45:477–481.

Jeletzky, J. A. 1965. Taxonomy and phylogeny of fossil Coleoidea (= Dibranchiata). *Geol. Surv. Papers Canada* 65-2:72–76.

Johannes, R. E., and L. Tepley. 1974. Examination of feeding of the reef coral *Porites lobata in sutu* using time lapse photography. *Proc. Second Intern. Coral Reef Symp.* 1:127–131.

John, D. M., and G. W. Lawson. 1974. Observations on the marine algal ecology of Gabon. *Botanica Marina* 17:249–254.

John, D. M., and W. Pople. 1973. The fish grazing of rocky shore algae in the Gulf of Guinea. *J. Exp. Mar. Biol. Ecol.* 11:81–90.

Johnson, R. K., and M. A. Barnett. 1975. An inverse correlation between meristic characters and food supply in mid-water fishes: evidence and possible explanations. *Fish. Bull.* 73:284–298.

Johnson, W. E., H. L. Carson, K. Y. Kaneshiro, W. W. M. Steiner, and M. M. Copper. 1975. Genetic variation in Hawaiian *Drosophila* II. Allozymic differentiation in the *D. planitibia* subgroup. In *Isozymes*, ed. C. L. Markert. New York: Academic Press, vol. 4, pp. 563–584.

Jones, D. A. 1972. Aspects of the ecology and behaviour of *Ocypode ceratophthalmus* (Pallas) and *O. kuhlii* de Haan (Crustacea: Ocypodidae). *J. Exp. Mar. Biol. Ecol.* 8:31–43.

Jones, N. S. 1948. Observations and experiments on the biology of *Patella vulgata* at Port St. Mary, Isle of Man. *Proc. Trans. Liverpool Biol. Soc.* 56:60–77.

Jones, R. S. 1968. Ecological relationships in Hawaiian and Johnston Island Acanthuridae (surgeonfishes). *Micronesica* 4:309–361.

Jung, P., and R. T. Abbott. 1967. The genus *Terebellum* (Gastropoda: Strombidae). *Indo-Pacific Mollusca* 1 (7):445–454.

Kauffman, E. G. 1973. Cretaceous bivalvia. In *Atlas of palaeobiogeography*, ed. A. Hallam. Amsterdam: Elsevier, pp. 358–383.

Kauffman, E. G., and N. F. Sohl. 1974. Structure and evolution of Antillean Cretaceous rudist frameworks. *Verhandl. Naturf. Ges. Basel* 84:399–467.

Kaufman, L. 1977. The three spot damselfish: effects on benthic biota of Caribbean coral reefs. *Proc. Third Intern. Coral Reef Symp.* 1:559–564.

Kawaguti, S. 1950. Observations on the heart shell, *Corculum cardissa* (L.), and its associated zooxanthellae. *Pac. Sci.* 4:43–49.

Kay, E. A. 1967. The composition and relationships of marine molluscan fauna of the Hawaiian Islands. *Venus* 25:94–104.

———. 1971. The littoral marine molluscs of Fanning Island. *Pac. Sci.* 25: 260–281.

Keast, A. 1969. Evolution of mammals on southern continents. VII. Comparisons of the contemporary mammalian faunas of the southern continents. *Quart. Rev. Biol.* 44:121–167.

———. 1970. Adaptive evolution and shifts in niche occupation in island birds. *Biotropica* 2:61–75.

Keen, A. M. 1971. *Seashells of tropical West America* (2nd ed.). Palo Alto, Calif.: Stanford University Press.

Kempf, M. 1970. Notes on the benthic bionomy of the N-NE Brazilian shelf. *Mar. Biol.* 5:213–224.

Kempf, M., and J. Laborel. 1968. Formations de vermets et d'algues calcaires sur les côtes du Brésil. *Rec. Trav. Sta. Mar. Endoume Bull.* 43:9–23.

Kennedy, W. J., J. D. Taylor, and A. Hall. 1969. Environmental and biological controls on bivalve shell mineralogy. *Biol. Revs.* 44:499–530.

Kenny, R., and N. Hayson. 1962. Ecology of rocky shore organisms at Mac-Quarie Island. *Pac. Sci.* 16:245–263.

Kent, B. W. 1978. Interoceanic and latitudinal patterns in spiny lobster mandible size. *Crustaceana*, in press.

Kier, P. M. 1974. Evolutionary trends and their functional significance in the post-Paleozoic echinoids. *J. Paleont. 48, Paleont. Soc. Mem.* 5, II:1–95.

Kilburn, R. N., 1975. A new species of *Turbinella* (Mollusca: Gastropoda: Turbinellidae) from northern Mocambique waters. *Durban Mus. Novitates* 10:221–225.

Kinzie, R. A. III. 1968. The ecology of the replacement of *Pseudosquilla ciliata* (Fabricius) by *Gonodactylus falcatus* (Forskal) (Crustacea; Stomatopoda) recently introduced into the Hawaiian Islands. *Pac. Sci.* 22:465–475.

———. 1973. The zonation of West Indian gorgonians. *Bull. Mar. Sci.* 23:93–155.

Kitching, J. A., and F. J. Ebling. 1961. The ecology of Lough Ine. XI. The control of algae by *Paracentrotus lividus* (Echinoidea). *J. Anim. Ecol.* 30: 373–383.

Kitching, J. A., and J. Lockwood. 1974. Observations on shell form and its ecological significance in thaisid gastropods of the genus *Lepsiella* in New Zealand. *Mar. Biol.* 28:131–144.

Kitching, J. A., J. F. Sloan, and F. J. Ebling. 1959. The ecology of Lough Ine. VIII. Mussels and their predators. *J. Anim. Ecol.* 28:331–341.

Kitching, J. A., L. Muntz, and F. J. Ebling. 1966. The ecology of Lough Ine. XV. The ecological significance of shell and body forms in *Nucella*. *J. Anim. Ecol.* 35:113–126.

Knight, J. B., L. R. Cox, A. M. Keen, R. L. Batten, E. L. Yochelson, and R. Robertson. 1960. Systematic descriptions. In *Treatise on invertebrate paleontology*, ed. R. C. Moore. Lawrence: University of Kansas Press, pt. 1, pp. I147–I351.

Knox, G. A. 1954. Intertidal flora and fauna of the Chatham Islands. *Nature* 174:871–873.

———. 1960. Littoral ecology and biogeography of the southern oceans. *Proc. Roy. Soc. London (B)* 152:567–624.

———. 1963. The biogeography and intertidal ecology of the Australasian coasts. *Oceanogr. Mar. Biol. Ann. Rev.* 1:341–404.

Kohn, A. J. 1959a. The ecology of *Conus* in Hawaii. *Ecol. Monogrs.* 29:47–90.

———. 1959b. The Hawaiian species of *Conus* (Mollusca: Gastropoda). *Pac. Sci.* 13:368–401.

Kohn, A. J., and J. W. Nybakken. 1975. Ecology of *Conus* on eastern Indian Ocean fringing reefs. *Mar. Biol.* 29:211–234.

Kohn, A. J., and A. C. Riggs. 1975. Morphometry of the *Conus* shell. *Syst. Zool.* 24:346–359.

Kohn, A. J., and V. Waters. 1966. Escape responses of three herbivorous gastropods to the predatory gastropod *Conus textile*. *Anim. Beh.* 14:340–345.

Kraeuter, J. N. 1974 (1976). Offshore currents, larval transport, and establishment of southern populations of *Littorina littorea* Linné along the U.S. Atlantic coast. *Thalassia Jugoslavica* 10:159–170.

Kramer, P. 1967. Beobachtungen zur Biologie und zum Verhalten der Klippenkrabbe *Grapsus grapsus* (Brachyura Grapsidae) auf Galapagos und am ecuadorianischen Festland. *Z. Tierpsychol*. 24:385–402.

Krebs, C. J., M. S. Gaines, B. L. Keller, J. H. Myers, and R. H. Tamarin. 1973. Population cycles in small rodents. *Science* 179:35–41.

Kühnemann, O. 1972. Bosquejo fitogeografico de la vegetacion marina del litoral Argentino. *Physis* 31:117–142, 295–325.

Laborel, J. 1969. Les peuplements de madréporaires des côtes tropicales du Brésil. *Ann. Univ. Abidjan (E)* 2 (3):7–261.

———. 1974. West African reef corals: an hypothesis on their origin. *Proc. Second Intern. Coral Reef Symp.* 1:425–443.

Ladd, H. S. 1960. Origin of the Pacific Island molluscan fauna. *Amer. J. Sci.* 258:137–150.

———. 1966. Chitons and gastropods (Haliotidae through Adeorbidae) from the Western Pacific Islands. *U.S. Geol. Surv. Prof. Paper* 531:1–98.

———. 1972. Cenozoic fossil mollusks from Western Pacific Islands; gastropods (Turritellidae through Strombidae). *U.S. Geol. Survey Prof. Paper* 532:1–79.

———. 1977. Cenozoic fossil mollusks from Western Pacific Islands; gastropods (Eratoidae through Harpidae). *U.S. Geol. Survey Prof. Paper* 533:1–84.

Landers, W. S. 1954. Notes on the predation of the hard clam, *Venus mercenaria*, by the mud crab, *Neopanope texana*. *Ecology* 35:422.

Lang, J. 1971. Interspecific aggression by scleractinian corals. I. The rediscovery of *Scolemia cubensis* (Milne Edwards and Haime). *Bull. Mar. Sci.* 21:952–959.

———. 1973. Interspecific aggression by scleractinian corals. II. Why the race is not only to the swift. *Bull. Mar. Sci.* 23:260–279.

Lang, J. 1971. Interspecific aggression by scleractinian corals. I. The rediscovery of *Scolemia cubensis* (Milne Edwards and Haime). *Bull. Mar. Sci.* 21: 952–959.

Lavallard, R., G. Balas, et R. Schlenz. 1969. Contribution a l'étude de la croissance relative chez *Mytilus perna* L. *Zool. Biol. Mar.* (São Paulo) 26 (324):19–31.

Lawson, G. W. 1955. Rocky shore zonation in the British Cameroons. *West Afr. Sci. Assoc. J.* 1:78–88.

———. 1956. Rocky shore zonation on the Gold Coast. *J. Ecol.* 44:153–170.

———. 1957. Some features of the intertidal ecology of Sierre Leone. *West Afr. Sci. Assoc. J.* 3:166–174.

———. 1966. The littoral ecology of West Africa. *Oceanogr. Mar. Biol. Ann. Rev.* 4:405–448.

———. 1969. Some observations on the littoral ecology of rocky shores in East Africa (Kenya and Tanzania). *Trans. Roy. Soc. S. Afr.* 38:329–339.

Lawson, G. W., and T. A. Norton. 1971. Some observations on the littoral and sublittoral zonation at Teneriffe (Canary Islands). *Bot. Mar.* 14:116–120.

Laxton, J. H. 1970. Shell growth in some New Zealand Cymatiidae (Gastropoda: Prosobranchia). *J. Exp. Mar. Biol. Ecol.* 4:250–260.

———. 1974. Aspects of the ecology of the coral-eating starfish *Acanthaster planci*. *Biol. J. Linn. Soc.* 6:19–45.

Lee, M. A. B. 1974. Distribution of native and invader plant species on the island of Guam. *Biotropica* 6:158–164.

Levin, D. A. 1975. Pest pressure and recombination systems in plants. *Amer. Nat.* 109:437–451.

——. 1976. Alkaloid-bearing plants: an ecogeographic perspective. *Amer. Nat.* 110:261–284.

Levinton, J. S. 1970. The paleoecological significance of opportunistic species. *Lethaia* 3:69–78.

——. 1973. Genetic variation in a gradient of environmental variability: marine Bivalvia (Mollusca). *Science* 180:75–76.

——. 1975. Levels of genetic polymorphism at two enzyme encoding loci in eight species of the genus *Macoma* (Mollusca: Bivalvia). *Mar. Biol.* 33:41–47.

Levinton, J. S., and R. K. Bambach. 1970. Some ecological aspects of bivalve mortality patterns. *Amer. J. Sci.* 268:97–112.

——. 1975. A comparative study of Silurian and Recent deposit-feeding bivalve communities. *Paleobiology* 1:97–124.

Levinton, J. S., and D. S. Fundiller. 1975. An ecological and physiological approach to the study of biochemical polymorphisms. *Proc. Ninth Eur. Mar. Biol. Symp.*:165–178.

Lewinsohn, C., and L. B. Holthuis. 1964. New records of decapod Crustacea from the Meriterranean coast of Israel and the Eastern Mediterranean. *Zool. Meded.* 40:45–63.

Lewis, J. B. 1960. The fauna of rocky shores of Barbados, West Indies. *Canad. J. Zool.* 38:391–435.

——. 1963. Environmental and tissue temperatures of some tropical intertidal marine animals. *Biol. Bull.* 124:277–284.

Lewis, J. R. 1954. Observations on a high-level population of limpets. *J. Anim. Ecol.* 23:85–100.

——. 1955. The mode of occurrence of the universal intertidal zone in Great Britain. *J. Ecol.* 43:270–286.

——. 1964. *The ecology of rocky shores*. London: English Universities Press.

Lewis, J. R., and R. S. Bowman. 1975. Local habitat-induced variations in the population dynamics of *Patella vulgata* L. *J. Exp. Mar. Biol. Ecol.* 17:165–203.

Lewontin, R. C. 1974. *The genetic basis of evolutionary change*. New York: Columbia University Press.

Lipkin, Y. 1972. Marine algal and sea-grass flora of the Suez Canal. *Israel J. Zool.* 21:405–446.

——. 1975. Food of the Red Sea dugong (Mammalia: Sirenia) from Sinai. *Israel J. Zool.* 24:81–98.

Lipkin, Y., and U. Safriel. 1971. Intertidal zonation on rocky shores at Mikhmoret (Mediterranean, Israel). *J. Ecol.* 59:1–30.

Lipps, J. H. 1970. Plankton evolution. *Evolution* 24:1–22.

Lipps, J. H., and E. Mitchell. 1976. Trophic model for the adaptive radiations and extinctions of pelagic marine mammals. *Paleobiology* 2:147–155.

Long, E. R. 1974. Marine fouling studies off Oahu, Hawaii. *Veliger* 17:23–36.

Longhurst, A. R. 1958. An ecological survey of the West African marine benthos. *Colonial Office Fish. Publ.* 11:1–102.

Lowenstam, H. A. 1954. Factors affecting the aragonite/calcite ratios in carbonate-secreting marine organisms. *J. Geol.* 62:285–322.

Lowenstam, H. A., and D. P. Abbott. 1975. Vaterite: a mineralization product of the hard tissues of a marine organism (Ascidiacea). *Science* 188:363–365.

Lowry, L. F., and J. S. Pearse. 1973. Abalones and sea urchins in an area inhabited by sea otters. *Mar. Biol.* 23:213–219.

Loya, Y. 1972. Community structure and species diversity of hermatypic corals at Eilat, Red Sea. *Mar. Biol.* 13:100–123.

Loya, Y., and L. B. Slobodkin. 1971. Coral reefs of Eilat (Gulf of Eilat, Red Sea). In *Regional variations in Indian Ocean coral reefs*, ed. D. R. Stoddard and C. M. Yonge. New York: Academic Press, pp. 117–138.

Lubchenco, J., and B. A. Menge. 1978. Community development and persistence in a low rocky intertidal zone. *Ecol. Monogrs.* 48:in press.

Lucas, J. S., and M. M. Jones. 1976. Hybrid crown-of-thorns starfish *(Acanthaster planci* x *A. brevispinus)* reared to maturity in the laboratory. *Nature* 263:409–412.

Luckens, P. A. 1970. Predation and intertidal zonation at Asamushi. *Bull. Mar. Biol. Sta. Asamushi* 14:33–52.

———. 1975. Predation and intertidal zonation of barnacles at Leigh, New Zealand. *N.Z. J. Mar. Fresh-W. Res.* 9:355–378.

Lunz, G. R., Jr. 1947. *Callinectes* versus *Ostrea. J. Elisha Mitchell Sci. Soc.* 63:81.

MacArthur, R. H. 1965. Patterns of species diversity. *Biol. Revs.* 40:510–533.

———. 1972. *Geographical ecology: patterns in the distribution of species.* New York: Harper and Row.

MacArthur, R. H., and E. O. Wilson. 1967. *The theory of island biogeography.* Princeton, N.J.: Princeton University Press.

MacArthur, R. H., J. M. Diamond, and J. R. Karr. 1972. Density compensation in island faunas. *Ecology* 53:330–342.

McCosker, J. E., and C. E. Dawson. 1975. Biotic passage through the Panama Canal, with particular reference to fishes. *Mar. Biol.* 30:343–351.

McDermott, J. J. 1960. The predation of oysters and barnacles by crabs of the family Xanthidae. *Proc. Penn. Acad. Sci.* 34:199–211.

Macdonald, K. B. 1969. Quantitative studies of salt marsh faunas from the North American Pacific coast. *Ecol. Monogrs.* 39:33–66.

MacGeachy, J. K., and C. W. Stearn. 1976. Boring by macro-organisms in the coral *Montastrea annularis* on Barbados reefs. *Int. Rev. Ges. Hydrobiol.* 61:715–745.

MacIntire, I. G., and O. H. Pilkey. 1969. Tropical reef corals: tolerance of low temperatures on the North Carolina continental shelf. *Science* 166:374–375.

McLaughlin, J. J. A., and P. A. Zahl. 1966. Endozoic algae. In *Symbiosis*, ed. S. M. Henry. New York: Academic Press, vol. 1, pp. 257–297.

McLean, R. B., and R. N. Mariscal. 1973. Protection of a hermit crab by its symbiotic sea anemone *Calliactis tricolor. Experientia* 29:128–130.

McNab, B. K. 1971. On the ecological significance of Bergmann's rule. *Ecology* 52:845–854.

Macnae, W. 1968. A general account of the flora and fauna of the mangrove swamps and forests in the Indo-West-Pacific region. *Adv. Mar. Biol.* 6:73–270.

Madsen, H. 1940. A study of the littoral fauna of northwest Greenland. *Meded. Gronland* 100 (3):1–24.

Maes, V. O. 1967. Radulae of two species of *Pleuroploca* (Fasciolariidae) from the Indo-Pacific. *Nautilus* 81:48–54.

Maes, V. O., and D. Raeigle. 1975. Systematics and biology of *Thala floridana* (Gastropoda: Vexillidae). *Malacologia* 15:43–67.

Magalhaes, H. 1948. An ecological study of snails of the genus *Busycon* at Beaufort, North Carolina. *Ecol. Monogrs.* 18:377–409.

Mann, K. H. 1973. Seaweeds: their productivity and strategy of growth. *Science* 182:975–981.

Manning, R. B. 1967. Review of the genus *Odontodactylus* (Crustacea: Stomatopoda). *Proc. U.S. Nat. Mus.* 123 (3606):1–35.

———. 1969. Stomatopod Crustacea of the Western Atlantic. *Stud. Trop. Oceanogr., Univ. Miami* 8:1–380.

Marcus, E., and E. Marcus. 1959. Studies on "Olividae." *Bol. Fac. Fil. Cienc. Letr. São Paulo* 232 (Zool. 22):99–188.

Margolin, A. S. 1964a. The mantle response of *Diodora aspera*. *Animal Behavior* 12:187–194.

———. 1964b. A running response of *Acmaea* to seastars. *Ecology* 45:191–193.

Marincovich, L., Jr. 1973. Intertidal mollusks of Iquique, Chile. *Nat. Hist. Mus. Los Angeles County Sci. Bull.* 16:1–49.

———. 1977. Cenozoic Naticidae (Mollusca: Gastropoda) of the northeastern Pacific. *Bulls. Amer. Paleont.* 70 (294):169–494.

Marshall, A. G., and Lord Medway. 1976. A mangrove community in the New Hebrides, south-west Pacific. *Biol. J. Linn. Soc.* 8:319–336.

Massé, H. 1975. Étude de l'alimentation de *Astropecten arantiacus* Linné. *Cah. Biol. Mar.* 16:495–510.

Mauzey, K. P., C. Birkeland, and P. K. Dayton. 1968. Feeding behavior of asteroids and escape responses of their prey in the Puget Sound region. *Ecology* 49:603–619.

May, V., I. Bennett, and T. E. Thompson. 1970. Herbivore-algal relationships on a coastal rock platform (Cape Banks, N.S.W.). *Oecologia (Berlin)* 3:1–14.

Mayr, E. 1963. *Animal species and evolution*. Cambridge, Mass.: Harvard University Press.

Menge, B. A. 1972a. Foraging strategy of a starfish in relation to actual prey availability and environmental predictability. *Ecol. Monogrs.* 42:25–50.

———. 1972b. Competition for food between two intertidal starfish species. *Ecology* 53:635–644.

———. 1973. Effect of predation and environmental patchiness on the body size of a tropical pulmonate limpet. *Veliger* 16:87–92.

———. 1974. Effect of wave action and competition on brooding and reproductive effort in the seastar *Leptasterias hexactis*. *Ecology* 55:84–93.

———. 1976. Organization of the New England rocky intertidal community: role of predation, competition, and environmental heterogeneity. *Ecol. Monogrs.* 46:355–393.

Menge, B. A., and J. P. Sutherland. 1976. Species diversity gradients: synthesis of the roles of predation, competition, and temporal heterogeneity. *Amer. Nat.* 110:351–369.

Menzel, R. W., and S. H. Hopkins. 1955. Crabs as predators of oysters in Louisiana. *Proc. Nat. Shell Fish. Assoc.* 46:177–184.

Menzel, R. W., and F. E. Nichy. 1958. Studies of the distribution and feeding habits of some oyster predators in Alligator Harbor, Florida. *Bull. Mar. Sci. Gulf Caribbean* 8:129–145.

Menzies, R. J., J. Imbrie, and C. Heezen. 1961. Further considerations regarding the antiquity of the abyssal fauna with evidence of a changing abyssal environment. *Deep Sea Res.* 8:79–94.

Meyer, D. L. 1973a. Distribution and living habits of comatulid crinoids near Discovery Bay, Jamaica. *Bull. Mar. Sci.* 23:244–259.

――――. 1973b. Feeding behavior and ecology of shallow-water unstalked crinoids (Echinodermata) in the Caribbean Sea. *Mar. Biol.* 22:105–129.

Meyer, D. L., and D. B. Macurda. 1977. Adaptive radiation of the comatulid crinoids. *Paleobiology* 3:74–82.

Mienis, H. K. 1973. *Vasum turbinellus* from the Mediterranean. *Argamon, Israel J. Malacol.* 4:6.

Miller, B. A. 1975. The biology of *Terebra gouldi* Deshayes, 1859, and a discussion of life history similarities among other terebrids of similar proboscis type. *Pac. Sci.* 29:227–241.

Miller, R. J., and K. H. Mann. 1973. Ecological energetics of the seaweed zone in a marine bay on the Atlantic coast of Canada. III. Energy transformations by sea urchins. *Mar. Biol.* 18:99–114.

Mooney, H. A., and E. L. Dunn. 1970. Convergent evolution of the Mediterranean-climate evergreen sclerophyll shrubs. *Evolution* 24:292–303.

Moore, H. B. 1972. Aspects of stress in the tropical marine environment. *Adv. Mar. Biol.* 10:217–269.

Moore, H. B., and B. F. McPherson. 1965. A contribution to the study of the productivity of the urchins *Tripneustes esculentus* and *Lytechinus variegatus*. *Bull. Mar. Sci.* 15:855–871.

Moore, I. A., and J. W. Moore. 1974. Food of shorthorn sculpin, *Myxocephalus scorpius*, in the Cumberland Sound area of Baffin Island. *J. Fish. Res. Bd. Canada* 31:355–359.

Moore, M. M., Jr. 1975. Foraging of the Western gull *Larus occidentalis* and its impact on the chiton *Nuttallina californica*. *Veliger* 18 (suppl.):51–53.

Morrison, J. P. E. 1972. Mediterranean *Siphonaria*: west and east—old and new. *Argamon, Israel J. Malacol.* 3:51–62.

Morse, D. H. 1974. Niche breadth as a function of social dominance. *Amer. Nat.* 108:818–830.

――――. 1975. Ecological aspects of adaptive radiation in birds. *Biol. Revs.* 50:167–214.

Morton, J. E. 1955. The evolution of the Ellobiidae with a discussion on the origin of the Pulmonata. *Proc. Zool. Soc. London* 125:127–168.

Morton, J. E., and M. Miller. 1968. *The New Zealand seashore*. London: Collins.

Muntz, L., F. J. Ebling, and J. A. Kitching. 1965. The ecology of Lough Ine. XIV. Predatory activity of large crabs. *J. Anim. Ecol.* 35:315–329.

Murphy, D. J. 1977a. Metabolic and tissue solute changes associated with changes in the freezing tolerance of the bivalve mollusc *Modiolus demissus*. *J. Exp. Biol.* 69:1–12.

――――. 1977b. A calcium-dependent mechanism responsible for increasing the freezing tolerance of the bivalve mollusc *Modiolus demissus*. *J. Exp. Biol.* 69:13–21.

Murphy, D. J., and S. K. Pierce. 1975. The physiological basis of changes in the freezing tolerance of intertidal molluscs. I. Response to subfreezing temperatures and the influence of salinity and temperature acclimation. *J. Exp. Zool.* 193:313–321.

Nations, J. D. 1975. The genus *Cancer* (Crustacea: Brachyura): systematics, biogeography and fossil record. *Nat. Hist. Mus. Los Angeles County Sci. Bull.* 23:1–104.

Nesis, K. N. 1965. Ecology of *Cyrtodaria siliqua* and history of the genus *Cyrtodaria* (Bivalvia: Hiatellidae). *Malacologia* 3:197–210.

Neudecker, S. 1977. Transplant experiments to test the effect of fish grazing on coral distribution. *Proc. Third Intern. Coral Reef Symp.* 1:317–323.

Newcombe, C. L. 1935. A study of the community relationships of the sea mussel, *Mytilus edulis* L. *Ecology* 16:234–243.

Newell, N. D. 1971. An outline history of tropical organic reefs. *Amer. Mus. Novitates* 2465:1–37.

Newell, R. C. 1970. *Biology of intertidal animals.* New York: Elsevier.

Newman, W. A. 1960. The paucity of intertidal barnacles in the tropical Western Pacific. *Veliger* 2:89–94.

Nicol, D. 1964. Lack of shell-attached pelecypods in Arctic and Antarctic waters. *Nautilus* 77:92–93.

———. 1965. Ecological implications of living pelecypods with calcareous spines. *Nautilus* 78:109–116.

———. 1966. Size of pelecypods in Recent marine faunae. *Nautilus* 79:109–113.

———. 1967. Some characteristics of cold-water marine pelecypods. *J. Paleont.* 41:1330–1340.

Nicotri, M. E. 1977. Grazing effects of four marine intertidal herbivores on the microflora. *Ecology* 58:1020–1032.

Nielsen, C. 1975. Observations on *Buccinum undatum* L. attacking bivalves and on prey responses, with a short review on attack methods of other prosobranchs. *Ophelia* 13:87–108.

Norton-Griffiths, M. 1967. Some ecological aspects of the feeding behaviour of the oystercatcher *Haematopus ostralegus* on the edible mussel *Mytilus edulis. Ibis* 109:412–424.

Nybakken, J. 1968. Notes on the food of *Conus dalli* Stearns, 1873. *Veliger* 11:50.

Ockelmann, K. W. 1965. Developmental types in marine bivalves and their distribution along the Atlantic coast of Europe. *Proc. First Eur. Malacol. Congr.* (1962):25–35.

Odum, E. P. 1969. Strategy of ecosystem development. *Science* 164:162–170.

Old, W. E., Jr. 1975. Living *Conus* from the New World, with special reference to "twin species." *Bull. Amer. Malacol. Union*:52.

Olivier, S. R., and P. E. Penchaszadeh. 1968. Observaciones sobre la ecología y biología de *Siphonaria (Pachysiphonaria) lessoni* (Blainville, 1824) (Gastropoda: Siphonariidae) en el litoral rocoso de Mar del Plata (Buenos Aires). *Cah. Biol. Mar.* 9:469–491.

Olivier, S. R., I. Kreibohm de Paternoster, and R. Bastida. 1968. Estúdios biocenóticos en las costas de Chubut (Argentina). I. Zonación biocenológica de Puerto Pardelas (Golfo Nuevo). *Bol. Inst. Biol. Mar. (Mar del Plata)* 10:1–77.

Olla, B. L., A. J. Bejda, and A. D. Martin. 1974. Daily activity, movements, feeding, and seasonal occurrence in the tautog, *Tautoga onitis*. *Fish. Bull.* 72:27–35.

Olney, P. J. S. 1963. The food and feeding habits of tufted duck *Aythya fuligula*. *Ibis* 105:55–62.

———. 1965. The food and feeding habits of sheld duck *Tadorna tadorna*. *Ibis* 107:527–533.

Olney, P. J. S., and D. H. Mills. 1963. The food and feeding habits of goldeneye *Bucephala clangula* in Great Britain. *Ibis* 105:293–300.

Olsson, A. A. 1956. Studies on the genus *Olivella*. *Proc. Acad. Sci. Philadelphia* 108:155–225.

———. 1961. *Mollusks of the tropical Eastern Pacific particularly from the southern half of the Panamic-Pacific faunal province (Panama to Peru) (Pelecypoda)*. Ithaca, N.Y.: Paleontological Research Institute.

———. 1972. Origin of the existing Panamic molluscan biotas in terms of their geologic history and their separation by the isthmian land bridge. *Bull. Biol. Soc. Washington* 2:117–124.

Olsson, A. A., and L. E. Crovo. 1968. Observations on aquarium specimens of *Oliva sayana*, Ravenel. *Veliger* 11:31–32.

Olsson, A. A., and R. E. Petit. 1968. Notes on *Siphocypraea*. *Bull. Amer. Paleont.* 54:279–289.

Onuf, C. P., J. M. Teal, and I. Valiela. 1977. Interactions of nutrients, plant growth and herbivory in a mangrove ecosystem. *Ecology* 58:514–526.

Orians, G. H., and O. T. Solbrig. 1977. A cost-income model of leaves and roots with special reference to arid and semiarid areas. *Amer. Nat.* 111:677–690.

Ormond, R. F. G., N. J. Hanscomb, and D. H. Beach. 1976. Food selection and learning in the crown-of-thorns starfish, *Acanthaster planci* (L.). *Mar. Behav. Physiol.* 4:93–105.

Ott, B., and J. B. Lewis. 1972. The importance of the gastropod *Coralliophila abbreviata* (Lamarck) and the polychaete *Hermodice carunculata* (Pallas) as coral reef predators. *Canad. J. Zool.* 50:1651–1656.

Owen, G. 1958. Shell form, pallial attachment and ligament in the Bivalvia. *Proc. Zool. Soc. London* 131:637–648.

Paine, R. T. 1962. Ecological diversification in sympatric gastropods of the genus *Busycon*. *Evolution* 16:515–523.

———. 1963a. Trophic relationships of eight sympatric predatory gastropods. *Ecology* 44:63–73.

———. 1963b. Food recognition and predation on opisthobranchs by *Navanax inermis* (Gastropoda: Opisthobranchia). *Veliger* 6:1–9.

———. 1966a. Food web complexity and species diversity. *Amer. Nat.* 100:65–75.

———. 1966b. Function of labial spines, composition of diet, and size of certain marine gastropods. *Veliger* 9:17–24.

———. 1969. The *Pisaster-Tegula* interaction: prey patches, predator food preference, and intertidal community structure. *Ecology* 50:950–961.

———. 1971. A short-term experimental investigation of resource partitioning in a New Zealand rocky intertidal habitat. *Ecology* 52:1096–1106.

———. 1974. Intertidal community structure: experimental studies on the relationship between a dominant competitor and its principal predator. *Oecologia (Berlin)* 15:93–120.

————. 1976a. Biological observations on a subtidal *Mytilus californianus* bed. *Veliger* 19:125–130.

————. 1976b. Size-limited predation: an observational and experimental approach with the *Mytilus-Pisaster* interaction. *Ecology* 57:858–873.

Paine, R. T., and A. R. Palmer. 1978. *Sycyases sanguineus*: a unique trophic generalist from the Chilean intertidal. *Copeia* 1:75–80.

Paine, R. T., and R. L. Vadas. 1969a. Effects of grazing by the sea urchins, *Strongylocentrotus* spp., on benthic algal populations. *Limnol. Oceanogr.* 14:710–719.

————. 1969b. Calorific values of benthic marine algae and its postulated relation to invertebrate food preference. *Mar. Biol.* 4:79–86.

Palmer, A. R. 1977. Function of shell sculpture in marine gastropods: hydrodynamic destabilization in *Ceratostoma foliatum*. *Science* 197:1293–1295.

————. 1978. Fish predation as an evolutionary force molding gastropod shell form: a tropical-temperate comparison. *Evolution:* in press.

Palmer, J. B., and P. W. Frank. 1974. Estimates of growth of *Cryptochiton stelleri* (Middendorff, 1846) in Oregon. *Veliger* 16:301–304.

Pannella, G., and C. MacClintock. 1968. Biological and environmental rhythms reflected in molluscan shell growth. *J. Paleont.* 42, *Paleont. Soc. Mem.* II:64–80.

Parker, R. H., 1961. Speculations on the origin of the invertebrate faunas of the lower continental slope. *Deep Sea Res.* 8:286–293.

Parsons, D. J., and P. R. Moldenke. 1975. Convergence in vegetation structure along analogous climatic gradients in California and Chile. *Ecology* 56: 950–957.

Penrith, M. L., and B. F. Kensley. 1970a. The constitution of the intertidal fauna of rocky shores of South West Africa. I. Luderitzbucht. *Cimbebasia* (A) 1:191–239.

————. 1970b. The constitution of the intertidal fauna of rocky shores of South West Africa. II. Rocky Point. *Cimbebasia* (A) 1:245–268.

Percharde, P. L. 1974. Underwater observations on *Conus ermineus* Born 1778 in Trinidad and Tobago. *Verhandl. Naturf. Ges. Basel* 84:501–507.

Pérès, J. M., and J. Picard. 1955. Biotopes et biocénoses de la Mediterranée occidentale comparés à ceux de l'Atlantique nord-oriental. *Arch. Zool. Exp. Gen.* 92:1–72.

Peterson, C. H. 1975. Stability of species and of community of the benthos of two lagoons. *Ecology* 56:958–965.

Pettitt, C. 1975. A review of the predators of *Littorina*, especially those of *L. saxatilis* (Olivi) (Gastropoda: Prosobranchia). *J. Conchol.* 28:343–357.

Petuch, E. J. 1976. An unusual molluscan assemblage from Venezuela. *Veliger* 18:322–325.

Phillips, B. F., N. A. Campbell, and B. R. Wilson. 1973. A multivariate study of geographic variation in the whelk *Dicathais*. *J. Exp. Mar. Biol. Ecol.* 11: 27–69.

Phillips, D. W. 1976. The effect of a species-specific avoidance response to predatory starfish on the intertidal distribution of two gastropods. *Oecologia* (*Berlin*) 23:83–94.

Picard, J. 1957. Note sommaire sur les equivalences entre la zonation marine de la côte atlantique du Portugal et des côtes de Mediterranée occidentale. *Rec. Trav. Sta. Mar. Endoume Bull.* 12:22–27.

Pichon, M. 1971. Comparative study of the main features of some coral reefs of Madagascar, La Réunion, and Mauritius. In *Regional variation in Indian Ocean coral reefs*, ed. D. R. Stoddard and C. M. Yonge. New York: Academic Press, pp. 185–215.

Pilson, M. E. Q., and P. B. Taylor. 1961. Hole drilling by *Octopus*. *Science* 134:1366–1368.

Pimentel, D., and A. C. Bellotti. 1976. Parasite-host population systems and genetic stability. *Amer. Nat.* 110:877–888.

Pitelka, L. F. 1977. Energy allocation in annual and perennial lupines (*Lupinus*: Leguminosae). *Ecology* 58:1055–1065.

Plante, R. 1964. Contribution à l'étude des peuplements des hauts niveaux sur substrats solides nonrécifaux dans la region de Tuléar. *Rec. Trav. Sta. Mar. Endoume-Marseille*, suppl. 2:207–315.

Ponder, W. F. 1970. The morphology of *Alcithoe arabica* (Gastropoda: Volutidae). *Malacol. Revs.* 3:127–165.

———. 1973. The origin and evolution of the Neogastropoda. *Malacologia* 12:295–338.

Por, F. D. 1971. One hundred years of Suez Canal—a century of Lessepsian migration: retrospect and viewpoints. *Syst. Zool.* 20:138–159.

———. 1975. Pleistocene pulsation and preadaptation of biotas in mediterranean seas: consequences for Lessepsian migration. *Syst. Zool.* 24:72–78.

Porter, J. W. 1972a. Ecology and species diversity of coral reefs on opposite sides of the Isthmus of Panama. *Bull. Biol. Soc. Washington* 2:89–116.

———. 1972b. Patterns of species diversity in Caribbean reef corals. *Ecology* 53:745–748.

———. 1974a. Community structure and coral reefs on opposite sides of the Isthmus of Panama. *Science* 186:543–545.

———. 1974b. Zooplankton feeding by the Caribbean reef-building coral *Montastrea cavernosa*. *Proc. Second Intern. Coral Reef Symp.* 1:111–125.

———. 1976. Autotrophy, heterotrophy, and resource partitioning in Caribbean reef-building corals. *Amer. Nat.* 110:731–742.

Porter, K. G. 1973. Selective grazing and differential digestion of algae by zooplankton. *Nature* 244:179–180.

Powell, A. W. B. 1964. The family Turridae in the Indo-Pacific. I. The subfamily Turrinae. *Indo-Pacific Mollusca* 1 (5):227–345.

Purchon, R. D., and I. Enoch. 1954. Zonation of the marine fauna and flora on a rocky shore near Singapore. *Bull. Raffles Mus.* 25:47–65.

Quast, J. C. 1968. Observations on the food of the kelp-bed fishes. *Calif. Dept. Fish Game Bull.* 139:109–142.

Radwin, G. E. 1969. A Recent molluscan fauna from the Caribbean coast of southeastern Panama. *Trans. San Diego Soc. Nat. Hist.* 15:229–236.

———. 1977. The family Columbellidae in the Western Atlantic. *Veliger* 19:403–417.

Radwin, G. E., and A. D'Attilio. 1976. *Murex shells of the world*. Palo Alto, Calif.: Stanford University Press.

Radwin, G. E., and H. W. Wells. 1968. Comparative radular morphology and feeding habits of muricid gastropods from the Gulf of Mexico. *Bull. Mar. Sci.* 18:72–85.

Randall, J. E. 1961. Overgrazing of algae by herbivorous marine fishes. *Ecology* 42:812.

———. 1964. Contribution to the biology of the queen conch, *Strombus gigas*. *Bull. Mar. Sci. Gulf Caribbean* 14:246–295.

———. 1965. Grazing effect on sea grasses by herbivorous reef fishes in the West Indies. *Ecology* 46:255–260.

———. 1967. Food habits of reef fishes of the West Indies. *Stud. Trop. Oceanogr., Univ. Miami* 5:665–847.

———. 1974. The effect of fishes on coral reefs. *Proc. Second Intern. Coral Reef Symp.* 1:159–166.

Randall, J. E., and W. D. Hartman. 1968. Sponge-feeding fishes of the West Indies. *Mar. Biol.* 1:216–225.

Randall, J. E., and G. L. Warmke. 1967. The food habits of the hogfish (*Lachnolaimus maximus*), a labrid fish from the Western Atlantic. *Carib. J. Sci.* 3–4:141–144.

Ranson, G. 1952. Notes sur la cause probable de l'absence de recifs coralliens aux îles Marquises et de l'activité reduite des coraux recifaux à Tahiti, aux Tuamotu, aux Hawaii, etc. *C. R. Sommaire Sci. Soc. Biogeogr.* 248–249:3–11.

Rao, K. P. 1953. Rate of water propulsion in *Mytilus californianus* as a function of latitude. *Biol. Bull.* 104:173–181.

Ray, C. 1960. The application of Bergmann's and Allen's Rules to poikilotherms. *J. Morphol.* 106:85–108.

Ray, C. E. 1976. Geography of phocid evolution. *Syst. Zool.* 25:391–406.

Reaka, M. L. 1978. The evolutionary ecology of life history patterns in stomatopod Crustacea. In *Reproductive ecology of marine invertebrates*, Belle Baruch Inst. Mar. Biol. Symp., ed. S. Stancyk. Columbia: University of South Carolina Press, forthcoming.

Rebach, S. 1974. Burying behavior in relation to substrate and temperature in the hermit crab, *Pagurus longicarpus*. *Ecology* 55:195–198.

Recher, H. F. 1966. Some aspects of the ecology of migrant shorebirds. *Ecology* 47:393–407.

Reeder, W. G. 1951. Stomach analysis of a group of shorebirds. *Condor* 53:43–45.

Rehder, H. A. 1973. The family Harpidae of the world. *Indo-Pacific Mollusca* 3 (16):207–274.

Rehder, H. A., and J. E. Randall. 1975. Ducie Atoll: its history, physiography and biota. *Atoll Res. Bull.* 183:1–40.

Reimer, A. A. 1976. Description of a *Tetraclita stalactifera panamensis* community on a rocky intertidal Pacific shore of Panama. *Mar. Biol.* 35:225–238.

Reiswig, H. M. 1973. Population dynamics of three Jamaican Demospongiae. *Bull. Mar. Sci.* 23:191–226.

Remington, C. L. 1968. Suture-zones of hybrid interaction between recently joined biotas. *Evol. Biol.* 2:321–428.

Repenning, C. A. 1976a. *Enhydra* and *Enhydriodon* from the Pacific coast of North America. *J. Res. U.S. Geol. Surv.* 4:305–315.

———. 1976b. Adaptive evolution of sea lions and walruses. *Syst. Zool.* 25:375–390.

Revelle, A., and R. Fairbridge. 1957. Carbonates and carbon dioxide. *Geol. Soc. Amer. Mem.* 67 (1):239–296.

Rex, M. A. 1973. Deep-sea species diversity: decreased gastropod diversity at abyssal depths. *Science* 181:1051–1053.

———. 1976. Biological accommodation in the deep-sea benthos: comparative evidence on the importance of predation and productivity. *Deep Sea Res.* 23:975–987.

Rex, M. A., and K. J. Boss. 1976. Open coiling in Recent gastropods. *Malacologia* 15:289–297.

Reyment, R. A. 1967. Paleoethology and fossil drilling gastropods. *Kansas Acad. Sci. Trans.* 70:33–50.

Reynolds, W. W., and L. J. Reynolds. 1977. Zoogeography and the predator-prey "arms race": a comparison of *Eriphia* and *Nerita* species from three faunal regions. *Hydrobiologia* 56:63–68.

Rhoads, D. C., and J. W. Morse. 1971. Evolutionary and ecologic significance of oxygen-deficient marine basins. *Lethaia* 4:413–428.

Rhoads, D. C., and G. Pannella. 1970. The use of molluscan shell growth patterns in ecology and paleoecology. *Lethaia* 2:143–161.

Richards, P. W. 1952. *The tropical rain forest: an ecological study.* London: Cambridge University Press.

———. 1973. Africa, the "odd man out." In *Tropical forest ecosystems in Africa and South America: a comparative review*, ed. B. J. Meggers, E. S. Ayensu, and W. D. Duckworth. Washington, D.C.: Smithsonian Press, pp. 21–26.

Ricketts, E. F., and J. Calvin. 1968. *Between Pacific tides* (rev. J. W. Hedgpeth). Palo Alto, Calif.: Stanford University Press.

Ricklefs, R. E., and G. W. Cox. 1972. Taxon cycles in the West Indian avifauna. *Amer. Nat.* 106:195–219.

Ricklefs, R. E., and K. O'Rourke. 1975. Aspect diversity in moths: a temperate-tropical comparison. *Evolution* 29:313–324.

Roberts, T. R. 1972. Ecology of fishes in the Amazon and Congo basins. *Bull. Mus. Comp. Zool. Harvard Univ.* 143:117–147.

Robertson, R. 1957. The subgenus *Halopsephus* Rehder, with notes on the Western Atlantic species of *Turbo* and the subfamily Bothropomatinae Thiele. *J. Wash. Acad. Sci.* 47:316–319.

———. 1963. Wentletraps (Epitoniidae) feeding on sea anemones and corals. *Proc. Malacol. Soc. London* 35:51–63.

———. 1970. Review of the predators and parasites of stony corals, with special reference to symbiotic prosobranch gastropods. *Pac. Sci.* 24:43–54.

———. 1975. Faunal affinities of the Architectonicidae in the Eastern Pacific. *Bull. Amer. Malacol. Union:* 51.

———. 1976. *Heliacus trochoides:* an Indo-West-Pacific architectonicid newly found in the Eastern Pacific (mainland Ecuador). *Veliger* 19:13–18.

Robilliard, G. A. 1971. Predation by the nudibranch *Dirona albolineata* on three species of prosobranchs. *Pac. Sci.* 25:429–435.

Ropes, J. W. 1968. The feeding habits of the green crab, *Carcinus maenas* (L.). *Fish. Bull.* 67:183–203.

Rosen, B. R. 1971. The distribution of reef coral genera in the Indian Ocean. In *Regional variation in Indian Ocean coral reefs*, ed. D. R. Stoddard and C. M. Yonge. New York: Academic Press, pp. 263–299.

Rosenblatt, R. H. 1963. Some aspects of speciation in marine shore fishes. In *Speciation in the sea*, ed. J. P. Harding and N. Tebble. London: Systematics Assoc. Publ. 5, pp. 171–180.

———. 1967. The zoogeographic relationships of the marine shore fishes of tropical America. *Stud. Trop. Oceanogr., Univ. Miami* 5:579–587.

Rosenthal, R. J. 1971. Trophic interaction between the sea star *Pisaster giganteus* and the gastropod *Kelletia kelleti*. *Fish. Bull.* 69:669–679.

Rosenzweig, M. L. 1973. Evolution of the predator isocline. *Evolution* 27:84–94.

Rosewater, J. 1965. The family Tridacnidae in the Indo-Pacific. *Indo-Pacific Mollusca* 1 (6):347–394.

———. 1970. The family Littorinidae in the Indo-Pacific. I. The subfamily Littorininae. *Indo-Pacific Mollusca* 2 (11):417–506.

———. 1972. The family Littorinidae in the Indo-Pacific. II. Subfamilies Tectariinae and Echininae. *Indo-Pacific Mollusca* 2 (12):507–533.

Rossi, A. C., and V. Parisi. 1973. Experimental studies of predation by the crab *Eriphia verrucosa* on both snail and hermit crab occupants of conspecific gastropod shells. *Boll. Zool.* 40:117–135.

Rubinoff, I. 1968. Central American sea-level canal: possible biological effects. *Science* 161:857–861.

Rubinoff, I., and C. Kropach. 1970. Differential reactions of Atlantic and Pacific predators to sea snakes. *Nature* 228:1288–1290.

Rubinoff, R. W., and I. Rubinoff. 1968. Interoceanic colonization of a marine goby through the Panama Canal. *Nature* 217:476–478.

———. 1971. Geographic and reproductive isolation in Atlantic and Pacific populations of Panamanian *Bathygobius*. *Evolution* 25:88–97.

Rudwick, M. J. S. 1962. Notes on the ecology of brachiopods in New Zealand. *Trans. Roy. Soc. N.Z.* 25:327–335.

———. 1964. The function of zigzag deflexions in the commissures of fossil brachiopods. *Palaeontology* 7:135–171.

Runnegar, B. 1974. Evolutionary history of the bivalve subclass Anomalodesmata. *J. Paleont.* 48:904–940.

Runnegar, B., and P. A. Jell. 1976. Australian Middle Cambrian molluscs and their bearing on early molluscan evolution. *Alcheringa* 1:109–138.

Sacchi, C. F. 1970. Les epibiontes animaux de *Littorina obtusata* (L.) et de *Littorina mariae* Sacchi et Rast. (Gastropoda, Prosobranchia). *Cah. Biol. Mar.* 11:43–56.

Safriel, U., and Y. Lipkin. 1975. Patterns of colonization of the Eastern Mediterranean intertidal zone by Red Sea immigrants. *J. Ecol.* 63:61–63.

Saito, T. 1976. Geologic significance of coiling direction in the planktonic Foraminifera *Pullenatina*. *Geology* 4:305–309.

Sale, P. F. 1977. Maintenance of high diversity in coral reef fish communities. *Amer. Nat.* 111:337–359.

Salvat, B. 1970. Études quantitatives (comptages et biomasses) sur les mollusques récifaux de l'atoll de Fangataufa (Tuamotu—Polynesie). *Cah. Pac.* 14:1–58.

————. 1971. Mollusques lagunaires et récifaux de l'île de Raevavae (Australes, Polynesie). *Malacol. Revs.* 4:1–15.

Salvat, B., and J. P. Ehrhardt. 1970. Mollusques de l'île Clipperton. *Bull. Mus. Nat. Hist. Natur. Paris* (2) 42:223–231.

Sammarco, P. W., J. S. Levinton, and J. C. Ogden. 1974. Grazing and control of coral reef community structure by *Diadema antillarum* Philippi (Echinodermata: Echinoidea): a preliminary study. *J. Mar. Res.* 32:47–53.

Sanders, H. L. 1968. Marine benthic diversity: a comparative study. *Amer. Nat.* 102:243–282.

Sasekumar, A. 1974. The distribution of macrofauna on a Malayan mangrove shore. *J. Anim. Ecol.* 43:51–69.

Scheltema, R. S. 1971. Larval dispersal as a means of genetic exchange between geographically separated populations of shallow-water benthic marine gastropods. *Biol. Bull.* 140:284–322.

————. 1977. Dispersal of marine invertebrate organisms: paleobiogeographic and biostratigraphic implications. In *Concepts and methods of biostratigraphy*, ed. E. G. Kauffman and J. E. Hazel. Stroudsburg, Penn.: Dowden, Hutchinson, and Ross, pp. 73–108.

Scholander, P. F., W. Flagg, V. Walters, and L. Irving. 1953. Climatic adaptation in Arctic and tropical poikilotherms. *Physiol. Zool.* 26:67–92.

Seapy, R. R., and W. J. Hoppe. 1973. Morphological and behavioral adaptations to desiccation in the intertidal limpet *Acmaea* (*Collisella*) *strigatella*. *Veliger* 16:181–188.

Sebens, K. P. 1976. Ecology of Caribbean sea anemones in Panama: utilization of space on a coral reef. In *Coelenterate ecology and behavior*, ed. G. O. Mackie. New York: Plenum Press, pp. 67–77.

Seed, R. 1968. Factors influencing shell shape in the mussel *Mytilus edulis*. *J. Mar. Biol. Assoc. U.K.* 48:561–584.

————. 1969a. The ecology of *Mytilus edulis* L. (Lamellibranchiata) on exposed rocky shores. I. Breeding and settlement. *Oecologia (Berlin)* 3:277–316.

————. 1969b. The ecology of *Mytilus edulis* (Lamellibranchiata) on exposed rocky shores. II. Growth and mortality. *Oecologia (Berlin)* 3:317–350.

————. 1971. A physiological and biochemical approach to the taxonomy of *Mytilus edulis* and *M. galloprovincialis* Lmk. from s.w. England. *Cah. Biol. Mar.* 12:291–322.

Segal, E. 1956a. Microgeographic variation as thermal acclimation in an intertidal mollusc. *Biol. Bull.* 111:129–152.

————. 1956b. Adaptive differences in water-holding capacity in an intertidal gastropod. *Ecology* 37:174–178.

Segal, E., K. P. Rao, and T. W. James. 1953. Rate of activity as a function of intertidal height within populations of some littoral molluscs. *Nature* 172:1108–1109.

Segerstråle, S. H. 1957. Baltic Sea. *Geol. Soc. Amer. Mem.* 67 (1):751–800.

Seilacher, A. 1973. Fabricational noise in adaptive morphology. *Syst. Zool.* 22:451–465.

Selander, R. K., S. Y. Yang, R. C. Lewontin, and W. E. Johnson. 1970. Genetic variation in the horseshoe crab (*Limulus polyphemus*), a phylogenetic "relic." *Evolution* 24:402–414.

Shepherd, S. A. 1973. Studies on southern Australian abalone (genus *Haliotis*). I. Ecology of the five sympatric species. *Austr. J. Mar. Fresh-W. Res.* 24:217–257.

Shotwell, J. A. 1950. Distribution of volume and relative linear measurement changes in *Acmaea*, the limpet. *Ecology* 31:51–61.

Shoup, J. B. 1968. Shell opening by crabs of the genus *Calappa*. *Science* 160:887–888.

Shuto, T. 1974. Larval ecology of prosobranch gastropods and its bearing on biogeography and paleontology. *Lethaia* 7:239–256.

Simpson, G. G. 1953. *The major features of evolution*. New York: Columbia University Press.

Simpson, R. D. 1976. Physical and biotic factors limiting the distribution and abundance of littoral molluscs on Macquarie Island (Sub-Antarctic). *J. Exp. Mar. Biol. Ecol.* 21:11–49.

Slobodkin, L. B., and H. L. Sanders. 1969. On the contribution of environmental predictability to species diversity. *Brookhaven Symp. Biol.* 22:82–95.

Smith, D. A. S. 1975. Polymorphism and selective predation in *Donax faba* Gmelin (Bivalvia: Tellinacea). *J. Exp. Mar. Biol. Ecol.* 17:205–219.

Snyder, N. F. R., and H. A. Snyder. 1969. A comparative study of mollusc predation by limpkins, Everglade kites, and boat-tailed grackles. *Living Bird* 8:177–223.

———. 1971. Defenses of the Florida apple snail *Pomacea paludosa*. *Behaviour* 40:175–215.

Sohl, N. F. 1969. The fossil record of shell boring by snails. *Amer. Zool.* 9:725–734.

Soot-Ryen, T. 1955. A report on the family Mytilidae (Pelecypoda). *Allan Hancock Pac. Exped.* 20(1):1–176.

Sourie, R. 1954. Contribution à l'étude écologique des côtes rocheuses du Sénégal. *Mém. Inst. Franc. Afr. Noir* 38:7–342.

Southward, A. J. 1958a. The zonation of plants and animals on rocky sea shores. *Biol. Revs.* 33:137–177.

———. 1958b. Note on the temperature tolerances of some intertidal animals in relation to environmental temperatures and geographical distribution. *J. Mar. Biol. Assoc. U.K.* 37:49–66.

———. 1975. Intertidal and shallow water Cirripedia of the Caribbean. *Stud. Curaçao Other Carib. Islands* 150:1–53.

Southward, A. J., and E. C. Southward. 1967. On the biology of an intertidal chthamalid (Crustacea, Cirripedia) from the Chukchi Sea. *Arctic* 20:8–20.

Spight, T. M. 1973. Ontogeny, environment, and shape of a marine snail *Thais lamellosa* Gmelin. *J. Exp. Mar. Biol. Ecol.* 13:215–228.

———. 1976a. Censuses of rocky shore prosobranchs from Washington and Costa Rica. *Veliger* 18:309–317.

———. 1976b. Ecology of hatching size for marine snails. *Oecologia (Berlin)* 24:183–294.

———. 1976c. Colors and patterns of an intertidal snail, *Thais lamellosa*. *Res. Popul. Ecol.* 17:176–190.

Spight, T. M., and J. Emlen. 1976. Clutch sizes of two marine snails with a changing food supply. *Ecology* 57:1162–1178.

Spight, T. M., and A. Lyons. 1974. Development and functions of the shell sculpture of the marine snail *Ceratostoma foliatum*. *Mar. Biol.* 24:77–83.

Spight, T. M., C. Birkeland, and A. Lyons. 1974. Life histories of large and small murexes (Prosobranchia: Muricidae). *Mar. Biol.* 24:229–242.

Squires, H. J. 1970. Lobster *(Homarus americanus)* fishery and ecology in Port au Port Bay, 1960-65. *Proc. Nat. Shell Fish. Assoc.* 60:22–39.

Staiger, H. 1957. Genetical and morphological variation in *Purpura lapillus* with respect to local and regional differentiation of population groups. *Année Biol.* 33:252–258.

Stanley, S. M. 1968. Post-Paleozoic adaptive radiation of infaunal bivalve molluscs—a consequence of mantle fusion and siphon formation. *J. Paleont.* 42:214–229.

———. 1969. Bivalve mollusk burrowing aided by discordant shell ornamentation. *Science* 166:634–635.

———. 1970. Relation of shell form to life habits of the Bivalvia (Mollusca). *Geol. Soc. Amer. Mem.* 125:1–296.

———. 1972. Functional morphology and evolution of byssally attached bivalve molluscs. *J. Paleont.* 46:165–212.

———. 1973. Effects of competition on rates of evolution, with special reference to bivalve mollusks and mammals. *Syst. Zool.* 22:486–506.

———. 1975a. A theory of evolution above the species level. *Proc. Nat. Acad. U.S.A.* 72:646–650.

———. 1975b. Why clams have the shape they have: an experimental analysis of burrowing. *Paleobiology* 1:48–58.

———. 1977. Trends, rates, and patterns of evolution in the Bivalvia. In *Patterns of evolution as illustrated by the fossil record*, ed. A. Hallam. Amsterdam: Elsevier, pp. 209–250.

Starmühlner, F. 1969. Zur Molluskenfauna des Felslitorals bei Rovinj (Istrien). *Malacologia* 9:217–242.

Stasek, C. R. 1961. The form, growth, and evolution of the Tridacnidae (giant clams). *Arch. Zool. Exp. Gen.* 101:1–40.

Stearns, S. C. 1976. Life-history tactics: a review of the ideas. *Quart. Rev. Biol.* 51:3–47.

Stehli, F. G., and J. W. Wells. 1971. Diversity and age patterns in hermatypic corals. *Syst. Zool.* 20:115–126.

Stehli, F. G., A. McAlester, and C. E. Helsley. 1967. Taxonomic diversity of Recent bivalves and some implications for geology. *Geol. Soc. Amer. Bull.* 78:455–465.

Stehli, F. G., R. D. Douglas, and N. D. Newell. 1969. Generation and maintenance of gradients in taxonomic diversity. *Science* 164:947–949.

Stehli, F. G., R. G. Douglas, and I. A. Kafescioglu. 1972. Models for the evolution of planktonic Foraminifera. In *Models in paleontology*, ed. T. J. M. Schopf. San Francisco: Freeman, Cooper, pp. 116–128.

Stephenson, T. A. 1944. The constitution of the intertidal fauna and flora of South Africa—II. *Ann. Natal Mus.* 10:261–358.

———. 1948. The constitution of the intertidal fauna and flora of South Africa—III. *Ann. Natal Mus.* 11:207–324.

Stephenson, T. A., and A. Stephenson. 1949. The universal features of zonation between tide-marks on rocky coasts. *J. Ecol.* 37:289–305.

———. 1950. Life between tide-marks in North America. I. Florida Keys. *J. Ecol.* 38:354–402.

———. 1952. Life between tide-marks in North America. II. Northern Florida and the Carolinas. *J. Ecol.* 40:1–49.

———. 1954a. Life between tide-marks in North America. III A. Nova Scotia and Prince Edward Island: description of the region. *J. Ecol.* 42:14–45.

———. 1954b. Life between tide-marks in North America. III B. Nova Scotia and Prince Edward Island: the geographical features of the region. *J. Ecol.* 42:46–70.

———. 1972. *Life between tidemarks on rocky shores.* San Francisco: Freeman.

Stephenson, W. 1959. Evolution and ecology of portunid crabs, with especial reference to Australian species. In *The evolution of living organisms (Symp. Roy. Soc. Victoria)* pp. 311–327.

———. 1961. Experimental studies on the ecology of intertidal environments at Heron Island. II. The effect of substratum. *Austr. J. Mar. Fresh-W. Res.* 12:164–176.

———. 1972. An annotated check list and key to the Indo-West-Pacific swimming crabs (Crustacea: Decapoda: Portunidae). *Roy. Soc. N.Z. Bull.* 10:1–64.

Stephenson, W., and R. B. Searles. 1960. Experimental studies on the ecology of intertidal environments at Heron Island. I. Exclusion of fish from beach rock. *Austr. J. Mar. Fresh-W. Res.* 11:241–267.

Stephenson, W., R. Endean, and I. Bennett. 1958. An ecological survey of the marine fauna of Low Isles, Queensland. *Austr. J. Mar. Fresh-W. Res.* 9:261–318.

Stimson, J. 1970. Territorial behavior of the owl limpet, *Lottia gigantea*. *Ecology* 51:113–118.

———. 1973. The role of the territory in the ecology of the intertidal limpet *Lottia gigantea* (Gray). *Ecology* 54:1020–1030.

Stott, R. S., and D. P. Olson. 1973. Food-habitat relationship of sea ducks on the New Hampshire coast. *Ecology* 54:996–1007.

Struhsaker, J. W. 1968. Selection mechanisms associated with intraspecific shell variation in *Littorina picta* (Prosobranchia: Mesogastropoda). *Evolution* 22:459–480.

Stump, T. E. 1975. Pleistocene molluscan paleoecology and community structure of the Puerto Libertad region, Sonora, Mexico. *Paleogeogr., Paleoclimatol., Paleoecol.* 17:177–226.

Sutherland, J. P. 1970. Dynamics of high and low populations of the limpet, *Acmaea scabra* (Gould). *Ecol. Monogrs.* 40:169–188.

———. 1974. Multiple stable points in natural communities. *Amer. Nat.* 108:859–873.

Tagatz, M. E. 1968. Biology of the blue crab, *Callinectes sapidus* Rathbun, in the St. John's River, Florida. *Fish. Bull.* 67:17–33.

Taylor, J. D. 1968. Coral reef and associated invertebrate communities (mostly molluscan) around Mahé, Seychelles. *Phil. Trans. Roy. Soc. London* (B) 254:130–206.

———. 1970. Feeding habits of predatory gastropods in a Tertiary (Eocene) molluscan assemblage from the Paris Basin. *Palaeontology* 13:255–260.

———. 1971. Reef associated molluscan assemblages in the western Indian Ocean. In *Regional variation in Indian Ocean coral reefs*, ed. D. R. Stoddard and C. M. Yonge. New York: Academic Press, pp. 501–534.

———. 1976. Habitats, abundance and diets of muricacean gastropods at Aldabra Atoll. *Zool. J. Linn. Soc.* 59:155–193.

Taylor, J. D., and M. S. Lewis. 1970. The flora, fauna and sediments of the marine grass beds of Mahé, Seychelles. *J. Nat. Hist.* 4:199–220.

Thayer, C. W. 1971. Fish-like crypsis in swimming Monomyaria. *Proc. Malacol. Soc. London* 39:371–376.

Thomas, R. D. K. 1976. Gastropod predation on sympatric Neogene species of *Glycymeris* (Bivalvia) from the eastern United States. *J. Paleont.* 50:488–499.

Thompson, T. E. 1960. Defensive adaptations of opisthobranchs. *J. Mar. Biol. Assoc. U.K.* 39:123–134.

———. 1969. Acid secretion in Pacific Ocean gastropods. *Austr. J. Zool.* 17:755–764.

Thorson, G. 1950. Reproductive and larval ecology of marine bottom invertebrates. *Biol. Revs.* 25:1–45.

———. 1957. Bottom communities (sublittoral or shallow shelf). *Geol. Soc. Amer. Mem.* 67 (1):461–534.

———. 1961. Length of pelagic larval life in marine bottom invertebrates as related to larval transport by ocean currents. In *Oceanography*, ed. M. Sears. Amer. Assoc. Advancement Sci. Publ. 67:455–474.

Topp, R. W. 1969. Interoceanic sea-level canal: effects on the fish faunas. *Science* 165:1324–1327.

Trevallion, A., R. R. C. Edwards, and J. H. Steele. 1970. Dynamics of a benthic bivalve. In *Marine food chains*, ed. J. H. Steele. Los Angeles: University of California Press, pp. 285–295.

Tsuda, R. T., and J. E. Randall. 1971. Food habits of the gastropods *Turbo argyrostoma* and *T. setosus*, reported as toxic from the tropical Pacific. *Micronesica* 7:153–162.

Turner, R. D. 1954. The family Pholadidae in the Western Atlantic and the Eastern Pacific. I—Pholadinae. *Johnsonia* 3 (33):1–64.

———. 1955. The family Pholadidae in the Western Atlantic and the Eastern Pacific. II—Martesiinae, Jouannetiinae and Xylophaginae. *Johnsonia* 3 (34):65–160.

Turner, R. D., and J. Rosewater. 1958. The family Pinnidae in the Western Atlantic. *Johnsonia* 3 (38):285–326.

Underwood, A. J. 1974. The reproductive cycles and geographical distribution of some common eastern Australian prosobranchs (Mollusca: Gastropoda). *Austr. J. Mar. Fresh-W. Res.* 25:63–88.

Valentine, J. W. 1969. Niche diversity and niche size pattern in marine fossils. *J. Paleont.* 43:905–915.

———. 1973. *Evolutionary paleoecology of the marine biosphere*. Englewood Cliffs, N.J.: Prentice Hall.

———. 1976. Genetic strategies of adaptation. In *Molecular evolution*, ed. F. J. Ayala. Sunderland, Mass.: Sinauer Associates, pp. 78–94.

Valentine, J. W., and F. J. Ayala. 1976. Genetic variability in crill. *Proc. Nat. Acad. Sci. U.S.A.* 73:658–660.

Van Cleave, H. J. 1936. Reversal of symmetry in *Campeloma rufum*, a fresh-water snail. *Amer. Nat.* 70:567–573.

Van Dolah, R. F. 1978. Factors regulating the distribution and population dynamics of the amphipod *Gammarus palustris* in an intertidal salt marsh community. *Ecol. Monogr.*, in press.

Van Valen, L. 1971. Group selection and the evolution of dispersal. *Evolution* 25:591–598.

———. 1974. Two modes of evolution. *Nature* 252:298–300.

———. 1976. Energy and evolution. *Evol. Theory* 1:179–229.

Vanzolini, P. E. 1973. Paleoclimates, relief, and species multiplication in equatorial forests. In *Tropical forest ecosystems in Africa and South America: a comparative review*, ed. B. J. Meggers, E. S. Ayensu, and W. D. Duckworth. Washington, D. C.: Smithsonian Institution Press, pp. 255–258.

Vegas, M. D. 1963. Contribución al conocimiento de la zona de *Littorina* en la costa peruana. *Anales Cientificos, Dept. Publ. Univ. Agr. (La Molina)* 1 (2):174–193.

Vermeij, G. J. 1969. Observations on the shells of some fresh-water neritid gastropods from Hawaii and Guam. *Micronesica* 5:155–162.

———. 1971a. Temperature relationships of some tropical Pacific intertidal gastropods. *Mar. Biol.* 10:308–314.

———. 1971b. Substratum relationships of some tropical Pacific intertidal gastropods. *Mar. Biol.* 10:315–320.

———. 1972a. Endemism and environment: some shore molluscs of the tropical Atlantic. *Amer. Nat.* 106:89–101.

———. 1972b. Intraspecific shore-level size gradients in intertidal molluscs. *Ecology* 53:693–700.

———. 1973a. Morphological patterns in high intertidal gastropods: adaptive strategies and their limitations. *Mar. Biol.* 20:319–346.

———. 1973b. West Indian molluscan communities in the rocky intertidal zone: a morphological approach. *Bull. Mar. Sci.* 23:351–386.

———. 1973c. Molluscs in mangrove swamps: physiognomy, diversity, and regional differences. *Syst. Zool.* 22:609–624.

———. 1974a. Regional variations in tropical high intertidal gastropod assemblages. *J. Mar. Res.* 32:343–357.

———. 1974b. Marine faunal dominance and molluscan shell form. *Evolution* 28:656–664.

———. 1975. Evolution and distribution of left-handed and planispiral coiling in snails. *Nature* 254:419–420.

———. 1976. Interoceanic differences in vulnerability of shelled prey to crab predation. *Nature* 260:135–136.

———. 1977a. Patterns in crab claw size: the geography of crushing. *Syst. Zool.* 26:138–151.

———. 1977b. The architectural geography of some gastropods. In *Historical biogeography, plate tectonics and the changing environment*, ed. J. Gray and A. J. Boucot. Corvallis: Oregon State University Press.

———. 1977c. The Mesozoic marine revolution: evidence from gastropods, predators, and grazers. *Paleobiology* 3:245–258.

Vermeij, G. J., and A. P. Covich. 1978. Co-evolution of freshwater gastropods and their predators. *Amer. Nat.* 112, in press.

Vermeij, G. J., and J. W. Porter. 1971. Some characteristics of the dominant intertidal molluscs from rocky shores in Pernambuco, Brazil. *Bull. Mar. Sci.* 21:440–454.

Vermeij, G. J., and J. A. Veil. 1978. A latitudinal pattern in bivalve shell gaping. *Malacologia* 17:57–61.

Vernberg, F. J. 1962. Comparative physiology: latitudinal effects on physiological properties of animal populations. *Ann. Rev. Physiol.* 24:517–546.

Veron, J. E. N. 1974. Southern geographic limits to the distribution of Great Barrier Reef hermatypic corals. *Proc. Second Intern. Coral Reef Symp.* 1:465–473.

Verrill, A. E. 1908. Decapod Crustacea of Bermuda; I—Brachyura and Anomura. *Trans. Connecticut Acad. Arts Sci.* 13:299–474.

Vince, S., I. Valiela, N. Backus, and J. M. Teal. 1976. Predation by the salt marsh killifish *Fundulus heteroclitus* (L.) in relation to prey size and habitat structure: consequences for prey distribution and abundance. *J. Exp. Mar. Biol. Ecol.* 23:255–266.

Vine, P. J. 1974. Effects of algal grazing and aggressive behaviour of the fishes *Pomacentrus lividus* and *Acanthurus sohal* on coral-reef ecology. *Mar. Biol.* 24:131–136.

Vivien, M. L. 1973. Contribution à la connaissance de l'éthologie alimentaire de l'ichtyofaune du platier interne des récifs coralliens de Tuléar (Madagascar). *Téthys* suppl. 5:221–308.

Vohra, F. C. 1971. Zonation on a tropical sandy shore. *J. Anim. Ecol.* 40:679–708.

Vokes, E. H. 1964. The genus *Turbinella* (Mollusca, Gastropoda) in the New World. *Tulane Studs. Geol.* 2:39–68.

———. 1966. The genus *Vasum* (Mollusca: Gastropoda) in the New World. *Tulane Studs. Geol.* 5:1–36.

———. 1967. The genus *Vitularia* (Mollusca: Gastropoda) discovered in the Miocene of southern Florida. *Tulane Studs. Geol.* 5:90–92.

———. 1971. The geologic history of the Muricinae and the Ocenebrinae. *Echo* 4:37–54.

———. 1975a. Eastern Pacific–Western Atlantic faunal affinities—Muricinae and Muricopsinae. *Bull. Amer. Malacol. Union*: 54.

———. 1975b. Notes on the Chipola Formation—XVIII. Some new or otherwise interesting members of the Calyptraeidae (Mollusca: Gastropoda). *Tulane Studs. Geol. Paleontol.* 11:163–172.

Vroman, M. 1968. The marine algal vegetation of St. Martin, St. Eustatius and Saba (Netherlands Antilles). *Studs. Flora Curaçao Other Carib. Islands* 2:1–120.

Vuilleumier, B. S. 1971. Pleistocene changes in the fauna and flora of South America. *Science* 173:771–780.

Wade, B. A. 1967. Studies on the biology of the West Indian beach clam, *Donax denticulatus* Linné. I. Ecology. *Bull. Mar. Sci.* 17:149–174.

Wagner, F. J. E. 1977. Paleoecology of marine Pleistocene Mollusca, Nova Scotia. *Canad. J. Earth Sci.* 14:1305–1323.

Walker, A. J. M. 1972. Introduction to the ecology of the Antarctic limpet *Patinigera polaris* (Hombron and Jacquinot) at Signy Island, South Orkney Islands. *Brit. Antarctic Serv. Bull.* 8:49–69.

Walker, K. R., and L. B. Alberstadt. 1975. Ecological succession as an aspect of structure in fossil communities. *Paleobiology* 1:238–257.

Wallace, L. R. 1972. Some factors affecting vertical distribution and resistance to desiccation in the limpet, *Acmaea testudinalis* (Müller). *Biol. Bull.* 142:186–193.

Waller, T. R. 1972. The functional significance of some shell microstructures in the Pectinacea (Mollusca: Bivalvia). Intern. Geol. Congr., 24th Sess., Montreal, Canada, Sec. 7, Paleontology:48–56.

Warmke, G. L., and D. S. Erdman. 1963. Records of marine mollusks eaten by bonefish in Puerto Rican waters. *Nautilus* 76:115–120.

Warner, G. F. 1970. Behaviour of two species of grapsid crab during intraspecific encounters. *Behaviour* 36:9–19.

Webb, S. D. 1976. Mammalian faunal dynamics of the great American interchange. *Paleobiology* 2:220–234.

Webster, J. D. 1941. Feeding habits of the black oystercatcher. *Condor* 43:175–179.

Welch, W. R. 1968. Changes in abundance of the green crab, *Carcinus maenas* (L.), in relation to recent temperature changes. *Fish. Bull.* 67:337–345.

Wells, B., D. H. Steele, and A. V. Tyler. 1973. Intertidal feeding of winter flounder *Pseudopleuronectes americanus* in the Bay of Fundy. *J. Fish. Res. Bd. Canada* 30:1374–1378.

Wells, H. W. 1958a. Feeding habits of *Murex fulvescens. Ecology* 39:556–558.

––––––. 1958b. Predation of pelecypods and gastropods by *Fasciolaria hunteria* (Perry). *Bull. Mar. Sci. Gulf Carib.* 8:152–166.

Wells, H. W., and I. E. Gray. 1960. Seasonal occurrence of *Mytilus edulis* on the Carolina coast as a result of transport around Cape Hatteras. *Biol. Bull.* 119:550–559.

Wells, J. W. 1957. Coral reefs. *Geol. Soc. Amer. Mem.* 67 (1):609–631.

Weyl, P. K. 1968. The role of the oceans in climatic change: a theory of the ice ages. *Meteorol. Monogrs.* 8 (30):37–62.

Whitney, R. R. 1961. *Bairdiella, Bairdiella icistius* (Jordan and Gilbert). In *The Ecology of the Salton Sea, California, in relation to the sportfishery*, ed. B. W. Walker. Calif. Dept. Fish and Game Bull. 113:105–151.

Wiebe, W. J., R. E. Johannes, and K. L. Webb. 1975. Nitrogen fixation in a coral-reef community. *Science* 188:257–259.

Williams, A. B. 1974. The swimming crabs of the genus *Callinectes* (Decapoda: Portunidae). *Fish. Bull.* 72:685–798.

Williams, G. C. 1975. *Sex and evolution*. Princeton, N.J.: Princeton University Press.

Wilson, J. W. III. 1974. Analytical zoogeography of North American mammals. *Evolution* 28:124–140.

Wodinsky, J. 1969. Penetration of the shell and feeding from gastropods by *Octopus. Amer. Zool.* 9:997–1010.

––––––. 1973. Mechanism of hole boring in *Octopus vulgaris. J. Gen. Psychol.* 88:179–183.

Wolcott, T. G. 1973. Physiological ecology and intertidal zonation in limpets (*Acmaea*): a critical look at "limiting factors." *Biol. Bull.* 145:389–422.

Wommersley, H. B. S., and S. J. Edmonds. 1958. A general account of the intertidal ecology of South Australian coasts. *Austr. J. Mar. Fresh-W. Res.* 9:217–260.

Wonders, J. B. W. 1977. The role of benthic algae in the shallow reef of Curaçao (Netherlands Antilles). III. The significance of grazing. *Aquatic Botany* 3:357–390.

Woodin, S. A. 1974. Polychaete abundance patterns in a marine soft-sediment environment: importance of biological interactions. *Ecol. Monogrs.* 44:171–187.

———. 1976. Adult-larval interactions in dense faunal assemblages: patterns of abundance. *J. Mar. Res.* 34:25–41.

———. 1978. Refuges, disturbance, and community structure: a soft-bottom example. *Ecology* 59: in press.

Woodring, W. P. 1966. The Panama land bridge as a sea barrier. *Amer. Phil. Soc. Proc.* 110:425–433.

———. 1970. Geology and paleontology of Canal Zone and adjoining parts of Panama: description of Tertiary mollusks (gastropods: Eulimidae, Marginellidae to Helminthoglyptidae). *U.S. Geol. Survey Prof. Paper* 306-D:299–452.

———. 1973a. Affinities of Miocene molluscan faunas on Pacific side of Central America. *Publ. Geol. Inst. Centroamer. Invest. Tecnol. Industr.* 4:179–188.

———. 1973b. Geology and paleontology of Canal Zone and adjoining parts of Panama: description of Tertiary mollusks (additions to gastropods, scaphopods, pelecypods: Nuculidae to Malleidae). *U.S. Geol. Surv. Prof. Paper* 306-E:453–539.

———. 1974. The Miocene Caribbean Faunal Province and its subprovinces. *Verhandl. Naturf. Ges. Basel* 84:209–213.

Woodward, B. B. 1892. On the mode of growth and the structure of the shell in *Velates conoideus*, Lamk., and other Neritidae. *Proc. Zool. Soc. London* (1892):528–540.

Wray, J. L. 1971. Algae in reefs through time. *Proc. North Amer. Paleontol. Conv., 1969,* (2):1358–1373.

Wulff, B. L., E. M. T. Wulff, B. H. Robison, J. K. Lowry, and H. J. Humm. 1968. Summer marine algae of the jetty at Ocean City, Maryland. *Chesapeake Sci.* 9:56–60.

Yaldwyn, J. C. 1965. Antarctic and subantarctic decapod Crustacea. In *Biogeography and ecology in Antarctica,* ed. P. van Oye and J. van Mieghem. The Hague: Dr. W. Junk, pp. 324–332.

Yamaguchi, M. 1975a. Coral-reef asteroids of Guam. *Biotropica* 7:12–23.

———. 1975b. Sea level fluctuations and mass mortalities of reef animals in Guam, Mariana Islands. *Micronesica* 11:227–243.

Yaron, I. 1972. Further additions to the Mediterranean molluscan fauna of Israel. *Argamon, Israel J. Malacol.* 3:33–36.

———. 1976. *Minolia nedyma* Melvill 1897—an Indo-Pacific trochid in the Mediterranean. *Argamon, Israel J. Malacol.* 5:53–60.

Yonge, C. M. 1952. Studies on Pacific coast mollusks. IV. Observations on *Siliqua patula* Dixon and on evolution within the Solenidae. *Univ. Calif. Publ. Zool.* 55:421–438.

———. 1953. Form and habit in *Pinna carnea* Gmelin. *Phil. Trans. Roy. Soc. London* (B) 237:355–374.

———. 1967. Form, habit and evolution in the Chamidae (Bivalvia) with reference to conditions in the rudists (Hippuritacea). *Phil. Trans. Roy. Soc. London* (B) 252:49–105.

Zaneveld, J. S. 1969. Factors controlling the delimitation of littoral benthic marine algal zonation. *Amer. Zool.* 9:367–391.

Zaret, T. M. 1972. Predator-prey interaction in a tropical lacustrine ecosystem. *Ecology* 53:248–257.

Zaret, T. M., and W. C. Kerfoot. 1975. Fish predation on *Bosmina longirostris*: body size selection versus visibility selection. *Ecology* 56:232–237.

Zaret, T. M., and R. T. Paine. 1973. Species introduction in a tropical lake. *Science* 182:449–455.

Zeigler, R. F., and H. C. Porreca. 1969. *Olive shells of the world*. Rochester, N.Y.: Rochester Polychrome Press.

Zipser, E., and G. J. Vermeij. 1978. Crushing behavior of tropical and temperate crabs. *J. Exp. Mar. Biol. Ecol.* 31:155–172.

INDEX

Acanthaster: as predator of molluscs, 55, 79; as predator of corals, 101, 141-142; distribution, 142, 188, 228, 256; as potential Panama migrant, 261
Acanthina, 77, 137
Acanthocarpus, 61
Acanthocyclus, 106
Acanthopleura, 137, 151
Acanthuridae, 92, 156, 259
Acila, 67
Acmaea: rate compensation, 9; mucus, 28; water balance, 33; allometry, 32-33; predation on, 36-37, 54; as grazer, 94-95, 145; distribution, 137, 189; living on other shells, 143; cognate species, 264
Acochlidioid opisthobranchs, 21
Acropora: predation on, 99-101, 142; distribution, 141, 152-153, 200; competition, 165
Acrosiphonia, 92
Acteon, 53
Adamussium, 17, 227
Adaptation, definition, 3, 25
Aetobatis, 187
Agaricia, 141
Agaronia, cognate species, 213, 234
Alcyonacean soft corals, distribution, 155-156
Algae: latitudinal gradients, 85-90; regional variations, 140, 156; ephemeral, 175
Algal ridges, distribution, 87, 150
Alima, distribution, 187
Allometry, 12-15
Alpheus: association with corals, 101, 141-142, 194; cognate species, 217
Amblyrhynchus, as grazer, 96
Anachis, distribution, 216
Anadara, distribution, 217-218, 236
Anaerobiosis, and calcification, 16-18
Ancilla, distribution, 233
Angiosperms, genetic system, 173-174
Aniculus, 99, 142
Anisotremus, as predator of molluscs, 54
Anodontia, distribution, 233
Anomia, 67
Anomuran crustaceans: distribution, 16, 124; as predators of molluscs, 39-40; as grazers, 95-96, 99; predation on, 164

Antarctic biota, 15-17, 227
Anthopleura, distribution, 86
Anthus, as predator of molluscs, 52
Aplodinotus, as predator of molluscs, 75
Aporrhais, distribution, 225
Aragonite, 18-19
Aramus, as predator of molluscs, 39
Aratus, distribution, 217
Arca, distribution, 126, 189
Archaeocyatha, 185
Architectonica, distribution, 187, 253
Arcopsis, 215
Arctic Ocean, migration through, 257-258
Arenaeus, 213
Arenaria, as predator of molluscs, 52
Argentina, biota, 138, 224
Arothron, distribution, 187-188, 256
Articulate brachiopods, predation on, 104-105, 177
Ascophyllum, 221
Astacid crayfishes, 64
Astarte, 181
Astartidae, 19, 67
Asterias: as predator, 80; distribution, 138, 226
Asterina, as Suez migrant, 252
Asteroids, as predators, 101, 104, 162-163, 261. *See also* Forcipulate asteroids
Astraea, distribution, 216
Astreopora, history, 229
Astrocoenia, history, 229
Astropecten, as predator of molluscs, 51, 81-82
Astropyga, 126
Atactodea, 131
Atergatis, as Suez migrant, 252
Atrina, distribution, 237
Aulacomya, distribution, 137
Austrotrophon, 136
Autochthonous taxa, 237-240
Avrainvillea, grazing on, 90, 92
Azteca, evolution, 194

Bailya, distribution, 215
Bairdiella, introduction, 192
Balanus, distribution, 106
Balistidae: as predators of molluscs, 50; as grazers, 92, 97; distribution, 142
Baltic Sea biota, 225-226